Study Guide and Workbook for

PHYSICS

A General Introduction

ALAN VAN HEUVELEN

Prepared by

LOIS M. KIEFFABER
UNIVERSITY OF NEW MEXICO

Little, Brown and Company

BOSTON   TORONTO

ISBN 0-316-897124

9 8 7 6 5 4 3 2 1

SEM

Published simultaneously in Canada
by Little, Brown & Company (Canada) Limited

Printed in the United States of America

Preface

"I understand the material, but I can't work the problems" is probably the most frequent lament a physics teacher hears from students. The role of problem solving in physics is monumental. It is the means whereby one learns whether or not one really does "understand the material." And it is almost universally the means by which a student is asked to demonstrate his understanding of the subject matter. When asked what would be most helpful to them in learning physics, the most common student request by far is to see more example problems.

This *Study Guide,* therefore, places primary emphasis on problem-solving techniques. It is intended for use with Alan Van Heuvelen's *Physics: A General Introduction.* Each chapter opens with a list of terms serving to review the chapter contents and an equation review that consolidates the quantitative relationships of the chapter and saves a lot of flipping around in the textbook to track them down. From then on, there are problems, problems, and more problems, each with detailed solutions including comments on the results and on problem-solving techniques. Difficult concepts as well as fine points are stressed in the framework of solving problems.

Each chapter ends with a list of helpful hints and several drill problems, particularly designed for before-test review. The section entitled "Avoiding Pitfalls" calls specific attention to common sources of student error, from fine points in mathematical calculations to errors that arise from basic misconceptions. The Appendix contains complete solutions, not just answers, to the Drill Problems, so that at no time is the student left wondering how to solve any of the problems.

An attempt has been made to frame the problems in situations that are of natural interest to the student, drawn from the everyday world, with a minimum of "objects" and "black boxes." The problems are specifically chosen to clarify the main points of the chapter. Diagrams and illustrative sketches abound.

There is no duplication of textbook end-of-chapter problems, so the teacher's choice of problem assignments from the text need not depend on whether some or all students have study guides.

Any physics teacher's collection of ideas for physics questions and problems results from a wide variety of sources, starting with texts, teachers, and tests encountered as a student to the wide variety of literature in the field that is assimilated over many years of teaching. These ideas grow, change and intertwine in such a way that it would be impossible to detail the history of all of them. However, certain ideas for problems can be traced to certain individuals and in that regard, I would like to express particular appreciation to L. Dwight Farringer, Manchester College; Victor H. Regener, the University of New Mexico; and Alan W. Peterson, the University of New Mexico. I am also grateful to Alan Van Heuvelen and Alan Peterson for their careful reading of the entire manuscript and their many helpful comments and suggestions, and to Richard Elston, the University of New Mexico, for his painstaking check of all solutions. Of great help also were the unfailing interest and encouragement of Ian Irvine, Sponsoring Editor at Little, Brown and the expertise of himself and Elizabeth Schaaf in turning the manuscript into a book. Finally, my sincere thanks to my own physics students, past and present, for the part they have played in their own education: they have asked the right questions—those which reveal where the trouble spots are; they have demanded unambiguous, physically realistic problems to solve; they have had a critical interest in learning physics as well and as efficiently as possible. They are, after all, the primary motivation for the writing of physics textbooks and study guides such as these.

To the Student

The material in this *Study Guide* is presented in chapters corresponding to the chapters of Alan Van Heuvelen's *Physics: A General Introduction*. Each chapter contains the following subsections:

Terms. Here you should be able to define or describe briefly the new terms encountered in the chapter. Each term is keyed to the appropriate chapter section if you need help. This exercise is intended both as a review of the subject matter and to give practice in verbal expression of ideas.

Equation Review. Important equations are listed, identified by the number given them in the textbook, for ready reference in solving the problems that follow. For each equation you should be able to state the situation to which the equation applies, what quantity each symbol represents, and the units for measuring each quantity. Actually doing this for each equation will help you choose the right equation for a given situation, rather than having to try out several before finding the appropriate one. It will reduce the number of incorrect substitutions of quantities for symbols. Finally, becoming familiar with and keeping track of the units of quantities provides an excellent check of your algebra; an answer with an incorrect unit is a good indicator of an error in the solution.

Problems with Solutions and Discussion. This section occupies the largest number of pages in each chapter. Although you may wish to try working the problems before studying the solutions, they are intended as *example* problems, to supplement those given in the text. They are chosen to be similar to the end-of-chapter problems which will ordinarily be assigned as homework. By studying them, you will find help with all aspects of problem-solving, such as translating verbal statements into mathematical relationships, identifying the equations that apply to the situation, techniques for actually solving equations, unit conversions and consistency, checks of numerical results, and discussion of the physical meaning of your results.

Avoiding Pitfalls. These items are intended to call your attention to some common student errors, in the hope that you will then be able to recognize and avoid them. They also include explicit techniques other students have found useful in avoiding such errors. It is best studied after you have finished your study of the textbook chapter and example problems and have completed (or at least are actively working on) your homework problems. It is also good to peruse this section just before the test covering the chapter.

Drill Problems. These problems are intended to be similar to those you might encounter on a test. They require numerical solutions and often involve putting together more than one main concept from the chapter. Use them as a practice test after you have completed your study of the chapter. The solutions are in an Appendix to encourage you to try to work them before looking at the solutions.

No one learns physics well without working many problems. Solving problems gives you confidence in what you do know and calls your attention to what you don't know. In the words of a very famous physics teacher, Arnold Sommerfeld, to a very famous physics student, Werner Heisenberg, "Just do the exercises diligently, then you will know what you have understood, and what you have not."

Contents

Scalars and Vectors

1

Terms

Define or describe briefly what is meant by the following terms. If you have difficulty, refer to the textbook section given in parentheses.

basic quantity (1.2)

derived quantity (1.2)

system of units (1.2)

scalar (1.3)

vector (1.3)

resultant (1.5)

components of a vector (1.6)

Equation Review

For each equation, be able to state the situation to which it applies, what quantity each symbol represents, and the units for measuring each quantity in some consistent set of units.

$$F_x = \pm F \cos \theta \tag{1.3}$$

$$F_y = \pm F \sin \theta \tag{1.4}$$

$$R_x = F_{1x} + F_{2x} + F_{3x} + \ldots \tag{1.5}$$

$$R_y = F_{1y} + F_{2y} + F_{3y} + \ldots \tag{1.6}$$

$$R = \sqrt{R_x^2 + R_y^2} \tag{1.7}$$

$$\tan \theta = |R_y/R_x| \tag{1.8}$$

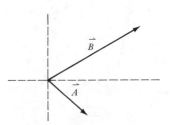

(a)

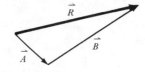

(b)

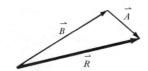

(c)

FIG. 1.1

Problems with Solutions and Discussion

PROBLEM: Sketch $\vec{R}$, the vector sum (resultant) of the two vectors $\vec{A}$ and $\vec{B}$ (Fig. 1.1a).

Solution: Only a sketch is asked for. Vectors $\vec{A}$ and $\vec{B}$ are placed head to tail, preserving their length and direction. Their resultant is the vector pointing from the tail of $\vec{A}$ to the

1

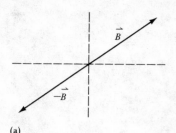

(a)

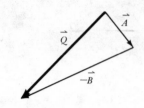

(b)

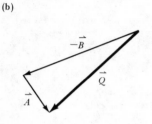

(c)

FIG. 1.2

head of $\vec{B}$. Note that the magnitude and direction of $\vec{R}$ in (b) is unchanged if the order of the vectors is reversed (c).

PROBLEM: For the vectors $\vec{A}$ and $\vec{B}$ of the previous problem, sketch $\vec{Q}$, the vector difference $\vec{B} - \vec{A}$.

Solution: We can treat this as a vector addition problem of the two vectors $\vec{A}$ and ($-\vec{B}$). $-\vec{B}$ is equal in magnitude (length) and opposite in direction to $\vec{B}$ (Fig. 1.2a). Thus $\vec{A} - \vec{B}$ is obtained by the head-to-tail method, (b) or (c).

PROBLEM: A baseball player hits a line drive. He runs 90 ft northeast to first base, then 90 ft northwest to second base, and has run 30 ft southwest toward third when he is tagged out. Determine graphically his resultant displacement from home plate when he is tagged.

Solution: Choosing an appropriate scale, the displacement vectors are constructed head to tail with their proper directions as shown in Fig. 1.3. Careful measurement with ruler and protractor yield a resultant of magnitude 108 ft, at an angle of 80° from the axis pointing west.

PROBLEM: Forces $\vec{A}$, $\vec{B}$, and $\vec{C}$ have magnitudes and directions as shown in Fig. 1.4a. Find the x and y components of each force and then determine their resultant by the method of components.

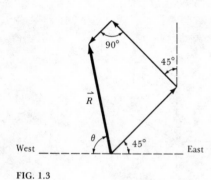

FIG. 1.3

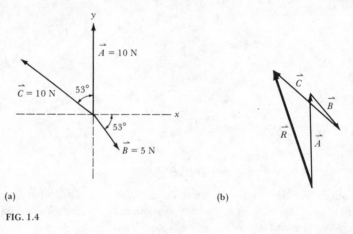

(a)

(b)

FIG. 1.4

Solution: A simple sketch (b) using the head-to-tail method shows quickly what result to expect and provides a rough check of the answer obtained. We expect a resultant lying in the second quadrant. (Note that when vectors are added graphically by the head-to-tail method, they may cross over each other.)

The components of $\vec{A}$ are (Fig. 1.5a):

$$A_x = +A \cos 90° = 10 \text{ N } (0) = 0$$
$$A_y = +A \sin 90° = 10 \text{ N } (1) = +10 \text{ N}$$

The components of $\vec{B}$ are (Fig. 1.5b):

$$B_x = +B \cos 53° = 5 \text{ N } (0.6) = +3 \text{ N}$$
$$B_y = -B \sin 53° = -5 \text{ N } (0.8) = -4 \text{ N}$$

Here we choose the negative sign for B_y to indicate that it lies along the negative y axis.

Vector $\vec{C}$ makes an angle of 53° with the y axis (Fig. 1.5c). The angle with the x axis is just 90 − 53 or 37°. Therefore,

2

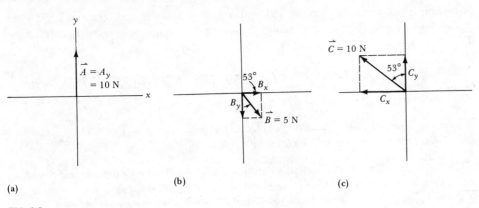

(a)

(b)

(c)

FIG. 1.5

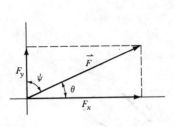

FIG. 1.6

$$C_x = -C \cos 37° = -10 \text{ N } (0.8) = -8 \text{ N}$$
$$C_y = +C \sin 37° = 10 \text{ N } (0.6) \quad = 6 \text{ N}$$

Note: If you know trigonometry, you may recall that the angle with the y axis, call it ψ (Fig. 1.6), can be used directly to calculate components:

$$C_x = \pm C \sin \psi$$
$$C_y = \pm C \cos \psi$$

All vectors are shown resolved into their components (Fig. 1.7). The original vectors are scratched out to indicate they have been replaced by their components.

Adding all vectors on the x axis:

$$R_x = A_x + B_x + C_x = 0 + 3 \text{ N} - 8 \text{ N} = -5 \text{ N}$$

On the y axis:

$$R_y = A_y + B_y + C_y = 10 \text{ N} - 4 \text{ N} + 6 \text{ N} = 12 \text{ N}$$

When R_x and R_y are sketched on an x-y coordinate system, the resultant, $\vec{R}$, is the diagonal of the completed rectangle (Fig. 1.8). Note that $\vec{R}$ is a second quadrant vector as expected. The magnitude of $\vec{R}$ is

$$R = \sqrt{R_x^2 + R_y^2} = \sqrt{(-5 \text{ N})^2 + (12 \text{ N})^2}$$
$$= \sqrt{25 \text{ N}^2 + 144 \text{ N}^2} = \sqrt{169} \text{ N} = \underline{13 \text{ N}}$$

FIG. 1.7

The angle labelled θ is obtained by

$$\tan \theta = \left| \frac{R_y}{R_x} \right| = \left| \frac{12 \text{ N}}{5 \text{ N}} \right| = 2.4$$

$$\theta = \underline{67°}$$

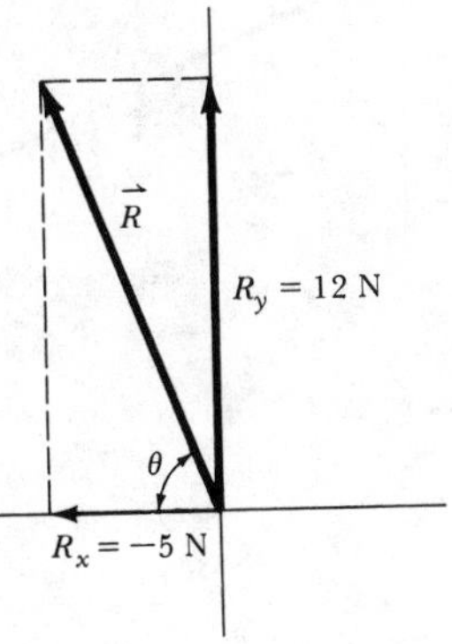

Discussion: In this problem, the method of components began by replacing 3 vectors ($\vec{A}$, $\vec{B}$, and $\vec{C}$) by six vectors (their rectangular components). Have we not therefore complicated the problem? No, because the six vectors lie along one of only two axes. This means that on each axis, the directional nature of the vector can be handled by a mere + or − sign. The increase in number of vectors and the fact that two additions are required (one for each axis) are more than compensated for by the ease of the additions, which have become simple algebraic additions with no angles involved. The final construction of $\vec{R}$ from R_x and R_y does involve a triangle, but it is a right triangle so the geometry is simple.

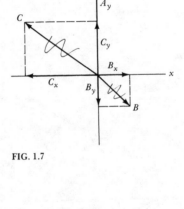

FIG. 1.8

PROBLEM: Velocities are vectors also. An airplane is flying with a ground speed of 500 mi/hr due west in the presence of a 150 mi/hr wind blowing due south (Fig. 1.9a). What is the resultant velocity of the plane, which is the sum of the two velocities?

3

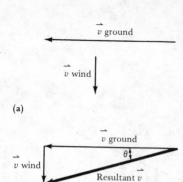

(a)

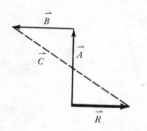

(b)

FIG. 1.9

Solution: The resultant $\vec{v}$ is the vector sum of the ground speed vector and the wind speed vector. A graphical solution shows $\vec{v}$ to be the hypotenuse of a right triangle, so the Pythagorean theorem is applicable (Fig. 1.9b).

$$v = \sqrt{v_{ground}^2 + v_{wind}^2} = \sqrt{(500 \text{ mi/hr})^2 + (150 \text{ mi/hr})^2}$$
$$= \underline{522 \text{ mi/hr}}$$
$$\tan \theta = \left| \frac{v_{wind}}{v_{ground}} \right| = \left| \frac{150}{500} \right| = 0.3, \quad \text{so} \quad \theta = \underline{17°}$$

Note that although we did not actually define velocity, we know how to add velocities if we know velocity is a vector. Similarly, we now know how to add *any* physical quantity which is a vector, regardless of whether we understand what that quantity is. If *skong* is a vector quantity, we know how to add *skongs* to find the resultant *skong*.

PROBLEM: Three forces, $\vec{A}$, $\vec{B}$, and $\vec{C}$, acting together, produce a resultant force of 10 N in the $+x$ direction. Force $\vec{A}$ is 10 N in the $+y$ direction. Force $\vec{B}$ is 10 N in the $-x$ direction. What is the magnitude and direction of force $\vec{C}$? Solve by the method of components.

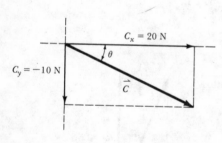

FIG. 1.10

Solution: A quick sketch of the graphical method shows the general properties of the unknown force: it is greater than any of the others and is a fourth quadrant vector (Fig. 1.10). The method of components is sometimes thought of exclusively as a method for finding a resultant, but it can equally well be used to find an unknown vector when the resultant is known. Resolving all vectors (including the resultant) into components,

$$R_x = 10 \text{ N} \quad A_x = 0 \quad B_x = -10 \text{ N}$$
$$R_y = 0 \quad A_y = 10 \text{ N} \quad B_y = 0$$

We may use Eq. (1.5) to solve for C_x:

$$R_x = A_x + B_x + C_x$$
$$10 \text{ N} = 0 - 10 \text{ N} + C_x$$
$$20 \text{ N} = C_x$$

and Eq. (1.6) to solve for C_y:

$$R_y = A_y + B_y + C_y$$
$$0 = 10 \text{ N} + 0 + C_y$$
$$-10 \text{ N} = C_y$$

Forming $\vec{C}$ from its two rectangular components as shown in Fig. 1.11,

$$C = \sqrt{(20 \text{ N})^2 + (-10 \text{ N})^2} = \underline{22.4 \text{ N}}$$
$$\tan \theta = \left| \frac{-10}{20} \right| = 0.5, \quad \text{so} \quad \theta = \underline{27°}$$

FIG. 1.11

Avoiding Pitfalls

1. When a vector is asked for, the correct answer includes both a magnitude and a direction, even if the direction is not explicitly requested.

2. Vectors may be moved from place to place, providing they retain their length and direction.

3. In many force diagrams, vectors to be added are shown tail-to-tail, since they are all thought of as acting at the same point. This is not the correct arrangement for a head-to-tail graphical method of vector addition, so the vectors must be rearranged.

4. When referring to vector components along a given axis, + and − indicate the two opposite directions along that axis. The positive sign may be applied to either direction, as long as that direction is considered positive throughout the problem.

5. Calculating components of a vector as $A_x = \pm A \cos \theta$ and $A_y = \pm A \sin \theta$ is correct only if θ is the angle the vector makes with the x axis. Therefore, if an angle ψ is given with respect to the y axis, the angle θ is its complement. Alternatively, when an angle ψ is measured from the y axis, the components are reversed: $A_x = \pm A \sin \psi$ and $A_y = \pm A \cos \psi$.

6. The method of components works for any frame of reference whose axes are mutually perpendicular. They need not be oriented horizontally and vertically (Fig. 1.12). In some problems (for example, an object on an inclined plane) the problem is simplified by choosing another reference frame (for example, parallel to and perpendicular to the inclined plane).

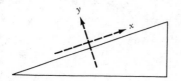

FIG. 1.12

7. The + and − signs have no directional meaning if there are more than two possible directions (that is, they have meaning only in a one-dimensional situation: to specify up or down, or to specify left or right, in accordance with an agreed-upon sign convention). For any two-dimensional situation, an angle is required to specify direction.

8. If an angle less than 90° is identified from $\tan \theta = |R_y/R_x|$, its location must be clearly specified, if it does not in fact lie in the first quadrant. This specification may be by means of a diagram, such as is shown in Fig. 1.13, or by words, such as "25° counterclockwise from the $-x$ axis."

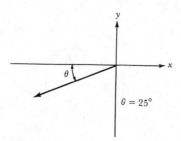

FIG. 1.13

Drill Problems
Answers in Appendix

1. For the three vectors shown in Fig. 1.14, convince yourself by a sketch that their resultant is independent of the *order* in which they are placed head-to-tail.

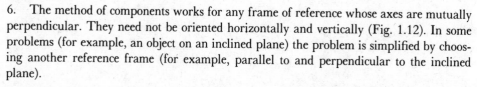

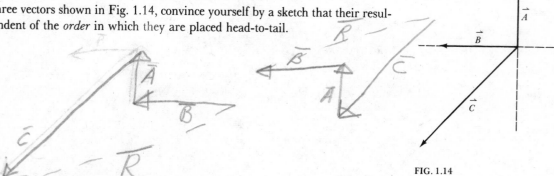

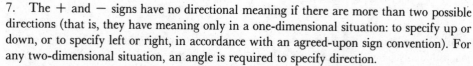

FIG. 1.14

2. Vector $\vec{A}$ is a displacement of 6 miles due north and vector $\vec{B}$ is a displacement of 8 miles due east. Find

(a) the vector $5\vec{A}$;

(b) the vector $-\vec{B}$;

(c) the resultant of $\vec{A}$ and $\vec{B}$;

(d) the vector $\vec{B} - \vec{A}$.

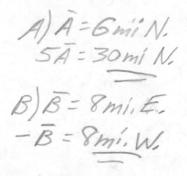

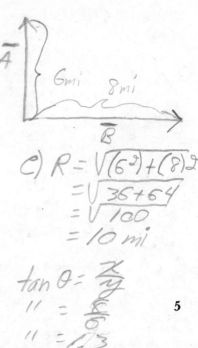

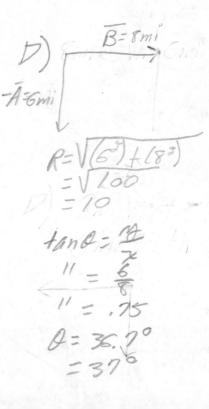

5

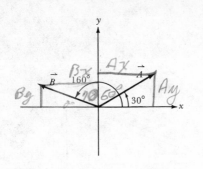

FIG. 1.15

3. The two forces $\vec{A}$ and $\vec{B}$ each have a magnitude of 10 N and are oriented as shown in Fig. 1.15. Their vector sum is $\vec{R}$. Determine the magnitude and direction of $\vec{R}$ by the method of components.

$A_x = A \cos 30° = (10)(.866) = 8.66$

$A_y = A \sin 30° = (10)(.5) = 5$

$B_x = -B \cos 20° = (-10)(.940) = -9.40$

$B_y = B \sin 20° = (10)(.342) = 3.42$

$R_x = A_x + B_x \qquad R_y = A_y + B_y$
$\quad = 8.66 - 9.4 \qquad = 5 + 3.42$
$\quad = -.74 \qquad\qquad = 8.42$

$R = \sqrt{(8.4)^2 + (-.74)^2}$

$R = \sqrt{70.56 + .49}$

$R = \sqrt{71.05}$

$R = 8.42$

$\tan \theta = \dfrac{8.42}{-.7}$

$\quad || = 12$

$\quad \theta = 85°$

4. A swimmer jumps into a large pool in front of the lifeguard's chair. He swims 15 m parallel to the pool's edge, then makes a 90° turn and swims 20 m toward the center of the pool. Suddenly he gets a cramp and yells for help. To reach the victim in the shortest time, at what angle with respect to the pool's edge should the lifeguard head, and how far must she swim?

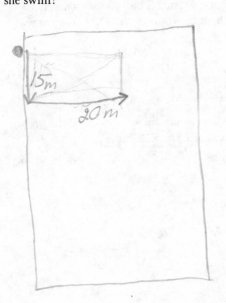

$R^2 = \sqrt{(20)^2 + (15)^2}$

$|| = \sqrt{400 + 225}$

$|| = \sqrt{625}$

$R = 25m$

$\tan \theta =$

$|| =$

$\theta =$

6

Statics

2

Terms

Define or describe briefly what is meant by the following terms. If you have difficulty, refer to the textbook section given in parentheses.

first condition of equilibrium (2.1)

second condition of equilibrium (2.4)

contact force (2.1)

tension force (2.2)

torque (2.3)

moment arm (2.3)

center of gravity (2.5)

Equation Review

For each equation, be able to state the situation to which it applies, what quantity each symbol represents, and the units for measuring each quantity in some consistent set of units.

$\Sigma F_x = 0$ $F_{1x} + F_{2x} + F_{3x} \cdots = 0$ **(2.2a)**

$\Sigma F_y = 0$ $F_{1y} + F_{2y} + F_{3y} \cdots = 0$ **(2.2b)**

$\Sigma \tau = 0$ $\tau_1 + \tau_2 + \tau_3 = 0$ **(2.5)**

$\tau = \pm Fl$ $\tau = F_1 l_1 + F_2 l_2 + F_3 l_3$ **(2.4)**

Problems with Solutions and Discussion

Before discussing numerical examples, let us get some practice in the second and third steps of solving statics problems, namely, choosing an object or objects to isolate, and drawing a diagram showing with arrows all forces acting on the isolated system.

PROBLEM: Show all the forces acting on a ladder leaning against a wall.

Solution: We isolate the ladder as our system (Fig. 2.1). The weight vector $\vec{w}$ may be attached to the geometric center of the ladder, assuming it to be symmetric. The only other two points where forces act are the ends. Looking first at the end touching the ground, the

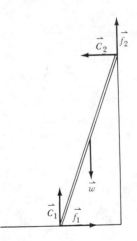

FIG. 2.1

7

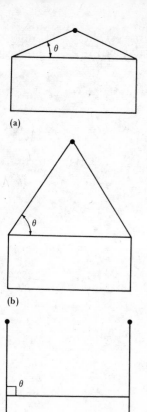

(a)

(b)

(c)

FIG. 2.2

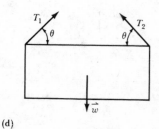

(d)

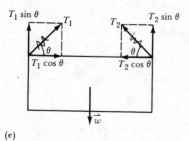

(e)

FIG. 2.2

force of the ground on the ladder may be thought of as resolved into two components, one perpendicular to and one parallel to the ground. (This is true for any two surfaces in contact. The perpendicular component is what we have referred to as the *contact* force. It is also commonly called the *normal* force, normal meaning "perpendicular." The parallel force component often turns out to be the *frictional* force.) We can determine that the horizontal force of the ground on the ladder is to the right rather than to the left by considering what the ladder would do if there were no force there. (It would slide down; a force to the left would make it slide even faster.)

The wall's force on the ladder also has in general a vertical and horizontal component. The horizontal component is the *contact* (or normal) force, being perpendicular to the surfaces in contact. The vertical component is the *frictional* force and must point upward to keep the ladder from slipping. So we have five forces: the ladder's weight, two contact forces, and two frictional forces as shown. (*Note:* the ladder, of course, exerts contact and frictional forces on the ground and on the wall. But in considering the equilibrium of the ladder, we need only concern ourselves with forces acting *on the ladder*.)

PROBLEM: Determine which method of hanging a picture, Fig. 2.2a, b, or c, results in the smallest tension in the wire.

Solution: The three situations shown differ in the size of the angle θ, so we need to determine how the tension in the wires depends on θ. The free-body diagram for the picture is shown in Fig. 2.2d. The picture's center of gravity is assumed to be its geometric center, which is true for most pictures. A further assumption is that the angles the wires make with the picture are equal, to be aesthetically pleasing.

Resolving the forces along horizontal and vertical axes (e), let us write the first equilibrium equation.

$$\Sigma F_x = 0: \quad T_1 \cos \theta - T_2 \cos \theta = 0$$

Solving for T_1 in terms of T_2, we find that

$$T_1 = T_2$$

The equality of the two angles requires that the two tensions be equal in magnitude; let us call them both T.

For equilibrium in the y direction,

$$\Sigma F_y = 0: \quad T \sin \theta + T \sin \theta - w = 0$$

Solving for T,

$$T = \frac{w}{2 \sin \theta}$$

We see that the smaller θ is, the larger T is. Thus the tension is greatest in (a) and least in (c). This can be understood by noting that as θ gets smaller, the horizontal components get larger, but they cancel each other and have no effect in supporting the picture's weight. The vertical components, on the other hand, get smaller, but they must always be great enough to support w. Therefore the smaller θ is, the larger T must be to insure enough vertical component to support w.

Let us now consider some numerical examples of statics problems.

PROBLEM: A 20-N weight is hung from the end of a uniform rod (Fig. 2.3a). It is kept in a horizontal position by balancing it at a point one-third of the way from that same end. Find the weight of the rod and the upward force C exerted on the rod at the balance point.

Solution: The rod is isolated and the forces on it are shown (b). The rod's center of gravity is its geometric center because it is uniform, so the weight vector can be considered as acting there. The x-y axes are taken as horizontal and vertical, respectively. Since there are no forces in the x direction, we proceed to the y direction.

$$\Sigma F_y = 0: \quad C - w - 20 = 0$$

$$C = w + 20$$

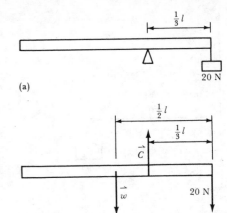

(a)

Taking torques about the fulcrum,

$$\Sigma \tau = 0: \quad w(\tfrac{1}{6}l) - 20(\tfrac{1}{3}l) = 0$$

Here we have expressed the moment arms as fractions of the length l of the rod. The 20-N weight is $\tfrac{1}{3}l$ from the fulcrum, and the vector $\vec{w}$, acting at the center of the rod, is $\tfrac{1}{2}l - \tfrac{1}{3}l = \tfrac{1}{6}l$ from the fulcrum. Note that the l's will cancel out, emphasizing a particularly important point: even though the torque equation contains lengths in the form of moment arms, it is not necessary to know these lengths explicitly if their *relative* magnitudes are known. When the above equation is solved, we find that

$$w = 20(\%) = \underline{40 \text{ N}}$$

From the y equation,

$$C = w + 20 = 40 + 20 = \underline{60 \text{ N}}$$

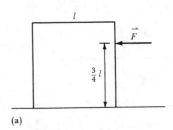

(b)

FIG. 2.3

Discussion: This problem demonstrates the power of the technique of attaching the entire weight vector of the center of gravity. Although the weight of the rod is in fact distributed throughout the rod, and each portion of the weight is capable of exerting a torque, the rod *does actually behave* as if all weight is concentrated at its center of gravity. One upward force there supports the entire weight. In addition, the torques exerted by all parts of the rod are exactly equivalent to the torque the entire weight would produce if it were located at the center of gravity. Treating the extended weight distribution as one weight acting at one point greatly simplifies the problem.

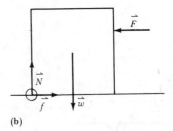

(a)

PROBLEM: What horizontal force is needed to tip over a cubical box weighing 60 lbs, if the force is applied at a point three-quarters of the way up the box on the centerline of that face (Fig. 2.4a)?

Solution: When the box is just beginning to tip, only the lower left edge will be in contact with the ground. The tipping is thus seen as a rotation about the lower left edge, and the horizontal and vertical components of the ground force will be applied at that edge. By treating this as an equilibrium problem, we can find the force needed to hold the box in that position. A slightly greater force will then cause the box to tip over.

The diagram (b) shows the forces acting on the box. Consideration of the first and second equilibrium equations will lead to the result that N must equal w and that F must equal f, but to get a value for F, we must write the torque equation. Taking torques about the lower left edge,

$$\Sigma \tau = 0: \quad F(\tfrac{3}{4}l) - 60(\tfrac{1}{2}l) = 0$$

$$F = 30(\%) = \underline{40 \text{ lb}}$$

(b)

FIG. 2.4

PROBLEM: A uniform plank, 8 m long and weighing 200 N, is supported at one end by a rope and at the other end by a rock, as shown in Fig. 2.5a. An 800-N man stands 2 m from the right end. The rope makes a 37° angle with the horizontal.

Show clearly on the diagram all forces acting on the plank. Then write three equations which are sufficient to express the equilibrium of the plank, putting in numbers where available.

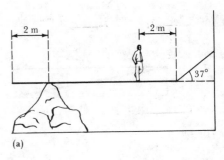

(a)

Solution: The forces are as shown in (b). The plank's weight acts upon its geometric center since it is uniform. The rock force and the rope tension are both shown resolved into their horizontal and vertical components. The direction of the rock's horizontal force $\vec{f}$ is chosen by considering what would happen if $\vec{f}$ were not present, or were in the opposite direction.

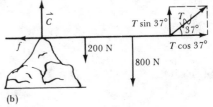

(b)

FIG. 2.5

9

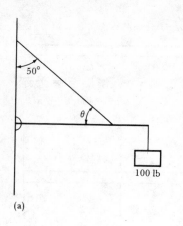

(a)

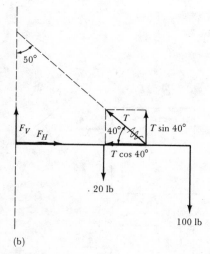

(b)

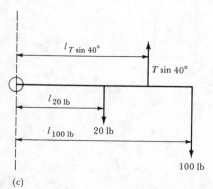

(c)

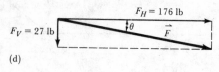

(d)

FIG. 2.6

The equilibrium equations are

$$\Sigma F_x = 0: \quad T\cos 37° - f = 0$$

$$T(0.8) = f \tag{1}$$

where forces to the left are taken as positive, forces to the right as negative.

$$\Sigma F_y = 0: \quad C + T\sin 37° - 200 - 800 = 0$$

$$C + T(0.6) = 1000 \tag{2}$$

Here up is chosen as positive and down as negative. Choosing the rock as the point about which to calculate torques,

$$\Sigma \tau_R = 0: \quad -(200\ \text{N})(2\ \text{m}) - (800\ \text{N})(4\ \text{m}) + (T\sin 37°)(6\ \text{m}) = 0$$

with counterclockwise torques taken as positive, clockwise torques as negative.

$$-400\ \text{N}\cdot\text{m} - 3200\ \text{N}\cdot\text{m} + T(0.6)(6\ \text{m}) = 0$$

$$3600\ \text{N}\cdot\text{m} = (3.6\ \text{m})T \tag{3}$$

Of course, different equations would result from different choices of points about which to calculate torques.

PROBLEM: A 20-lb uniform beam 4 ft long is supported horizontally by a cable and a hinge at a wall, as shown in Fig. 2.6a. The cable is attached 1 ft from the outer end of the beam and a 100-lb weight hangs from the end of the beam. Determine the force (magnitude and direction) of the hinge on the beam, and the tension in the cable.

Solution: Since the two unknown forces act on the beam, let us isolate it as our system. The free-body diagram is shown in (b). The force of the hinge on the beam is immediately resolved into a vertical and a horizontal component, labeled F_V and F_H respectively, since we do not know the angle of the force. We *do* know the direction of the tension in the cable, however. It *must* lie along the length of the cable, since the cable is not a rigid body and thus cannot sustain transverse forces.

The angle θ that T makes with the horizontal is 40°. (To see this, consider the right triangle made by wall, cable, and beam.) Horizontal-vertical x-y axes are chosen, since all forces except T are horizontal or vertical. T is then resolved as shown. We are now ready to write the equilibrium equations.

$$\Sigma F_x = 0: \quad F_H - T\cos 40° = 0$$

$$F_H = T(0.77)$$

$$\Sigma F_y = 0: \quad F_V + T\sin 40° - 20 - 100 = 0$$

$$F_V + T(0.64) = 120$$

There are two equations, but three unknowns, so a third equation is needed.

A good choice of points about which to calculate torques is the hinge, since two unknown forces, F_H and F_V, will then have no moment arms and thus not appear in the equation. Notice also that with this choice, $T\cos 40°$ has no moment arm, since its line of action when extended passes through the point about which we are calculating torques. The moment arms for the remaining forces are shown in the diagram (c).

$$\Sigma \tau_{\text{hinge}} = 0: \quad -(20\ \text{lb})(2\ \text{ft}) - (100\ \text{lb})(4\ \text{ft}) + (T\sin 40°)(3\ \text{ft}) = 0$$

This equation has only one unknown, T. Solving for T,

$$-20(2) - (100)(4) + T(0.64)(3) = 0$$

$$-40 - 400 + T(1.92) = 0$$

$$1.92T = 440$$

$$T = \underline{229\ \text{lb}}$$

10

The tension in the cable is 229 lb. F_H can now be obtained from the x equation:

$$F_H = T(0.77) = (229 \text{ lb})(0.77) = 176 \text{ lb}$$

F_V is obtained from the y equation:

$$F_V + (229 \text{ lb})(0.64) = 120 \text{ lb}$$

$$F_V = 120 - 147 = -27 \text{ lb}$$

The negative sign in the answer means the *direction* assumed for F_V was incorrect. F_V actually points down. (If the hinge were suddenly removed, the left end of the beam would try to slide *up* the wall because of the weight at the right end; F_V pulls down on the beam.) The *magnitude* of F_V, however, is correct.

Combining F_H and F_V to get the resultant force of the hinge on the beam, we now show F_V in its proper direction in (d). F is the composite force exerted by the hinge on the beam. Its magnitude is

$$F = \sqrt{(176)^2 + (27)^2} = \underline{178 \text{ lb}}$$

The angle θ made with the horizontal is given by

$$\tan \theta = {}^{27}\!/_{176} = 0.153; \quad \theta = \underline{8.7°}$$

Avoiding Pitfalls

1.   Once an object or objects have been isolated for application of the equilibrium conditions, make sure forces on other objects do not appear in the equations. If a box has been isolated, the force *on the floor* (albeit by the box) is excluded. If an object hanging from a wire is isolated, the force of the wire *on the wall* is excluded.

2.   If more than one object is included in the "isolated system," then the force of one such object on a second is an internal force, exactly balanced by the force of the second on the first. Their vector sum is zero; thus no "internal" forces of one part of the system on another part are ever included in the equilibrium equations.

3.   In assigning directions to tension forces, recall that ropes, cables, wires, etc. always *pull* on an object; they never push (which you already know if you've ever tried to push something with a rope).

4.   *Lengths* of ropes are not related to the *magnitudes* of forces. (Lengths never appear in force equations.) Lengths are often useful in determining the *angle* at which a tension force acts, because the tension pulls in the same direction as the length of the rope.

5.   The tension force exerted by a rope (or cable, wire, string, etc.) always lies along the rope, because ropes are flexible and cannot sustain transverse forces. But the force on a rigid body need not (and generally does not) lie along its length. For example, if a hinge exerts an unknown force on a beam, the force does not, in general, lie in the direction of the beam length. Such a force is best treated by determining its vertical and horizontal components separately, from the equilibrium conditions. Then, if desired, the composite force, magnitude, and direction, can be determined by vector addition of the components.

6.   If a body is symmetric and uniform, its center of gravity is the geometric center and the weight vector acts at that point. For nonuniform bodies, the center of gravity is not at the geometric center. But wherever it is located, this is the point at which the weight vector acts, both for force and torque considerations.

7.   Weight vectors always point toward the center of the earth. In most diagrams this direction is *vertically down.*

8.   If directions of forces are known, these directions must be included in the equilibrium equations for forces, using + and − signs. These + and − signs are *not* used with the forces appearing in the *torque* equation.

9.   For torques, + and − signs refer to whether or not the *torque* is counterclockwise or clockwise and *not* on whether the force producing the torque was originally called + or −.

10. If the direction of a force is not known, assume a direction and complete the solution. Any force which comes out *negative* will have the correct magnitude, but its actual direction will be opposite to that assumed. But since the original equations were based on the assumed directions, all forces must be continued through the solution with the + or − sign obtained.

11. When surfaces are in contact, the force exerted by one on the other has, in general, a component perpendicular to the surfaces (often referred to as the *contact* or *normal* force) and a component parallel to the surfaces (caused by friction).

12. If a body is in rotational equilibrium, it is true that the vector sum of the torques about any point is zero; thus any point may be used as an origin for a torque equation. However, the algebra of the problem can often be greatly simplified by choosing as an origin a point where one or more unknown forces are applied. Since those forces then have no moment arms, the number of unknowns in the torque equation is reduced.

Drill Problems
Answers in Appendix

1. A horizontal uniform plank 8 ft long and weighing 30 pounds is suspended by two ropes, each fastened one foot from an end of the plank (Fig. 2.7). What is the tension in each rope when a painter weighing 150 pounds stands 2 feet from one end of the plank?

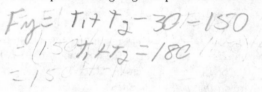

2. Solve equations (1), (2), and (3) of the plank-on-the-rock example problem to find the tension in the rope and the horizontal and vertical forces exerted on the plank by the rock.

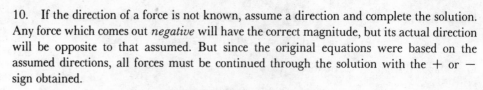

FIG. 2.7

3. A steel ball of mass 25 kg hangs at the end of a long wire. What horizontal force F is required to hold the ball at an angle of 30° from the vertical as shown in Fig. 2.8?

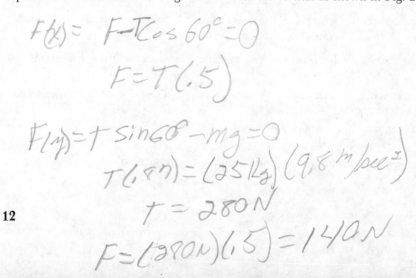

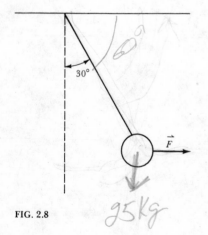

FIG. 2.8

12

Describing Motion: Kinematics

3

Terms

Define or describe briefly what is meant by the following terms. If you have difficulty, refer to the textbook section given in parentheses.

kinematics

translational motion (3.1)

rotational motion (3.1)

vibrational motion (3.1)

average speed (3.2)

average velocity (3.2)

instantaneous speed (3.3)

instantaneous velocity (3.3)

average acceleration (3.4)

instantaneous acceleration (3.4)

slope (3.5)

gravitational acceleration (3.7)

projectile (3.8)

Equation Review

For each equation, be able to state the situation to which it applies, what quantity each symbol represents, and the units for measuring each quantity in some consistent set of units.

$$\bar{v} = \frac{d}{t} \tag{3.1}$$

$$\bar{\vec{v}} = \frac{\vec{d}}{t} \tag{3.2}$$

$$v = \frac{\Delta x}{\Delta t} = \frac{x_2 - x_1}{t_2 - t_1} \tag{3.3}$$

$$\vec{v} = \frac{\Delta \vec{d}}{\Delta t} \tag{3.4}$$

$$\vec{a} = \frac{\vec{v}_2 - \vec{v}_1}{t_2 - t_1} \tag{3.5}$$

$$\vec{a} = \frac{\Delta \vec{v}}{\Delta t} \tag{3.6}$$

$$v = v_0 + at \tag{3.7}$$

$$x - x_0 = \left(\frac{v + v_0}{2}\right)t \tag{3.8}$$

$$x - x_0 = v_0 t + \tfrac{1}{2}at^2 \tag{3.9}$$

$$2a(x - x_0) = v^2 - v_o^2 \tag{3.10}$$

Problems with Solutions and Discussion

PROBLEM: Each day a jogger runs at 6 m/s for 2 min, and then walks at 2 m/s for 400 m. What is her average speed for the daily exercise period?

Solution: Average speed is defined as total distance divided by total time. Looking at the two legs of the trip separately:

$$\text{Running:} \quad t = 2 \text{ min} = 2 \text{ min}\left(\frac{60 \text{ s}}{1 \text{ min}}\right) = 120 \text{ s}$$

$$x = vt = (6 \text{ m/s})(120 \text{ s}) = 720 \text{ m}$$

$$\text{Walking:} \quad x' = 400 \text{ m}$$

$$t' = \frac{x'}{v'} = \frac{400 \text{ m}}{2 \text{ m/s}} = 200 \text{ s}$$

$$\text{Average speed:} \quad \bar{v} = \frac{\text{total distance}}{\text{total time}} = \frac{x + x'}{t + t'} = \frac{720 \text{ m} + 400 \text{ m}}{120 \text{ s} + 200 \text{ s}}$$

$$= \frac{1120 \text{ m}}{320 \text{ s}} = \underline{3.5 \text{ m/s}}$$

Notice that average speed is not an arithmetic average of the speeds of the two legs. That would yield 4 m/s, which is incorrect because it did not take into account the *time* each speed was maintained. Each leg must be separately analyzed for the distance and time so that the total distance can be divided by the total time for the correct average speed.

The next problem emphasizes the difference between *average speed* and *average velocity*.

PROBLEM: Andy drives 30 miles east in an hour and then 40 miles north in an hour. Determine his average speed and average velocity for the entire trip.

Solution:

$$\text{Average speed:} \quad \bar{v} = \frac{\text{distance traveled}}{\text{time}} = \frac{30 \text{ mi} + 40 \text{ mi}}{1 \text{ hr} + 1 \text{ hr}} = \frac{70 \text{ mi}}{2 \text{ hr}} = \underline{35 \frac{\text{mi}}{\text{hr}}}$$

His average speed is 35 mi/hr.

$$\text{Average velocity:} \quad \bar{\vec{v}} = \frac{\text{displacement}}{\text{time}} = \frac{\vec{d}}{t}$$

Displacement is a vector; we must add the two displacements as vectors to get the total $\vec{d}$ (Fig. 3.1).

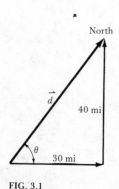

FIG. 3.1

$$d = \sqrt{30^2 + 40^2} = 50 \text{ mi}$$

14

Thus,

$$v = \frac{d}{t} = \frac{50 \text{ mi}}{2 \text{ hr}} = \underline{25 \text{ mi/hr}}$$

The direction of $\vec{v}$ is the same as the direction of $\vec{d}$; in a vector equation not only are the numbers on both sides equal, but also the directions are equal. The angle θ is given by

$$\tan \theta = {}^{40}\!/\!_{30} = 1.33 \quad \text{so} \quad \theta = \underline{53°}.$$

His average velocity is 25 mi/hr, at 53° north of east. This problem shows that average speed and average velocity do not necessarily have the same magnitude. They are equal only when distance and displacement are equal in magnitude.

PROBLEM: In nuclear physics, the *barn* is used as a unit of area for measuring nuclear cross sections. One barn is equal to 10^{-24} cm². Express a square meter in barns.

Solution:

$$1 \text{ m}^2 = 1 \text{ m}^2 \left(\frac{100 \text{ cm}}{1 \text{ m}} \right) \left(\frac{100 \text{ cm}}{1 \text{ m}} \right) \left(\frac{1 \text{ barn}}{10^{-24} \text{ cm}^2} \right) = \underline{1 \times 10^{28} \text{ barns}}$$

There are 10^{28} barns in one square meter. The conversion factors used are each fractions equal to one, since the numerator equals the denominator. Notice that without the second cm-to-m conversion factor, some units would be "left over." If you think of units as canceling algebraically like numbers, you can determine quickly which unit goes in the numerator and which in the denominator of the conversion factor. Also you can see quickly how many of a particular conversion factor are needed to make sure all the unwanted units cancel out.

PROBLEM: Convert a speed of 4 ft/sec to units of fathoms per day. One fathom is equal to 6 feet.

Solution:

$$4 \frac{\text{ft}}{\text{s}} = 4 \frac{\text{ft}}{\text{s}} \left(\frac{1 \text{ fathom}}{6 \text{ ft}} \right) \left(\frac{60 \text{ s}}{1 \text{ min}} \right) \left(\frac{60 \text{ min}}{1 \text{ hr}} \right) \left(\frac{24 \text{ hr}}{1 \text{ day}} \right) = 57,600 \frac{\text{fathoms}}{\text{day}}$$

PROBLEM: Friction can cause very large decelerations. A bullet traveling at 400 m/s can be stopped by a tree trunk in half a millisecond (1000 ms = 1 s). Determine the acceleration of the bullet in the wood.

Solution:

$$v_0 = 400 \text{ m/s}$$

$$v = 0$$

$$t = 0.5 \text{ ms} \left(\frac{1 \text{ s}}{1000 \text{ ms}} \right) = 5 \times 10^{-4} \text{ s}$$

Using the definition of acceleration,

$$a = \frac{v - v_0}{t} = \frac{0 - 400 \text{ m/s}}{5 \times 10^{-4} \text{ s}} = \underline{-8 \times 10^5 \text{ m/s}^2}$$

The acceleration is 800,000 m/s² in a direction opposite to that of the initial velocity which we took to be plus. We normally call this a *deceleration;* in physics the general term *acceleration* covers both an increase and a decrease in velocity as well as a change in the direction of the velocity vector.

PROBLEM: A boy sliding down a 12-foot-long sliding board reaches the bottom in 2 s. What is his acceleration (assuming it to be constant) and his speed at the bottom of the slide?

15

Solution:

$$x - x_0 = \cancel{v_0 t}^{\,0} + \tfrac{1}{2}at^2$$

where $(x - x_0) = 12$ ft, $v_0 = 0$, $t = 2$ s.

Solving for a,

$$a = \frac{2(x - x_0)}{t^2} = \frac{2(12 \text{ ft})}{(2 \text{ s})^2} = \underline{6 \text{ ft/s}^2}$$

Then, for the final speed,

$$v = v_0 + at = 0 + \left(6 \frac{\text{ft}}{\text{s}^2} \right)(2 \text{ s}) = \underline{12 \text{ ft/s}}$$

His acceleration is 6 ft/s² and his final velocity is 12 ft/s in the direction of the displacement. By putting in $(x - x_0)$ as a positive number, we have chosen the downward direction of the slide as positive. Thus, a and v come out positive, indicating they are in the same downward direction.

In this problem, as in many others, there is more than one correct method to get the required results. After obtaining a, the final velocity could also be calculated as follows:

$$v^2 - v_0^2 = 2a(x - x_0), \quad \text{where } v_0 = 0$$
$$v^2 = 2(6 \text{ ft/s}^2)(12 \text{ ft}) = 144 \text{ ft}^2/\text{s}^2$$
$$v = \underline{12 \text{ ft/s}}$$

Or the final velocity could be obtained first, using

$$x - x_0 = \left(\frac{v_0 + v}{2} \right)t, \quad \text{where } v_0 = 0$$
$$v = \frac{2(x - x_0)}{t} = \frac{2(12 \text{ ft})}{2 \text{ s}} = \underline{12 \text{ ft/s}}$$

Then a can be obtained from the defining equation for acceleration:

$$a = \frac{v - v_0}{t} = \frac{12 \text{ ft/s} - 0}{2 \text{ s}} = \underline{6 \text{ ft/s}^2}$$

Using an alternate method of solution is a good way to check your results.

PROBLEM: At a time $t = 0$, a car has a constant velocity of 20 m/s east. At the same time a motorcycle starts from rest in the same direction with an acceleration of 2 m/s². How long does it take the motorcycle to catch up with the car, and what distance have they traveled at that time?

Solution: The distance traveled by the car at any time t is given by

$$x_c = v_{c_0}t + \tfrac{1}{2}a_ct^2 = 20\,t \tag{1}$$

where $x_{c_0} = 0$ and $x_{m_0} = 0$ at $t = 0$, and $a_c = 0$. For the motorcycle,

$$x_m = v_{m_0}t + \tfrac{1}{2}a_mt^2 = 0 + \tfrac{1}{2}(2)t^2 = t^2 \tag{2}$$

At the time the motorcycle overtakes the car, both will have traveled the same distance, or

$$x_c = x_m$$
$$20t = t^2$$
$$t^2 - 20t = 0$$
$$t(t - 20) = 0$$

The equation has two roots. The first, $t = 0$, represents the original time when the car and motorcycle were together. The second root is given by $t - 20 = 0$, or $t = \underline{20 \text{ s}}$. After

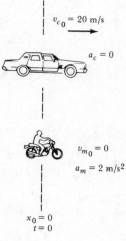

$v_{c_0} = 20$ m/s

$a_c = 0$

$v_{m_0} = 0$

$a_m = 2$ m/s²

$x_0 = 0$
$t = 0$

FIG. 3.2

20 s the motorcycle overtakes the car and they again have the same displacement. The distance both have travelled can be obtained from either (1) or (2).

$$\text{From (1),} \quad x_c = 20t = 20(20) = \underline{400 \text{ m}}$$
$$\text{From (2),} \quad x_m = t^2 = (20)^2 = \underline{400 \text{ m}}$$

PROBLEM: A marble rolls off a box 4 ft high at a speed of 10 ft/s. How far from the edge of the box does it strike the floor?

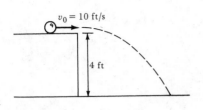

FIG. 3.3

Solution: The given quantities are

$$v_{x_0} = v_0 = 10 \text{ ft/s}, \quad v_{y_0} = 0$$
$$y - y_0 = 4 \text{ ft}, \quad a_y = g = 32 \text{ ft/s}^2, \quad a_x = 0$$

The quantity $(x - x_0)$ is what we wish to find. In the y direction:

$$y - y_0 = \cancel{v_{y_0}t}^{0} + \tfrac{1}{2} a_y t^2$$
$$t = \sqrt{\frac{2(y - y_0)}{g}} = \sqrt{\frac{2(4 \text{ ft})}{32 \text{ ft/s}^2}} = \sqrt{0.25 \text{ s}^2} = \underline{0.5 \text{ s}}$$

where down is taken as positive, so the y displacement and the accelaration of gravity are both positive.

In the x direction:

$$x - x_0 = v_{x_0}t + \frac{1}{2} \cancel{a_x t^2}^{0} = (10 \text{ ft/s})(0.5 \text{ s}) = \underline{5 \text{ ft}}$$

The marble strikes the floor 5 ft from the edge of the box. Note that from the very beginning of the problem, we looked at components and equations *either* for the y direction *or* for the x direction, never for both at the same time.

PROBLEM: In the previous problem, what is the marble's velocity (magnitude and direction) as it strikes the floor?

Solution: We need v_y and v_x to construct the final $\vec{v}$.

In the y direction, $v_y = v_{y_0} + a_y t = 0 + (32)(0.5) = 16 \text{ ft/s}$

In the x direction, $v_x = v_{x_0} + \cancel{a_x t}^{0} = 10 \text{ ft/s}$

Adding the x and y components of the velocity as vectors (and recalling that $+$ means *down* for the y direction),

$$v = \sqrt{v_x^2 + v_y^2} = \sqrt{(10)^2 + (16)^2} = \underline{19 \text{ ft/s}}$$
$$\tan \theta = {}^{16}\!/_{10} = 1.6, \quad \text{so } \theta = \underline{58°}$$

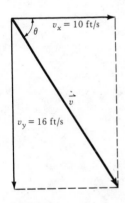

FIG. 3.4

As it strikes the floor, the marble's velocity is 19 ft/s at an angle of 58° below the horizontal.

PROBLEM: A diver jumps off a diving board, which is 4 m above the water level, giving herself an initial vertical upward speed of 5 m/s. (a) How much time has elapsed when she reaches the highest point above water? (b) How high does she rise? (c) With how large a speed does she hit the water?

Solution: Because the motion is entirely in the vertical direction, we will omit the y subscripts, with the understanding that all subsequent equations refer to the y direction.

(a) At the highest point, $v = 0$; the diver momentarily stops as she reverses direction. We divide the trip into legs at this point.

For Leg 1: $v = v_0 + at$, where $a = g = 9.8 \text{ m/s}^2$
$$0 = 5 \text{ m/s} + (-9.8 \text{ m/s}^2)t$$

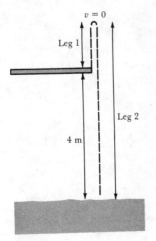

FIG. 3.5

17

where we have taken up as positive, so g, which points downward, is negative.

$$t = \frac{-5 \text{ m/s}}{-9.8 \text{ m/s}^2} = \underline{0.51 \text{ s}}$$

About half a second has elapsed when she reaches the top of her path.

(b) Again looking at Leg 1, for which we now know t,

$$y - y_0 = \left(\frac{v_0 + \cancel{v}^{\,0}}{2} \right) t = \frac{5 \text{ m/s}}{2} (0.51 \text{ s}) = \underline{1.275 \text{ m}}$$

She rises 1.275 m above the board, or 5.275 m above the water.

(c) Looking at Leg 2, and taking down as positive,

$$v_0 = 0, \quad y - y_0 = 5.275 \text{ m}, \quad a = 9.8 \text{ m/s}^2$$

$$2a(y - y_0) = v^2 - \cancel{v_0^2}^{\,0}$$

$$v^2 = 2(9.8 \text{ m/s}^2)(5.275 \text{ m}) = 103 \text{ m}^2/\text{s}^2$$

$$v = \pm 10 \text{ m/s}$$

We choose the positive square root to represent the downward velocity. She hits the water with a speed of 10 m/s.

Part (c) may also be worked by combining Legs 1 and 2 into one long leg, since acceleration is constant throughout. Taking up as positive,

$$v_0 = 5 \text{ m/s (up)} \quad y - y_0 = -4 \text{ m (down)} \quad a = -9.8 \text{ m/s}^2 \text{ (down)}$$

$$2a(y - y_0) = v^2 - v_0^2$$

$$2(-9.8)(-4) = v^2 - (5)^2$$

$$v^2 = 78.4 + 25 = 103.4 \quad \text{so} \quad v = -10 \text{ m/s}$$

We choose the negative square root to represent the downward velocity. This method is more concise, but we lose the information concerning the height to which the diver rose if we do not break the path above the water into legs.

Avoiding Pitfalls

1. *Velocity* is a vector quantity having magnitude and direction. The magnitude of the velocity vector is called the *speed*.

2. The words acceleration and deceleration are in common use when describing an increase or decrease in speed, respectively. In physics, the term *acceleration* is more general and includes both these situations and also a change in *direction* even if the speed remains constant.

3. When an acceleration vector points in the same direction as a velocity vector, regardless of whether that direction is called positive or negative, an increase in speed occurs. When an acceleration vector points in the opposite direction to the velocity vector, a decrease in speed occurs. Thus, a negative sign on an acceleration implies *de*celeration only if the velocity vector was taken as positive (which is usually the case).

4. The constant acceleration equations are applicable only when the acceleration is constant. If it changes, the motion must be broken up into "legs," within each of which the acceleration is constant.

5. Sometimes it is convenient to break up a motion into legs even when the acceleration does not change. For example, at the highest point of the path of a projectile, v_y is always zero. This makes it a convenient end point for the first leg or starting point for the second leg of the projectile's motion.

6. The constant acceleration equations are *vector* equations even when motion is one-dimensional. Displacement, velocity, and acceleration must be accompanied by + or − signs appropriate to their direction. These signs are entirely separate from + or − signs which might result from algebraic manipulation of an equation.

7. When a body is thrown, shot, or released into the air, it is considered a projectile only while the sole acceleration is gravitational. An initial velocity is the velocity the instant after

it is projected; the final velocity is the velocity just before it is brought to rest by some external force, such as the ground or a person's hand.

8. For a projectile, do not attempt to obtain a velocity from a single equation. The constant acceleration equations for projectile motion refer to the x axis, or to the y axis, but never to both simultaneously. If a velocity is given, resolve it into its x and y components immediately; if a velocity is asked for, get its components from the projectile motion equations, then construct the vector from the components.

9. In projectile motion, it is v_y which is zero at the top of the path, not v_x which continues unchanged, and certainly not the acceleration. The acceleration is always g, pointing downward.

10. For projectiles in the absence of air resistance, the velocity in the x direction *always continues unchanged* in both magnitude and direction, simply because there is no horizontal acceleration.

Drill Problems
Answers in Appendix

1. To demonstrate your ability to handle conversion of units, let a *moment* be a unit of time such that 1 moment = 4 s. Let a *hand* be a unit of length equal to 20 cm. Express an acceleration of 8 hands/moment^2 in the proper *mks* units for acceleration.

2. Johnny drives 80 km west to deliver a package. On the return trip, his truck breaks down halfway home, an hour and a half after he left home. Determine his average speed and his average velocity for the trip, in km/hr.

3. A driver traveling at a speed of 20 m/s on a foggy day suddenly sees a roadblock 50 m ahead. It takes the driver 0.7 s to react to the situation and apply her brakes. After she does, the car has a constant deceleration of 5 m/s^2.

   (a) Will the car hit the roadblock?
   (b) If so, what is its speed on impact?
   (c) What minimum deceleration would the car need to have to prevent the collision?

4. Two boys 100 ft apart are playing ball. When one boy throws the ball to the other, it spends 3 s in the air.

   (a) What was the initial velocity of the ball (magnitude and direction)?
   (b) How high did the ball go?

4

Newton's Laws of Motion

Terms

Define or describe briefly what is meant by the following terms. If you have difficulty, refer to the textbook section given in parentheses.

dynamics

Newton's first law of motion (4.1)

Newton's second law of motion (4.2)

Newton's third law of motion (4.5)

mass (4.2)

kilogram (4.3)

newton (4.3)

pound (4.3)

slug (4.3)

weight (4.4)

friction force (4.7)

static friction (4.7)

kinetic friction (4.7)

Equation Review

For each equation, be able to state the situation to which it applies, what quantity each symbol represents, and the units for measuring each quantity in some consistent set of units.

$$\Sigma \vec{F} = m\vec{a} \tag{4.1}$$

$$\vec{w} = m\vec{g} \tag{4.4}$$

$$F_s \leq \mu_s F_c \tag{4.9}$$

$$F_k = \mu_k F_c \tag{4.10}$$

Problems with Solutions and Discussion

PROBLEM: A 250-kg piano, initially at rest on a frictionless level surface, is pushed simultaneously by three people. One pushes with a force of 300 N north, another with 900 N west and the third with 500 N east. Find the magnitude and direction of the piano's acceleration.

Solution: From Newton's second law, $\Sigma \vec{F} = m\vec{a}$, we get

$$\vec{a} = \frac{\Sigma \vec{F}}{m}$$

where $\Sigma \vec{F}$ is the resultant of the forces. They are shown in Fig. 4.1a.

$$\Sigma \vec{F}_x = 500 \text{ N} - 900 \text{ N} = -400 \text{ N}$$

$$\Sigma \vec{F}_y = 300 \text{ N}$$

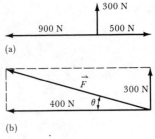

(a)

(b)

FIG. 4.1

The resultant force $\vec{F}$ therefore has magnitude

$$F = \sqrt{(-400)^2 + (300)^2} = 500 \text{ N}$$

and θ is given by

$$\tan \theta = \frac{300}{400} = 0.75, \quad \text{so} \quad \theta = 37°$$

Therefore,

$$a = \frac{\Sigma F}{m} = \frac{500 \text{ N}}{250 \text{ kg}} = 2 \frac{\text{kg} \cdot \text{m/s}^2}{\text{kg}} = \underline{2 \text{ m/s}^2}$$

where we have used the fact that $1 \text{ N} = 1 \text{ kg} \cdot \text{m/s}^2$ to make sure the answer has the proper *mks* units for acceleration. The direction of $\vec{a}$ must be the same as the direction of the resultant force $\vec{F}$, because $\vec{F} = m\vec{a}$ is a vector equation; not only are numbers and units on both sides equal, but so are *directions*. The piano's acceleration is 2 m/s² at 37° north of west.

PROBLEM: The highest average acceleration encountered in nature is 400 g's (1 g = 9.8 m/s²), experienced by the click beetle, a common British species, when jackknifing into the air to escape predators. What force is required to give a beetle of mass 4 milligrams this acceleration?

Solution: We will use Newton's second law, $F = ma$, where

$$m = 4 \text{ mg}\left(\frac{1 \text{ g}}{1000 \text{ mg}}\right)\left(\frac{1 \text{ kg}}{1000 \text{ g}}\right) = 4 \times 10^{-6} \text{ kg}$$

$$a = 400 \text{ } g = 400(9.8 \text{ m/s}^2) = 3920 \text{ m/s}^2$$

$$F = ma = (4 \times 10^{-6} \text{ kg})(3920 \text{ m/s}^2) = \underline{0.016 \text{ N}}$$

The force required is 0.016 N or about ¹⁄₁₆ oz. The beetle must push the ground with this force to generate a Newton's third law reaction force from the ground which then accelerates it upward.

PROBLEM: A 5-kg box is released from rest on a frictionless ramp inclined at an angle of 30°. How large is the box's acceleration and how long does it take for the box to attain a speed of 10 m/s?

Solution: The forces on the box are its weight $\vec{w}$, directed downward, and the contact force $\vec{F}_c$ of the ramp, directed perpendicular to the ramp, as shown in Fig. 4.2. The x axis is chosen parallel to the plane, since motion is confined to this direction. The weight $\vec{w}$ is resolved into components; note that the angle between $\vec{w}$ and the plane is the *complement* of the plane's angle. In the y direction, $\Sigma \vec{F}_y = 0$, because there is no acceleration in this direction. The box does not jump up from the plane or crash down through it.

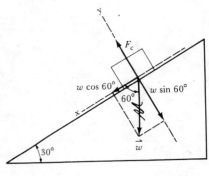

FIG. 4.2

21

In the x direction, there is a net unbalanced force, $w \cos 60°$, which will cause acceleration.

$$\Sigma \vec{F}_x = m\vec{a}$$

$$w \cos 60° = ma$$

$$a = \frac{w \cos 60°}{m} = \frac{mg \cos 60°}{m} = g \cos 60° = 9.8 \text{ m/s}^2(0.5) = \underline{4.9 \text{ m/s}^2}$$

The acceleration of the box is not g, as for a freely falling body, but $g \cos 60°$, since gravity is "diluted" by the inclined plane. But as the angle of inclination approaches 90°, the acceleration approaches g. Note that the acceleration is independent of mass, as it is for a freely falling body.

To get the time taken to attain a velocity of 10 m/s, a kinematics equation is needed:

$$v_0 = 0; \quad a = g \cos 60° = 4.9 \text{ m/s}^2; \quad v = 10 \text{ m/s}$$

$$v = \cancel{v_0}^{0} + at$$

$$t = \frac{v}{a} = \frac{10 \text{ m/s}}{4.9 \text{ m/s}^2} = \underline{2 \text{ s}}$$

After 2 s the box will have a speed of 10 m/s down the ramp.

PROBLEM: On Jupiter a stone falls 200 ft in 2.17 s when released from rest. Determine Jupiter's gravitational acceleration, g_J. If a person weighs 120 lb on earth, how much would she weigh on Jupiter?

Solution: The stone falls under Jupiter's gravitational acceleration in the same manner as on earth:

$$y - y_0 = v_0 t + \tfrac{1}{2}at^2, \quad \text{where } y - y_0 = 200 \text{ ft and } t = 2.17 \text{ s}$$

$$a = \frac{2(y - y_0)}{t^2} = \frac{2(200 \text{ ft})}{(2.17 \text{ s})^2} = \underline{84.9 \text{ ft/sec}^2}$$

g_J is 84.9 ft/s². This is $\frac{84.9}{32.2} = 2.64$ times earth's gravitational acceleration of 32.2 ft/s². And since $w = mg$, the weight increases in direct proportion to g. The 120-lb person would weigh $120 \times 2.64 = \underline{317 \text{ lb on Jupiter!}}$

PROBLEM: Andy and Gwendolyn, two mountain climbers on a glacier, are connected by a rope. Gwendolyn falls into a crevasse. Andy is pulled along the slippery glacier (no friction) whose angle of incline is 37°. Gwendolyn, whose mass is 40 kg, is observed to fall (from rest) 2 m in 2 s. Determine the rope's tension and Andy's mass (Fig. 4.3a). The rope has negligible weight.

Solution: Gwendolyn's acceleration can be obtained from kinematics. Choosing the x axis to be the direction of motion,

$$x - x_0 = v_0 t + \tfrac{1}{2}at^2$$

$$a = \frac{2(x - x_0)}{t^2} = \frac{2(2 \text{ m})}{(2 \text{ s})^2} = 1.0 \text{ m/s}^2$$

Her acceleration is not g; the upward tension of the rope prevents her from falling freely. Next we isolate Gwendolyn, draw in the forces acting on her (b) and write Newton's second law for her, taking the direction of a (down) as positive.

$$\Sigma \vec{F} = m\vec{a}$$

$$mg - T = ma$$

$$T = mg - ma = m(g - a) = (40)(9.8 - 1.0)$$

$$T = \underline{352 \text{ N}}$$

The tension in the rope is 352 N.

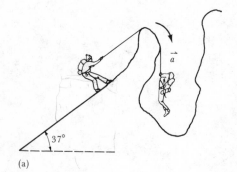

(a)

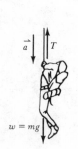

(b) Gwendolyn

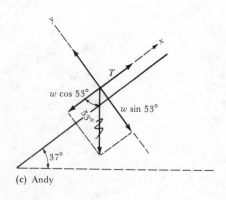

(c) Andy

FIG. 4.3

We now isolate Andy (c), draw all the forces on him, and write Newton's second law for him. Here again we choose the direction of the acceleration as positive to be consistent with our previous assignment of direction for a. For Andy,

$$\Sigma \vec{F} = M\vec{a}$$

$$T - w \cos 53° = Ma$$

$$T - Mg \cos 53° = Ma$$

Here we have used the relationship $w = Mg$. Putting in $a = 1$ m/s² and $T = 352$ N from above, we can solve for Andy's mass M.

$$M(a + g \cos 53°) = T$$

$$M = \frac{T}{a + g \cos 53°} = \frac{352}{1.0 + 9.8(0.6)} = \underline{51 \text{ kg}}$$

Andy's mass is 51 kg.

Notice the procedure carefully. Each body was separately isolated, the forces acting on it were drawn in, and Newton's second law for that body only was written. The equations for each body are then solved simultaneously for the unknowns.

In such problems we always assume there is one and only one tension T in a rope. And if the rope does not stretch, break, or buckle, bodies connected by the rope move as a system with the same acceleration; $a_{\text{Gwen}} = a_{\text{Andy}} = a$. As for the direction of $\vec{a}$, it changed in the problem—it went around a corner in passing over the glacier. This is one chief function of ropes (and also pulleys)—to change the direction of forces (tensions) and accelerations without affecting their magnitudes. If we assign the $+x$ axis to the direction of $\vec{a}$, no matter where it goes, then directions for forces, velocities, etc. can also be assigned accordingly: positive if they are in the acceleration direction, negative if opposite to it. Of course, the y axis is always perpendicular to the x axis and moves around corners right along with it.

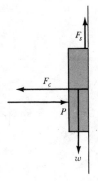

FIG. 4.4

PROBLEM: A man has to push with a horizontal force P of 15 lb to barely hold a 6-lb book against a wall without slipping. What is the coefficient of static friction between the book and the wall?

Solution: The forces on the book are shown in Fig. 4.4. They are the push $\vec{P}$, the contact force of the wall $\vec{F}_c$, the book's weight $\vec{w}$, and the static friction $\vec{F}_s$(max), parallel to the surfaces in contact and opposing the motion of falling. No acceleration occurs, so $\Sigma\vec{F} = 0$.

$$\Sigma\vec{F}_x = 0: \quad P - F_c = 0$$

$$F_c = P$$

$$\Sigma\vec{F}_y = 0: \quad F_s - w = 0$$

$$F_s = w$$

μ_s is the ratio of F_s(max) to F_c, so

$$\mu_s = \frac{F_s(\text{max})}{F_c} = \frac{w}{P} = \frac{6 \text{ lb}}{15 \text{ lb}} = \underline{0.4}$$

The coefficient of static friction is 0.4. Here we see (hopefully, once and for all) that the contact force on an object does not necessarily equal the object's weight. In fact, here the two forces are not even acting on the same axis; they are entirely independent of each other. The force $\vec{P}$ could push much harder, requiring $\vec{F}_c$ to increase, but the book's weight would be unchanged.

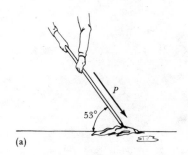

(a)

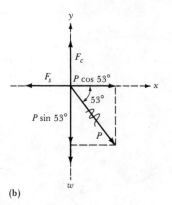

(b)

FIG. 4.5

PROBLEM: Brad is mopping the floor (Fig. 4.5a). The mop weighs 16 N. If he pushes at an angle of 53° with the floor, what force P must he exert along the mop handle for the mop to move with constant velocity? The coefficient of kinetic friction between the mop and the floor is 0.25.

Solution: Four forces act on the mop: the push $\vec{P}$, the contact force $\vec{F}_c$ of the floor, the mop's weight $\vec{w}$, and the kinetic friction $\vec{F}_k$. These are shown in (b) with x-y axes superimposed and the vector $\vec{P}$ placed so that its tail is at the origin. $\vec{P}$ is then resolved into components.

The mop moves at constant velocity, so the acceleration is zero, and the equilibrium equations apply.

$$\Sigma \vec{F}_x = 0: \quad P \cos 53° - F_k = 0$$

And; since $F_k = \mu_k F_c$,

$$P \cos 53° = \mu_k F_c \tag{1}$$

An expression for F_c can be obtained from the y equation:

$$\Sigma \vec{F}_y = 0: \quad F_c - w - P \sin 53° = 0$$

$$F_c = w + P \sin 53° \tag{2}$$

Putting (2) into (1) gives an equation which can be solved for P:

$$P \cos 53° = \mu_k(w + P \sin 53°)$$

$$P(\cos 53° - \mu_k \sin 53°) = \mu_k w$$

$$P = \frac{\mu_k w}{\cos 53° - \mu_k \sin 53°} = \frac{(0.25)(16\text{N})}{(0.6) - (0.25)(0.8)} = \frac{4\text{N}}{0.4} = \underline{10 \text{ N}}$$

For the mop to move at constant velocity, the force required is 10 N.

PROBLEM: A boy coasting on a sled has a speed of 15 ft/s when he encounters a 30° hill. The boy and sled have a total weight of 60 lb and the coefficient of kinetic friction between sled and hill is 0.1 (Fig. 4.6). What is the boy's acceleration up the hill and how far up does he go?

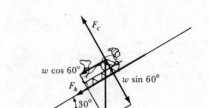

FIG. 4.6

Solution: There are three forces acting on the sled-and-boy: their weight $\vec{w}$, the contact force of the hill $\vec{F}_c$, and the kinetic friction $\vec{F}_k$, parallel to the hill and opposing the motion. The weight $\vec{w}$ is resolved into components parallel to and perpendicular to the hill. Taking the $+x$ axis *up* the plane, there is no acceleration in the y direction:

$$\Sigma \vec{F}_y = 0: \quad F_c - w \sin 60° = 0$$

$$F_c = w \sin 60° = (60)(0.866) = 52 \text{ lb} \tag{1}$$

In the x direction, there are two unbalanced forces down the hill:

$$\Sigma \vec{F}_x = ma$$

$$-F_k - w \cos 60° = ma$$

$$-\mu_k F_c - w \cos 60° = ma$$

The mass m is given by $m = \dfrac{w}{g} = \dfrac{60 \text{ lb}}{32 \text{ ft/s}^2} = 1.87$ slugs, and F_c is 52 lb, from Eq. (1).

$$a = \frac{-\mu_k F_c - w \cos 60°}{m} = \frac{-(0.1)(52) - (60)(0.5)}{1.87} = \underline{-19 \text{ ft/s}^2}$$

Here the negative sign means *down* the plane; being opposite in direction to the initial velocity, this is a deceleration of 19 ft/s².

To find how far up the hill the boy goes, we turn to kinematics.

$$v_0 = 15 \text{ ft/s}; \quad v = 0 \text{ (he stops)}; \quad a = -19 \text{ ft/s}^2; \quad x_0 = 0$$

$$\cancel{v^2}^{0} - v_0^2 = 2a(x - \cancel{x_0}^{0})$$

$$x = \frac{-v_0^2}{2a} = \frac{-(15 \text{ ft/s})^2}{2(-19 \text{ ft/s}^2)} = \frac{-225 \text{ ft}^2/\text{s}^2}{-38 \text{ ft/s}^2} = \underline{5.9 \text{ ft}}$$

The sled goes 5.9 ft up the hill.

Avoiding Pitfalls

1. Be sure you can identify a physical source for each force you put on your force diagram. A velocity is not a force, nor does the existence of a velocity mean a force necessarily acts in that direction. There can be no contact force if there are no surfaces in contact.

2. We tend to think a force must be acting if we see an object in motion. This is because we cannot easily get rid of friction; most moving objects stop if left alone (because they are not really "left alone"; frictional forces are acting on them.) So we "know" that forces are required to overcome friction and keep an object moving. *But if there is no friction,* an object in motion *does indeed* keep moving at constant velocity. We want to ask: What keeps it moving? A more appropriate question would be: What is there to stop its motion?

3. The term *equilibrium,* which means the absence of acceleration, applies equally well to two situations which seem quite different physically. The first is an object at rest, the second is an object in motion with constant velocity (no change in either speed or direction). In both cases, $\bar{a} = 0$, so $\Sigma \vec{F} = 0$; there are no net unbalanced forces acting on the object. The words *constant velocity* in a problem are your clue that the equilibrium equations apply.

4. To remember what a *newton* is, think of $F = ma$ as an equation of units. The right side has units of $\text{kg} \cdot \text{m/s}^2$ (for m and a respectively), so a newton must be equal to a $\text{kg} \cdot \text{m/s}^2$.

5. If coordinate axes are chosen such that the acceleration lies entirely on one of the two axes, the forces on the other axis will satisfy an equilibrium equation, $\Sigma \vec{F} = 0$, giving you at least one easy equation to work with.

6. In resolving the weight vector for an object on an inclined plane, recall that the angle between w and the plane is not the plane angle θ, but its *complement* $(90° - \theta)$.

7. The *directions* of frictional forces between two surfaces (1) are always parallel to the surfaces in contact, and (2) always oppose the motion or intended motion of the surfaces relative to one another. (So you must know the direction of motion or intended motion to assign the right direction to frictional forces.)

8. The equation $F_k = \mu_k F_c$ is *not* a vector equation. $\vec{F}_k$ and $\vec{F}_c$ are each vectors, but their directions are not the same; in fact, they are perpendicular to one another. So this equation relates their magnitudes only. The same is true for the equation $F_s \leq \mu_s F_c$.

9. F_c is not in general equal to an object's weight. If you have walked on thin ice, you know that the upward force of the ice does not have to be equal to your downward weight. If it isn't, you accelerate downward! If you have ridden an elevator, you know that the upward force of the elevator floor can be greater than your downward weight, causing you to accelerate upward.

10. The magnitude of the contact force $\vec{F}_c$ *is* equal to the weight of an object if the following very special conditions are satisfied: (1) the object is on a horizontal surface, so that $\vec{F}_c$ and $\vec{w}$ are opposite in direction; (2) $\vec{F}_c$ and $\vec{w}$ are the *only* two forces in that direction; and (3) there is no acceleration in that direction.

11. The magnitude of the contact force $\vec{F}_c$ is usually determined from a Newton's-second-law equation for the direction containing $\vec{F}_c$ and not by setting it equal to an object's weight.

12. When two or more bodies are connected by ropes, we consider this a system in which all bodies and ropes move as a unit and have the same speed and acceleration. (This is equivalent to assuming the ropes do not stretch or buckle.)

13. Ropes and pulleys with negligible weight and friction can be used to change the *direction* of forces and accelerations without changing their *magnitudes.* Thus, tension forces can "go around corners" via ropes and pulleys.

1. A human femur will break if it is compressed with a force of about 20,000 lb. What acceleration would cause the femur of a 160-lb person to break if she landed on one leg? How many *g*'s is this?

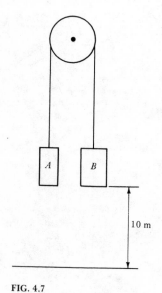

2. Two blocks, *A* (2 kg) and *B* (5 kg) are connected by a string which runs over a light frictionless pulley (Fig. 4.7). *A* and *B* are initially at rest at a height of 10 m above the ground. After the blocks are released, find

(a) the acceleration of the system;
(b) the tension in the string;
(c) the speed of block *B* when it hits the ground.

FIG. 4.7

3. The boy on the sled in the practice problem (page 24) went 5.9 ft up the hill before stopping. If the coefficient of static friction between snow and sled is 0.2, will he start back down again? If so, with what speed will he reach the bottom?

5

Momentum

Terms

Define or describe briefly what is meant by the following terms. If you have difficulty, refer to the textbook section given in parentheses.

momentum (5.1)

impulse of a force (5.2)

conservation of momentum principle (5.4)

Equation Review

For each equation, be able to state the situation to which it applies, what quantity each symbol represents, and the units for measuring each quantity in some consistent set of units.

$$\vec{p} = m\vec{v} \tag{5.1}$$

$$\text{impulse} = Ft \tag{5.2}$$

$$(\Sigma \vec{F})t = m\vec{v} - m\vec{v}_0 \tag{5.3}$$

$$t = \frac{2(x - x_0)}{v + v_0} \tag{3.8}$$

$$m_1\vec{v}_1 + m_2\vec{v}_2 = m_1\vec{v}_{1_0} + m_2\vec{v}_{2_0} \tag{5.6}$$

Problems with Solutions and Discussion

PROBLEM: How fast does a 2000-kg truck have to move in order to have as much momentum as a 200-g bullet traveling at 800 m/s?

Solution: Momentum is given by mv. If the truck and the bullet are to have equal momenta,

$$m_t v_t = m_b v_b$$

where $m_b = 200 \text{ g} \left(\dfrac{1 \text{ kg}}{1000 \text{ g}} \right) = 0.2$ kg, $v_b = 800$ m/s, and $m_t = 2000$ kg

$$v_t = \frac{m_b v_b}{m_t} = \frac{(0.2 \text{ kg})(800 \text{ m/s})}{2000 \text{ kg}} = \underline{0.08 \text{ m/s}}$$

The truck must have a speed of only 8 cm/s. A large momentum can arise from an object's large mass or its large speed or both.

PROBLEM: A sprinter weighing 160 lb runs 200 yd in 25 s. Assuming, for simplicity, that he runs at a constant speed, what is his momentum?

Solution:

$$p = mv$$

where $m = \dfrac{w}{g} = \dfrac{160\ \text{lb}}{32\ \text{ft/s}^2} = 5$ slugs; and $v = \dfrac{200\ \text{yd}}{25\ \text{s}} = 8\ \dfrac{\text{yd}}{\text{s}}\left(\dfrac{3\ \text{ft}}{1\ \text{yd}}\right) = 24$ ft/s.

$$p = (5\ \text{slugs})(24\ \text{ft/s}) = \underline{120\ \text{slug-ft/s}}$$

His momentum is 120 slug-ft/s. It is important to recall that mass, not weight, is needed to calculate momentum. Note also the proper units for momentum in the English system of units.

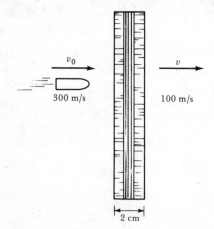

FIG. 5.1

PROBLEM: A 10-g bullet is shot through a 2-cm thick sheet of plywood. It enters with a speed of 300 m/s and comes out with a speed of 100 m/s. What force did the wood exert on the bullet as it passed through (Fig. 5.1)? (Assume this force to be constant.)

Solution: The impulse-momentum equation

$$\vec{F}t = m\vec{v} - m\vec{v}_0$$

can be used to obtain F if we can determine the time during which the momentum change occurred. From Eq. (3.8),

$$t = \frac{2(x - x_0)}{v + v_0} = \frac{2(0.02)}{100 + 300} = 1 \times 10^{-4}\ \text{s}$$

where $(x - x_0) = 2$ cm $= 0.02$ m, and the forward direction of v_0 and v is taken as positive. Now F can be determined from the impulse-momentum equation:

$$F = \frac{m(v - v_0)}{t} \quad \text{where } m = 10\ \text{g}\left(\frac{1\ \text{kg}}{1000\ \text{g}}\right) = 0.01\ \text{kg}$$

$$F = \frac{(0.01)(100 - 300)}{1 \times 10^{-4}} = \frac{-2}{1 \times 10^{-4}} = \underline{-2 \times 10^4\ \text{N}}$$

The wood exerts a force of 20,000 N on the bullet in the opposite direction to its velocity; the bullet decelerates.

By Newton's third law, the bullet also exerts a force of 20,000 N on the wood in the forward direction. Unless the wood is held securely in place by other forces, it will move as a result of the bullet passing through.

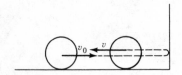

FIG. 5.2

PROBLEM: What is the change in momentum of a 0.2-kg ball when it is bounced against a wall with a speed of 10 m/s and rebounds in the opposite direction with the same speed?

Solution: Let $\Delta\vec{p}$ represent the change in momentum.

$$\Delta\vec{p} = m\vec{v} - m\vec{v}_0$$

If the direction of v_0 is taken as positive, the final v is negative.

$$\Delta p = (0.2\ \text{kg})(-10\ \text{m/s}) - (0.2\ \text{kg})(10\ \text{m/s})$$

$$\Delta p = (0.2\ \text{kg})(-20\ \text{m/s}) = \underline{-4\ \text{kg} \cdot \text{m/s}}$$

The momentum change is 4 kg·m/s, directed to the left.

PROBLEM: What force did the wall exert on the ball in the previous problem, if the collision time was 0.01 s? What force did the ball exert on the wall?

Solution: By the impulse-momentum equation,

$$\vec{F}t = m\vec{v} - m\vec{v}_0 = \Delta\vec{p} = -4 \text{ kg}\cdot\text{m/s}$$

$$\vec{F} = \frac{\Delta\vec{p}}{t} = \frac{-4 \text{ kg}\cdot\text{m/s}}{0.01 \text{ s}} = \underline{-400 \text{ N}}$$

The wall exerts a 400-N force to the left to reverse the direction of the ball. By Newton's third law the reaction force of the ball on the wall must be 400 N to the right.

PROBLEM: Suppose a 0.2-kg ball of putty struck a wall with a speed of 10 m/s and stuck. What would be the change in momentum of the putty and the force exerted by the wall in this case, if the collision time is again 0.01 s?

Solution: For the putty's change in momentum,

$$\Delta\vec{p} = \cancel{m\vec{v}}^{0} - m\vec{v}_0, \quad \text{since the putty is brought to rest.}$$

$$\Delta\vec{p} = -m\vec{v}_0 = -(0.2 \text{ kg})(10 \text{ m/s}) = \underline{-2 \text{ kg}\cdot\text{m/s}}$$

And for the wall's force,

$$\vec{F} = \frac{\Delta\vec{p}}{t} = \frac{-2 \text{ kg}\cdot\text{m/s}}{0.01 \text{ s}} = \underline{-200 \text{ N}}$$

The momentum change is 2 kg·m/s, to the left, caused by a force of 200 N from the wall in the same direction.

Compare the force exerted and the corresponding momentum change in bringing a moving object to rest (putty) and in reversing its velocity (rebounding ball). The force exerted and the change in momentum are twice as great for the ball as for the putty. To slow down and stop an object is easier than to slow down and stop it and then start it up again with the same speed in the opposite direction.

PROBLEM: A 5-kg rifle gives a 20-g bullet an initial speed of 600 m/s. With what velocity does the rifle recoil?

Solution: Since no external forces act on the bullet-and-rifle system, momentum is conserved during the firing (Fig. 5.3).

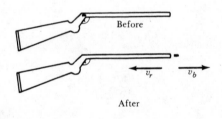

Before

After

FIG. 5.3

$$\text{Initial momentum} = \text{Final momentum}$$
$$0 \qquad = m_r\vec{v}_r + m_b\vec{v}_b$$

where $m_b = 20 \text{ g} = 0.02 \text{ kg}$; $\vec{v}_b = 600 \text{ m/s}$; $m_r = 5 \text{ kg}$; and the direction of the bullet's velocity is taken as positive.

$$\vec{v}_r = \frac{-m_b\vec{v}_b}{m_r} = \frac{-(0.02 \text{ kg})(600 \text{ m/s})}{5 \text{ kg}} = \underline{-2.4 \text{ m/s}}$$

The rifle recoils with a speed of 2.4 m/s in the opposite direction to the bullet's velocity. The rifle has a substantial kick.

PROBLEM: What is the recoil velocity if the rifle above is held firmly against the shoulder of a 60-kg person, so that the person and gun both recoil. Assume that half the person's mass is effective in sharing the recoil momentum.

Solution: Let M = effective recoil mass = $30 + 5 = 35$ kg. Again,

$$\text{Initial momentum} = \text{Final momentum}$$
$$0 \qquad = m_b\vec{v}_b + M\vec{v}_M$$

$$\vec{v}_M = \frac{-m_b\vec{v}_b}{M} = \frac{(0.02 \text{ kg})(600 \text{ m/s})}{35 \text{ kg}} = \underline{-0.34 \text{ m/s}}$$

29

The recoil velocity is 0.34 m/s backwards. When a rifle is braced against a shoulder, its kick is greatly reduced because the recoil momentum is absorbed by a much larger mass, reducing the velocity proportionately.

PROBLEM: A 2000-kg railroad car traveling 4 m/s on a straight track overtakes a car of mass 4000 kg traveling 1 m/s in the same direction. During the collision the cars couple together. What is their speed after the collision?

Solution: Conserving momentum in the direction of motion,

$$p_{before} = p_{after}$$
$$(2000 \text{ kg})(4 \text{ m/s}) + (4000 \text{ kg})(1 \text{ m/s}) = (2000 \text{ kg} + 4000 \text{ kg})v_f$$
$$v_f = \frac{12{,}000 \text{ kg} \cdot \text{m/s}}{6000 \text{ kg}} = 2 \text{ m/s}$$

The cars continue in the forward direction with a speed of 2 m/s.

PROBLEM: As Jane's sled moves across a level field and under a fence at 10 ft/s, her friend Kathy, waiting on the fence, drops down onto the sled. What happens to the sled's motion? Each girl weighs 64 lb, the sled weighs 16 lb. Assume negligible friction between the sled and snow.

Solution: Considering the two girls and the sled as a system, no external forces are exerted in the horizontal direction; call it the x axis. Thus,

$$\text{Momentum before} = \text{Momentum after} \quad (x \text{ direction})$$
$$(m_s + m_J)v_0 + m_K \cancel{v_K} = (m_s + m_J + m_K)v_f$$

where the subscripts s, J, and K represent the sled, Jane, and Kathy, respectively, and v_f is the final speed of all three.

$$m_K = m_J = \frac{w}{g} = \frac{64}{32} = 2 \text{ slugs}$$
$$m_s = \frac{16}{32} = 0.5 \text{ slug}$$
$$v_0 = 10 \text{ ft/s}$$

Putting in these values,

$$(2.5)(10) = (4.5)v_f$$
$$v_f = \frac{25}{4.5} = 5.6 \text{ ft/s}$$

The sled slows down from 10 ft/s to 5.6 ft/s.

In the y direction (vertical), external forces *do* act: the gravitational force causes Kathy to drop downward toward the sled and the earth's contact force stops her downward motion when she reaches the sled and also supports the weight of the entire system. In this direction momentum within the system was not conserved. The external forces caused changes in Kathy's y momentum, according to the impulse-momentum equation. Thus momentum may be conserved along one axis—if $\Sigma \vec{F} = 0$ in that direction—and not along the other axis.

PROBLEM: In an attempt to dislodge a frisbee caught in a tree, a boy shoots a suction arrow straight up at the frisbee. The arrow strikes the frisbee with a speed of 8 m/s and sticks. How high does the frisbee go above its original position? The frisbee's mass is 150 g and the arrow's mass is 100 g.

Solution: Since no net external forces act during the collision time (the downward weights are balanced by the upward contact force of the branches), momentum is conserved.

$$\vec{p}_{before} = \vec{p}_{after}$$
$$m_a\vec{v}_a = (m_a + m_f)\vec{v}_f$$

where m_a = arrow mass = 0.1 kg; m_f = frisbee mass = 0.15 kg; and $\vec{v}_a$ = arrow speed = 8 m/s.

$$v_f = \frac{m_a v_a}{m_a + m_f} = \frac{(0.1)(8)}{0.25} = 3.2 \text{ m/s}$$

The frisbee and arrow leave the collision with a speed of 3.2 m/s, in the same direction as the original direction of the arrow (upward). But as they rise from the branch, the gravitational force does act; we have a projectile motion problem where

$$v_0 = 3.2 \text{ m/s (up)} \quad \text{and} \quad a = -9.8 \text{ m/s}^2 \text{ (down)}$$

To find the maximum height, we set the final v equal to zero. From kinematics,

$$v^2 - v_0^2 = 2a(y - y_0)$$
$$y - y_0 = \frac{-v_0^2}{2a} = \frac{-(3.2 \text{ m/s})^2}{2(-9.8 \text{ m/s}^2)} = \underline{0.52 \text{ m}}$$

The 52-cm rise may or may not be sufficient to free the frisbee.

PROBLEM: A Volkswagen and a Thunderbird approach an intersection as shown in Fig. 5.4a. Their masses are 800 kg and 1200 kg respectively. After a collision the bumpers are locked together and the two cars are found some distance from the intersection at the angle shown. The T-bird's speed before the collision was 30 m/s. What was the VW's speed before the collision?

Solution: No external forces act on the system of VW-and-T-bird in the horizontal plane, so momentum is conserved. (This assumes that during the collision the frictional forces between the cars and the road are negligible compared to the forces the cars exert on each other, usually a good assumption.) The "before" and "after" sketches show the choice of x-y axes and the symbols assigned to the various masses and velocities. For two-dimensional problems we conserve momentum on the x and y axes separately. p_f is resolved into its x and y components. On each axis,

$$\vec{p}_{before} = \vec{p}_{after}$$

x axis: $p_1 = p_f \cos 40°$

$$m_1 v_1 = (m_1 + m_2)v_f \cos 40°$$
$$(1200)(30) = (1200 + 800)v_f(0.766)$$
$$v_f = \frac{36,000}{(2000)(0.766)} = 23.5 \text{ m/s}$$

y axis: $p_2 = p_f \sin 40°$

$$m_2 v_2 = (m_1 + m_2)v_f(0.643)$$

Substituting v_f from above and solving for v_2, we get:

$$v_2 = \frac{(2000)(23.5)(0.643)}{800} = \underline{38 \text{ m/s}}.$$

The VW's speed before the collision was 38 m/s.

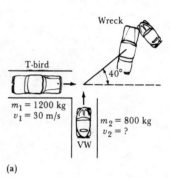

(a)

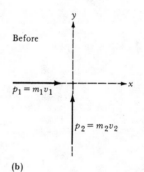

(b)

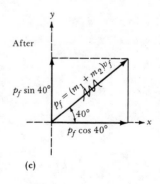

(c)

FIG. 5.4

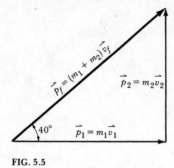

FIG. 5.5

Alternate Solution: Since momentum is a vector quantity, we may use graphical vector techniques to solve momentum problems. In this case, momentum conservation gives the vector relation:

$$\vec{p}_1 + \vec{p}_2 = \vec{p}_f$$

Thus if $\vec{p}_1$ and $\vec{p}_2$ are arranged head-to-tail, $\vec{p}_f$ is their resultant. This is shown in Fig. 5.5. A right triangle results, so

$$\tan 40° = \frac{p_2}{p_1} = \frac{m_2 v_2}{m_1 v_1}$$

Solving for v_2,

$$v_2 = \frac{m_1 v_1 \tan 40°}{m_2} = \frac{(1200)(30)(0.84)}{800} = \underline{38 \text{ m/s}}$$

The VW's speed before the collision was 38 m/s. Once v_2 is known, v_f can be determined by the Pythagorean theorem, although it was not asked for in this problem.

Avoiding Pitfalls

1. Momentum is conserved within a system only when no net unbalanced force acts on that system. If an unbalanced force acts, momentum *does* change, and the impulse-momentum equation applies.

2. The impulse-momentum equation is a vector equation. In applying it, be sure that v, v_0, and F are assigned algebraic signs consistent with their directions, if known. These signs are in addition to, and independent of, the minus sign appearing in the equation. In solving for a vector quantity, the sign of the solution indicates the direction of the vector.

3. Stating that $\Sigma \vec{F} = 0$, the condition for momentum conservation, may seem odd for a collision in which it is obvious that objects are exerting great forces on each other. The trick is to include all colliding objects in the system. Then the force of one object on a second is exactly balanced by the Newton's-third-law reaction force of the second object on the first, making the vector sum of the forces zero *for the system*.

4. In conserving momentum during collisions, the initial and final momenta refer to the instant before and the instant after the collision. Other forces such as friction may act on the bodies before and after the collision, but momentum conservation applies only when the sole unbalanced forces acting on a body are from other bodies within the system.

5. Momentum is a vector quantity. Its direction can be accounted for with positive and negative signs *only* in a one-dimensional problem.

6. Two-dimensional problems are most easily solved by graphical methods or by resolving all vectors into x and y components and conserving momentum separately along each axis.

7. Momentum may be conserved in one direction and not in another. If there is a direction in which no external unbalanced forces act on the system, momentum will be conserved in that direction.

8. In momentum conservation equations, each term appearing on both sides of the equation is a momentum term. Therefore the units of the two sides are not just equal, they are *identical*. In such cases, one may use any convenient velocity units, as long as they are the same on both sides. One may even put in weights in the mass positions and still get correct results, because $w = m/g$ and the g's will cancel out of each term:

$$m_1 v_1 = m_2 v_2$$

$$\frac{w_1}{g} v_1 = \frac{w_2}{g} v_2$$

$$w_1 v_1 = w_2 v_2$$

Drill Problems
Answers in Appendix

1. A softball of mass 0.15 kg and a speed of 30 m/s is hit by a bat, giving it a speed of 40 m/s in the opposite direction. If the time duration of the impact is 3×10^{-3} s, what average force did the bat exert on the ball?

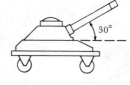

2. A cannon pointed at an angle of 30° with the horizontal is mounted on a car which runs on a straight frictionless track (Fig. 5.6). The cannon assembly has a mass of 20,000 kg. A cannonball of mass 100 kg is fired with a muzzle velocity of 200 m/s. With what velocity does the cannon recoil?

FIG. 5.6

3. A bowling ball of mass 6.0 kg and a speed of 3.0 m/s knocks a single pin straight ahead of it with a speed of 6.0 m/s. The pin's mass is 2.0 kg. What is the bowling ball's velocity after the collision?

Circular Motion at Constant Speed

Terms

Define or describe briefly what is meant by the following terms. If you have difficulty, refer to the textbook section given in parentheses.

arc-length coordinate (6.1)

angular-position coordinate (6.1)

radian (6.2)

tangential speed (6.3)

angular velocity (6.3)

centripetal acceleration (6.4)

centripetal force (6.5)

Newton's universal law of gravitation (6.6)

weightlessness (6.7)

Equation Review

For each equation, be able to state the situation to which it applies, what quantity each symbol represents, and the units for measuring each quantity in some consistent set of units.

$$\theta \text{ (in radians)} = \frac{s}{r} \tag{6.1}$$

$$360° = 1 \text{ rev} = 2\pi \text{ rad} \tag{6.2}$$

$$v = \frac{\Delta s}{\Delta t} \tag{6.3}$$

$$\omega = \frac{\Delta\theta}{\Delta t} \tag{6.4}$$

$$v = r\omega \tag{6.5}$$

$$a_c = \frac{v^2}{r} = r\omega^2 \tag{6.7}$$

$$a_c = r\omega^2 \qquad\qquad\qquad (6.8)$$

$$\Sigma F \text{ (in radial direction)} = \frac{mv^2}{r} = mr\omega^2$$

$$F = \frac{Gm_1m_2}{r^2} \qquad\qquad\qquad (6.9)$$

$$g = \frac{Gm_e}{r_e^2} \qquad\qquad\qquad (6.10)$$

Problems with Solutions and Discussion

PROBLEM: Consider a clock whose minute hand is 20 cm long. Through what angle, in degrees and in radians, does the minute hand travel during the time period from 7:30 A.M. to 11:15 A.M.? What corresponding linear distance is traveled by the tip of the minute hand?

Solution: From 7:30 to 11:15, the minute hand makes 3.75 revolutions.

$$\text{In degrees, } \theta = 3.75 \text{ rev} \left(\frac{360°}{1 \text{ rev}} \right) = \underline{1350°}$$

$$\text{In radians, } \theta = 3.75 \text{ rev} \left(\frac{2\pi \text{ rad}}{1 \text{ rev}} \right) = \underline{23.6 \text{ rad}}$$

The hand moved through an angle of 1350°, or 23.6 rad. (It would be correct to include a negative sign, if we wished to indicate that the angle was swept out in the clockwise direction.)

The corresponding linear distance traveled by the tip is along an arc,

$$s = r\theta$$

$$\text{where } r = 20 \text{ cm} \left(\frac{1 \text{ m}}{100 \text{ cm}} \right) = 0.2 \text{ m, and } \theta = 23.6 \text{ rad}$$

$$s = (0.2 \text{ m})(23.6) = \underline{4.72 \text{ m}}$$

The tip traveled a linear distance of 4.72 m. Note that θ must be expressed in radian measure (unitless) to obtain the proper distance unit of m.

PROBLEM: If a motor shaft is rotating 1800 revolutions per minute, what is its angular velocity in radians per second?

Solution:

$$\omega = 1800 \frac{\text{rev}}{\text{min}} \left(\frac{1 \text{ min}}{60 \text{ s}} \right) \left(\frac{2\pi \text{ rad}}{1 \text{ rev}} \right) = \underline{190 \text{ rad/s}}$$

Here we have converted an angular velocity in rev/min to rad/s.

PROBLEM: A highway is to be made so that a car can safely go around a certain curve at a speed of 30 m/s with its centripetal acceleration about 5 m/s². What should the radius of the curve be?

Solution: Equation (6.7) relates centripetal acceleration to speed and radius:

$$a_c = \frac{v^2}{r} \quad \text{so} \quad r = \frac{v^2}{a_c} = \frac{(30 \text{ m/s})^2}{5 \text{ m/s}^2} = \underline{180 \text{ m}}$$

The curve should have a radius of 180 m.

PROBLEM: A flea rides on a 33-rpm phonograph record at a distance of 6 inches from the center. What is the flea's centripetal acceleration?

Solution: Here we choose Eq. (6.8) instead of (6.7) to express centripetal acceleration because the information given lends itself more easily to using angular quantities like ω than linear quantities like v.

$$a_c = r\omega^2$$

$$\text{where } r = 6 \text{ in} \left(\frac{1 \text{ ft}}{12 \text{ in}} \right) = 0.5 \text{ ft}$$

$$\omega = 33 \frac{\text{rev}}{\text{min}} \left(\frac{1 \text{ min}}{60 \text{ s}} \right) \left(\frac{2\pi \text{ rad}}{1 \text{ rev}} \right) = 3.46 \text{ rad/s}$$

$$a_c = (0.5 \text{ ft})(3.46 \text{ rad/s})^2 = \underline{6 \text{ ft/s}^2}$$

The flea's centripetal acceleration is 6 ft/s².

FIG. 6.1

PROBLEM: A child weighing 64 lb is sitting on a swing 16 ft. long. The swing passes the lowest point of its path with a speed of 16 ft/s. (a) Show in a diagram the forces on the child. (b) How large is the force C with which the swing supports the child at the lowest point of the path?

Solution: The required diagram is shown in Fig. 6.1. The two forces on the child are her weight $\vec{w}$ and the upward contact force of the swing seat $\vec{C}$. To get C, we start with the centripetal force equation

$$\Sigma F = \frac{mv^2}{r}$$

since the path of the swinging child is the arc of a circle of radius $r = 16$ ft, the length of the swing. F, the centripetal force, represents the vector sum of all forces lying along the radius of the circle. Taking the direction of the centripetal acceleration (toward the circle's center) as positive, $\vec{C}$ is positive and $\vec{w}$ is negative.

$$C - w = \frac{mv^2}{r} \quad \text{where } m = \frac{w}{g} = \frac{64}{32} = 2 \text{ slugs}$$

Solving for C,

$$C = w + \frac{mv^2}{r} = 64 + \frac{(2)(16)^2}{16} = \underline{96 \text{ lb}}$$

The swing supplies an upward force of 96 lb which not only supports the child's weight, but also supplies the necessary centripetal force to cause the circular motion.

Discussion: A child's speed does not remain constant on a swing, nor does the child move around a complete circle. The centripetal force equation applies nevertheless. It describes the unbalanced force necessary to maintain any speed v along a path with any radius of curvature r. If v and r are unchanged, the motion is circular. If v and r are constantly changing, the path is not circular and the size of the centripetal force must be changing accordingly. But its direction is always along the instantaneous radius of curvature, perpendicular to the object's instantaneous velocity.

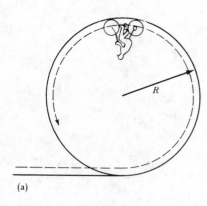

(a)

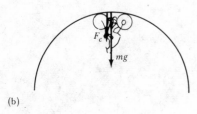

(b)

FIG. 6.2

PROBLEM: A person on a motorcycle performs a stunt on a vertical circular track (Fig. 6.2). Neglect the height of the motorcycle and driver compared to the radius of the track.

If the track radius R is 40 ft, what must the speed be at the top of the path so that the force from the track F_c is one-half the motorcycle and rider's weight?

Solution: The force diagram is shown in (b). Two forces, the weight and the track's contact force, act on the motorcycle. The sum of the forces along the radius supplies the centripetal force.

$$\Sigma F = \frac{mv^2}{R}$$

$$mg + F_c = \frac{mv^2}{R}$$

where the direction of the centripetal acceleration is taken as positive. We are given that $F_c = \frac{1}{2}w = \frac{1}{2}mg$.

$$\cancel{mg} + \frac{1}{2}\cancel{mg} = \frac{\cancel{m}v^2}{R}$$

$$v^2 = \frac{3}{2}gR = \frac{3}{2}(32 \text{ ft/s}^2)(40 \text{ ft}) = 1920 \text{ ft}^2/\text{s}^2$$

$$v = \underline{44 \text{ ft/s}}$$

The motorcycle's speed is 44 ft/s or about 30 mi/hr.

PROBLEM: What is the minimum speed the motorcycle in the previous problem could have without losing contact with the track at the top?

Solution: The equation of motion for the motorcycle in the topmost position is as before:

$$\cancel{m}g + F_c = \frac{\cancel{m}v^2}{R}$$

except that if contact is lost, $F_c = 0$. So

$$mg = \frac{mv^2}{R}$$

$$v^2 = Rg = (40 \text{ ft})(32 \text{ ft/s}^2) = 1280 \text{ ft}^2/\text{s}^2$$

$$v = \underline{35.8 \text{ ft/s}}$$

The minimum speed to maintain contact is about 36 ft/s or about 25 mi/hr.

PROBLEM: A 1-kg ball is attached to one end of a 50-cm string which will break if the tension is greater than 30 N. It is set in motion in a horizontal circle as shown in Fig. 6.3a. (This arrangement is sometimes called a *conical pendulum*.) (a) What is the maximum angle the string can make with the vertical without breaking? (b) What is the speed of the ball at this angle?

Solution: (a) Since this is a two-dimensional problem, we will consider the x and y axes separately. The force diagram for the ball is shown in (b). The two forces acting are the weight mg and the string tension T. The angle labeled ψ is the complement of θ. T is resolved into its x and y components. In the y direction, there is no motion up or down, so $\bar{a}_y = 0$.

$$\Sigma \bar{F}_y = 0: \quad T \sin \psi = mg$$

Since T can be no greater than 30 N,

$$(30) \sin \psi = (1)(9.8)$$

$$\sin \psi = \frac{9.8}{30} = 0.327$$

$$\text{so} \quad \psi = 19°, \quad \text{and} \quad \theta = 90 - \psi = \underline{71°}$$

The angle θ can get no larger than 71° or the tension required will exceed 30 N.

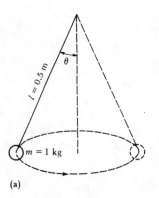

(a)

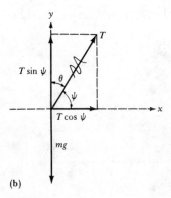

(b)

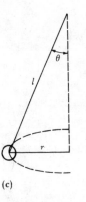

(c)

FIG. 6.3

37

(b) In the x direction, there must be a net unbalanced force pointing toward the circle's center to cause the circular motion. There is only one force in the x direction; it must be the centripetal force.

$$F_x = ma_c: \quad T\cos\psi = \frac{mv^2}{r}$$

Solving for v,

$$v^2 = \frac{T\cos\psi\, r}{m}$$

The radius of the circle is needed. It is *not* the length of the string, but it is related to the string's length l by a right triangle as shown in (c).

$$\sin\theta = \frac{r}{l} \quad \text{so} \quad r = l\sin\theta$$

Putting this in,

$$v^2 = \frac{(T\cos\psi)(l\sin\theta)}{m}$$

$$= \frac{(30\ \text{N})(\cos 19°)(0.5\ \text{m})(\sin 71°)}{1\ \text{kg}} = 13.4\ \text{m}^2/\text{s}^2$$

$$v = \underline{3.66\ \text{m/s}}$$

The speed of the ball is 3.66 m/s.

Discussion: The preceding problem may have looked difficult at first. Sometimes a person "just doesn't know where to start." The following technique almost always works, so why not memorize it? You can use it when you have "no idea what to do" and fool everyone.

1. Isolate the object of interest and draw in all forces acting on it. The centripetal force is not one of those forces; it is just a name for certain identifiable physical forces (or their components), such as weight, tension, friction, gravitational force, etc., which happen to point toward the center of a circular motion.

2. Choose an x-y axis and resolve all forces. If there is circular motion, one axis should lie along the radius of the circle, thereby confining the centripetal force and acceleration to that axis.

3. Write $\Sigma\vec{F} = m\vec{a}$, or $\Sigma\vec{F} = 0$ as appropriate for each axis and solve simultaneously for any unknowns.

By the time you've completed these steps, you've probably solved the problem. If not, you can almost invariably see what comes next—possibly a kinematics equation. Follow the application of this technique in the next problem.

PROBLEM: A very practical application of centripetal force ideas is the banking of curves on highways. At what angle should a curve of radius 1000 ft be banked, so that a car can go around it at 40 mi/hr without relying on friction to supply centripetal force (Fig. 6.4a)?

Solution: A cross section of the banked curve is shown in Fig. 6.4a and forces on the car are drawn in (b). They are the weight $\vec{w}$ and the contact force $\vec{F}_c$ of the road. The x axis is chosen to lie along the radius of the circle made by the car in rounding the curve. $\vec{F}_c$ is resolved into components. Notice that ψ is the complement of the banking angle θ.

On the y axis, we have equilibrium:

$$\Sigma\vec{F}_y = 0: \quad F_c\sin\psi - mg = 0$$

$$F_c\sin\psi = mg \tag{1}$$

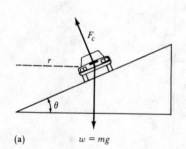

(a) $w = mg$

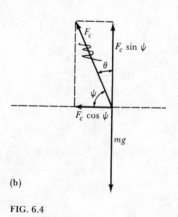

(b)

FIG. 6.4

38

On the x axis, $F_c \cos \psi$ is an unbalanced force; it must be the centripetal force.

$$F_x = ma_c: \quad F_c \cos \psi = \frac{mv^2}{r} \tag{2}$$

Solving (1) for F_c and putting it into (2):

$$F_c = \frac{mg}{\sin \psi}$$

$$\left(\frac{mg}{\sin \psi} \right) \cos \psi = \frac{mv^2}{r}$$

$$mg \cot \psi = \frac{mv^2}{r}$$

$$\cot \psi = \frac{v^2}{gr}$$

This equation can now be solved for ψ, and then the banking angle θ is $90 - \psi$. Or the equation can be expressed in terms of the banking angle θ by recalling that

$$\cot \psi = \tan(90 - \psi) = \tan \theta$$

Thus,

$$\tan \theta = \frac{v^2}{gr}$$

For $v = 40 \text{ mi/hr} \left(\frac{1.47 \text{ ft/s}}{1 \text{ mi/hr}} \right) = 59 \text{ ft/s}$, and $r = 1000 \text{ ft}$

$$\tan \theta = \frac{(59)^2}{(32)(1000)} = 0.109, \text{ and } \theta = 6.2°$$

The road should be banked at an angle of 6.2°. Interestingly enough, the result is independent of the mass of the vehicle. This is why the posted speed for a certain banked curve applies equally well to a bicycle or a freight truck.

PROBLEM: Planet X has 10 times the mass and 4 times the radius of earth. If you went there, with no change in mass, your weight would be how many times as great as it is now?

Solution: Since $w = mg$, weight increases in direct proportion to g. Let subscripts e and x represent Earth and Planet X respectively. Then

$$g_e = \frac{Gm_e}{r_e^2} \quad \text{and} \quad g_x = \frac{Gm_x}{r_x^2}$$

Now $m_x = 10m_e$ and $r_x = 4r_e$, so

$$g_x = \frac{G(10m_e)}{(4r_e)^2} = \frac{10}{16} \frac{Gm_e}{r_e^2} = \frac{10}{16} g_e = \frac{5}{8} g_e$$

Since g_x is ⅝g_e, on Planet X you would weigh ⅝ of your weight on earth. At first glance it might seem that g_x must surely be greater than g_e because X's mass was larger than earth's by a factor of 10 while its radius was only larger by a factor of 4. But g depends on the *square* of the radius and on only the first power of the mass.

PROBLEM: The planet Mercury has a radius of about 2400 km and a mass of about 3.3×10^{23} kg. Work out a formula for the speed of a satellite in a circular orbit at a distance of 3 Mercury radii from the center of Mercury. Determine the numerical value of this speed.

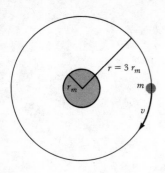

FIG. 6.5

Solution: The satellite is in a circular orbit, so it experiences a centripetal force; that force is supplied by the gravitational attraction between planet and satellite:

$$F_G = \frac{mv^2}{r}$$

where m is the satellite's mass, v its speed, and r the radius of its orbit. Using Newton's law of universal gravitation to express the gravitational force,

$$\frac{Gm_M m}{r^2} = \frac{mv^2}{r}, \quad \text{giving } v^2 = \frac{Gm_M}{r}$$

Now $r = 3r_M$, so

$$v = \sqrt{\frac{Gm_M}{3r_M}}$$

is the required formula. Putting in numerical values we get:

$$r = 2400 \text{ km} \left(\frac{1000 \text{ m}}{1 \text{ km}}\right) = 2.4 \times 10^6 \text{ m}$$

$$v = \sqrt{\frac{(6.67 \times 10^{-11} \text{ N}\cdot\text{m}^2/\text{kg}^2)(3.3 \times 10^{23} \text{ kg})}{3(2.4 \times 10^6 \text{ m})}} = \sqrt{3.06 \times 10^6} = \underline{1750 \text{ m/s}}$$

The satellite's orbital speed is 1750 m/s.

Avoiding Pitfalls

1. A change in velocity can be a change in speed, or a change in direction of an object's motion, or both. All are accelerations and all require the action of an unbalanced force.

2. Forces acting in the direction of an object's motion speed it up. Forces acting opposite to the direction of motion slow it down. Forces acting perpendicular to the direction of motion do not alter the speed, but change the *direction* of the motion.

3. If an object is moving at constant speed in a circle, there must be an unbalanced force acting on it which is constant in magnitude but whose direction keeps changing so that it always points toward the center of the circle.

4. In circular motion problems, choose the x-y axes so that one axis lies along the radius of the circle. This confines the centripetal force to one axis. On that axis, choose the positive direction to point toward the circle's center. The equation $\Sigma F = mv^2/r$ (or $= mr\omega^2$) applies on that axis; on the other axis, $\Sigma F = ma$ or $\Sigma F = 0$.

5. When drawing force diagrams for circular motion, do not draw in mv^2/r as an *additional* applied force. The centripetal force does not appear as such in the force diagram. It is already present as a component or the sum of several components of the physically identifiable forces such as tension, weight, or friction.

6. Any object moving in a circle requires a force pointing *inward* toward the circle's center. It is very tempting to want to say there is an equal and opposite force pushing the object *outward* so that the object is in equilibrium, going around the circle at constant speed. Not so! If that force existed and the object were indeed in equilibrium, it would have no net forces acting and thereby, from Newton's first law, would continue in a straight line at constant speed. The fact that it goes on a curved line means it is not in equilibrium; there must be an unbalanced force acting.

7. Don't worry if you don't immediately see a force pointing toward the center of a circular motion. The centripetal force may be a component of another force which, when resolved, will point in the proper direction.

8. Look carefully at the circular motion in determining its radius. It is not necessarily equal to any of the lengths given in the problem, and it may have to be determined geometrically, as in the case of the conical pendulum.

Drill Problems
Answers in Appendix

1. The second hand of a clock is 10 cm long. For the tip of the second hand, find the magnitudes of

  (a) the linear velocity;
  (b) the angular velocity; and
  (c) the centripetal acceleration.

2. How much centripetal force acts on a 60-kg person at latitude 40° due to the earth's rotation? The earth's radius is 6.38×10^6 m.

3. A spaceship travels around the moon in a circular orbit of radius R, requiring a time T for one complete revolution. Make use of what you know about centripetal force to work out an expression for the mass of the moon, in terms of R and T.

7 Rotational Motion with Angular Acceleration

Define or describe briefly what is meant by the following terms. If you have difficulty, refer to the textbook section given in parentheses.

angular acceleration (7.1)

tangential acceleration (7.2)

moment of inertia (7.5)

angular momentum (7.7)

conservation of angular momentum principle (7.7)

Equation Review

For each equation, be able to state the situation to which it applies, what quantity each symbol represents, and the units for measuring each quantity in some consistent set of units.

$$\alpha = \frac{\Delta\omega}{\Delta t} \tag{7.1}$$

$$\bar{\alpha} = \frac{\omega - \omega_0}{t} \tag{7.2}$$

$$a_t = \frac{\Delta v}{\Delta t} \tag{7.3}$$

$$a_t = r\alpha \tag{7.6}$$

$$v = r\omega \tag{6.5}$$

$$s = r\theta$$

$$\omega = \omega_0 + \alpha t \tag{7.7}$$

$$\theta - \theta_0 = \left(\frac{\omega + \omega_0}{2}\right) t \tag{7.8}$$

$$\theta - \theta_0 = \omega_0 t + \tfrac{1}{2}\alpha t^2 \tag{7.9}$$

42

$$2\alpha(\theta - \theta_0) = \omega^2 - \omega_0^2 \tag{7.10}$$

$$\tau = \pm Fl \tag{2.4}$$

$$\Sigma\tau = I\alpha \tag{7.12}$$

$$I = m_1 r_1^2 + m_2 r_2^2 + m_3 r_3^2 + \ldots \tag{7.13}$$

$$L = I\omega \tag{7.14}$$

$$I\omega = I_0\omega_0 \tag{7.16}$$

$$L = mrv \tag{7.18}$$

Problems with Solutions and Discussion

PROBLEM: An electric fan takes 3 s to come up to its operating speed of 1800 rpm when turned on. What is its angular acceleration (assumed constant) during that period, and how many revolutions does it make?

Solution: Angular acceleration is defined by Eq. (7.2):

$$\alpha = \frac{\omega - \omega_0}{t}, \text{ where } \omega_0 = 0$$

$$\omega = 1800 \frac{\text{rev}}{\text{min}} \left(\frac{2\pi \text{ rad}}{1 \text{ rev}}\right)\left(\frac{1 \text{ min}}{60 \text{ s}}\right) = 60\pi \frac{\text{rad}}{\text{s}} = 188 \frac{\text{rad}}{\text{s}}$$

$$\alpha = \frac{60\pi - 0}{3 \text{ s}} = 20\pi \frac{\text{rad}}{\text{s}^2} = \underline{63 \frac{\text{rad}}{\text{s}^2}}$$

The angular acceleration is 63 rad/s². The number of revolutions is a measure of angular displacement, $\theta - \theta_0$.

$$\theta - \theta_0 = \cancel{\omega_0 t}^{\,0} + \frac{1}{2}\alpha t^2 = \frac{1}{2}\left(20\pi \frac{\text{rad}}{\text{s}^2}\right)(3 \text{ s})^2 = 90\pi \text{ rad}$$

Converting to revolutions,

$$90\cancel{\pi} \text{ rad}\left(\frac{1 \text{ rev}}{2\cancel{\pi} \text{ rad}}\right) = \underline{45 \text{ rev}}$$

The fan makes 45 revolutions in coming up to speed. If a final radians-to-revolutions conversion will occur, the arithmetic is often simplified by leaving answers with π in them.

PROBLEM: An automobile traveling at 30 m/s has tires 30 cm in radius. (a) What is the angular velocity of the tires about the axle? (b) If the tires are brought to a stop uniformly in 25 turns, what is the angular acceleration? (c) How far does the car travel during this braking period?

Solution: (a) The tires' angular velocity is given by

$$\omega = \frac{v}{r} = \frac{30 \text{ m/s}}{0.3 \text{ m}} = \underline{100 \text{ rad/s}}$$

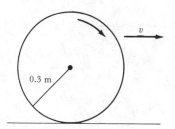

FIG. 7.1

Here the radian unit is added to remind us what angular measure the result is expressed in. (See the discussion following the problem if you wonder why we can put in the forward speed v of the *car*, 30 m/s, for the tangential speed of the *edge of the tire* rotating around the axle.)

$$\text{(b) } \theta - \theta_0 = 25 \text{ turns} = 25 \text{ rev}\left(\frac{2\pi \text{ rad}}{1 \text{ rev}}\right) = 157 \text{ rad}$$

From rotational kinematics Eq. (7.10),

43

$$2\alpha(\theta - \theta_0) = \omega^2 - \omega_0^2$$

$$\alpha = \frac{\omega^2 - \omega_0^2}{2(\theta - \theta_0)} = \frac{-(100 \text{ rad/s})^2}{2(157 \text{ rad})} = \underline{-32 \text{ rad/s}^2}$$

The angular deceleration is 32 rad/s².

(c) The linear distance the car advances is equal to the arc-length distance traveled by the edge of the tire, since it rolls without slipping on the road. This distance is related to the angular distance θ by

$$s = r\theta = (0.3 \text{ m})(157 \text{ rad}) = \underline{47 \text{ m}}$$

The car travels 47 m in stopping.

Or, α may be converted to a tangential acceleration which can then be used in the appropriate kinematics equation:

$$a = \alpha r = (-32 \text{ rad/s}^2)(0.3 \text{ m}) = -9.6 \text{ m/s}^2$$
$$v^2 - v_0^2 = 2a(x - x_0)$$
$$x - x_0 = \frac{-v_0^2}{2a} = \frac{-(30 \text{ m/s})^2}{2(-9.6 \text{ m/s}^2)} = 47 \text{ m}$$

which agrees with the previous result.

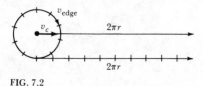

FIG. 7.2

Discussion: In the preceding problem the speed v of the car was used in the equation $\omega = v/r$, where v represents the speed of the outside edge of the tire. Are the two v's in fact equal? Yes, they are, when an object *rolls*. Consider a rolling wheel (Fig. 7.2). As the wheel turns once, a point on its edge moves a distance of one circumference, $2\pi r$. Simultaneously, the wheel moves along the surface a linear distance equal to its circumference because there is a one-to-one correspondence between points on its circumference and points on the surface as it rolls. The wheel's center of mass, always remaining above the point of contact between wheel and surface, also travels a distance $2\pi r$. So the wheel's center moving forward and the wheel's edge rotating around the center travel the same distance and in the same time: they have the same speed v. This equality, $v_{\text{edge}} = v_{\text{center of mass}}$, exists only in the case of rolling without slipping.

PROBLEM: Three 10-kg masses are joined by very light rods 1 m long to form an equilateral triangle, as shown in Fig. 7.3a. What is the moment of inertia of this system for rotation about an axis passing through the point O at the center of rod AB and perpendicular to the plane of the triangle?

Solution: The moment of inertia is given by Eq. (7.13):

$$I = m_1 r_1^2 + m_2 r_2^2 + m_3 r_3^2 + \ldots$$

so the distance of each mass m from the axis of rotation is needed. For A and B, $r = L/2$. For C, a little geometry is needed. Triangle OAC (b) is a right triangle; by the Pythagorean theorem:

$$(1 \text{ m})^2 = R^2 + (0.5 \text{ m})^2$$
$$R^2 = 1 \text{ m}^2 - 0.25 \text{ m}^2 = 0.75 \text{ m}^2$$
$$R = 0.867 \text{ m}$$

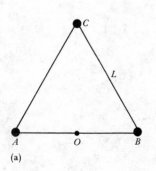

(a)

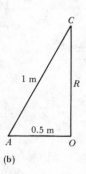

(b)

FIG. 7.3

Now I can be calculated:

$$I = m_A \left(\frac{L}{2}\right)^2 + m_B \left(\frac{L}{2}\right)^2 + m_c R^2$$
$$= 10 \text{ kg}(0.5 \text{ m})^2 + 10 \text{ kg}(0.5 \text{ m})^2 + 10 \text{ kg}(0.867 \text{ m})^2$$
$$I = 2.5 + 2.5 + 7.5 = \underline{12.5 \text{ kg} \cdot \text{m}^2}$$

The moment of inertia is 12.5 kg·m².

PROBLEM: A grindstone in the form of a solid cylinder has a diameter of 1 m and a mass of 50 kg. What torque will bring it from rest to an angular velocity of 300 rev/min in 10 s?

Solution: The rotational equivalent of Newton's second law is $\Sigma\tau = I\alpha$. So we need both I and α. From Table 7.1 in the text:

For a solid cylinder, $I = \frac{1}{2}mR^2 = \frac{1}{2}(50 \text{ kg})(0.5 \text{ m})^2 = 6.25 \text{ kg}\cdot\text{m}^2$

α is obtained from rotational kinematics:

$$\alpha = \frac{\omega - \omega_0}{t} \quad \text{where } \omega_0 = 0, \quad \omega = 300 \frac{\text{rev}}{\text{min}}\left(\frac{2\pi \text{ rad}}{1 \text{ rev}}\right)\left(\frac{1 \text{ min}}{60 \text{ s}}\right) = 31.4 \text{ rad/s}$$

$$\alpha = \frac{31.4 \text{ rad/s} - 0}{10 \text{ s}} = 3.14 \text{ rad/s}^2$$

So the torque is

$$\tau = I\alpha = (6.25 \text{ kg}\cdot\text{m}^2)(3.14 \text{ rad/s}^2) = \underline{19.6 \text{ N}\cdot\text{m}}$$

PROBLEM: A 3-kg bucket is attached to a cord wrapped around the windlass over a well (Fig. 7.4a). The windlass has an outer radius of 20 cm. When the bucket is released from rest, its downward acceleration is 2 m/s². Find the tension in the cord and the moment of inertia of the windlass.

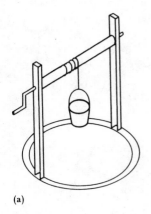
(a)

Solution: The standard technique for two-body problems—isolating each body and writing Newton's second law for each, then solving the resulting equations simultaneously—will work here. For the bucket,

$$\Sigma\vec{F} = m\vec{a}$$

$$mg - T = ma$$

where the forces are positive if aiding the acceleration and negative if opposing it. Solving for T,

$$T = mg - ma = m(g - a) = (3)(9.8 - 2) = 3(7.8) = \underline{23.4 \text{ N}}$$

The tension in the rope is 23.4 N.

For the body in rotational motion, the windlass, we write the rotational equivalent of Newton's second law:

$$\Sigma\tau = I\alpha$$

One torque acts, the tension T at moment arm R, so $\Sigma\tau = TR$.

$$TR = I\alpha$$

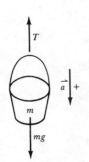
(b)

Solving for I,

$$I = \frac{TR}{\alpha}$$

Now α is the angular acceleration of the windlass. Its equivalent tangential acceleration a_t is related by $a_t = \alpha r$. This tangential acceleration and the acceleration of the cord and bucket are the same, since the cord doesn't slip. Thus:

$$\alpha = \frac{a_t}{R} = \frac{2 \text{ m/s}^2}{0.2 \text{ m}} = 10 \text{ rad/s}^2$$

(c)

FIG. 7.4

So

$$I = \frac{TR}{\alpha} = \frac{(23.4)(0.2)}{10} = \underline{0.47 \text{ kg}\cdot\text{m}^2}$$

The moment of inertia of the windlass is 0.47 kg·m². This suggests another method of determining the moment of inertia for bodies with unusual mass distributions or no useful symmetries. A known torque can be applied and the resulting angular acceleration measured. Then I is solved for in the equation $\tau = I\alpha$.

45

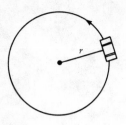

FIG. 7.5

PROBLEM: A 1200-kg car is traveling at 40 m/s (about 90 mi/hr) around a circular track. The length of one lap is 400 m (about ¼ mile). What is the car's angular momentum?

Solution: Angular momentum is given by $L = I\omega$. The car's mass is all at essentially the same distance from the center of the circular motion, so $I = mr^2$. Now r is the distance of the mass from that center; in this case, the radius of the circular track.

$$2\pi r = 400 \text{ m} \quad \text{so} \quad r = 400 \text{ m}/2\pi = 64 \text{ m}$$

This gives

$$I = mr^2 = 1200 \text{ kg}(64 \text{ m})^2 = 4.9 \times 10^6 \text{ kg} \cdot \text{m}^2$$

The car's angular velocity is related to its linear velocity by the equation $\omega = v/r$.

$$\omega = \frac{v}{r} = \frac{40 \text{ m/s}}{64 \text{ m}} = 0.63 \text{ rad/s}$$

The car's angular momentum is thus

$$L = I\omega = (4.9 \times 10^6 \text{ kg} \cdot \text{m}^2)(0.63 \text{ rad/s}) = \underline{3.1 \times 10^6 \text{ kg} \cdot \text{m}^2/\text{s}}$$

Alternate Solution: Equation (7.18) for angular momentum may also be used, in this case more readily:

$$L = mvr = (1200 \text{ kg})(40 \text{ m/s})(64 \text{ m}) = \underline{3.1 \times 10^6 \text{ kg} \cdot \text{m}^2/\text{s}}$$

which agrees with the result obtained previously.

PROBLEM: A girl of mass 60 kg stands on the edge of a merry-go-round making 10 turns per minute. The merry-go-round has a moment of inertia of 1200 kg·m² and a radius of 3 m. The girl walks to the center of the merry-go-round. Assuming no net external torques act, determine the angular velocity of the merry-go-round.

Solution:

$$L_{\text{initial}} = L_{\text{final}}$$
$$(I_M + I_{g0})\omega_0 = (I_M + I_g)\omega_f$$

Initially,

$$I_{g0} = mr^2 = (60 \text{ kg})(3 \text{ m})^2 = 540 \text{ kg} \cdot \text{m}^2$$
$$I_M = 1200 \text{ kg} \cdot \text{m}^2$$
$$\omega_0 = 10 \text{ rev/min for girl and merry-go-round}$$

Afterwards,

$$I_M = 1200 \text{ kg} \cdot \text{m}^2$$
$$I_g = mr^{2\;0} = 0$$

Putting in these values,

$$(1200 + 540)(10) = (1200 + 0)\omega_f$$
$$\frac{174,000}{1200} = \omega_f = \underline{14.5 \text{ rev/min}}$$

As the girl walks inward, the merry-go-round speeds up to 14.5 rev/min. Since angular momentum is conserved, the angular momentum lost by the girl as she walks inward is gained by the merry-go-round.

Discussion: When an equation is a relationship among angular quantities only, such as $\theta - \theta_0 = \omega_0 t + \frac{1}{2}\alpha t^2$, any convenient angular units may be used as long as they are

46

consistent. For example, θ may be expressed in rev, ω in rev/min, α in rev/min^2, and t in min. However, in an equation that relates *linear* and *angular* quantities, such as $v = r\omega$, the *radian* is the only unit that can be used in measuring angles and derived angular quantities. It is a *unitless* angular measure which allows the linear units on both sides of the equation to be equal.

Avoiding Pitfalls

1.   Angular acceleration α and tangential acceleration a_t are two ways of describing the *same* change. When a point moving in a circle changes its speed, its acceleration can be expressed in linear units of m/s^2 (a_t) or in angular units of rad/s^2 (α). The relationship between the two is $a_t = r\alpha$.

2.   Tangential acceleration a_t and centripetal acceleration a_c are ways of describing two *different* changes. Centripetal acceleration occurs whenever an object moves in a circle; this acceleration points toward the center of the circle and indicates that the object's *direction* is changing. Tangential acceleration occurs when the object's *speed* is changing, and it points parallel (or antiparallel) to the direction of motion.

3.   If an object is changing speed as it moves in a circle, it is experiencing both a tangential and a centripetal acceleration.

4.   In applying $\Sigma\tau = I\alpha$, recall that torques have direction; they can accelerate the angular motion or oppose the motion. They must have the appropriate positive or negative signs to indicate this.

5.   Angular momentum is conserved only if no net external torques act on a system. A frictional force at the axis of rotation does not exert a torque because its moment arm is zero.

Drill Problems
Answers in Appendix

1.   A flywheel is undergoing constant angular acceleration. At one time it is turning at 400 rpm and 2 s later it is turning at 500 rpm.

 (a) What is its angular acceleration, in rad/s^2?
 (b) How many revolutions has it made in reaching 500 rpm from rest?

2.   A rotating restaurant sign needs a neon tube replaced (Fig. 7.6). The sign is 1.6 m wide and has a moment of inertia of 200 kg·m^2. It turns 5 times per minute. The repairman turns off the motor and applies a perpendicular force which brings the sign to rest in ⅛ rotation. With what force did he push? (Assume negligible friction.)

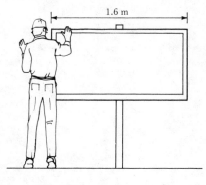

FIG. 7.6

3. A drum majorette is twirling a baton.

   (a) Calculate its moment of inertia by assuming it to be a long thin rod of length 80 cm and mass 0.2 kg (see Table 7.1) with a 0.2-kg mass at each end.

   (b) What is its angular momentum if twirled at a rate of 2 turns per second?

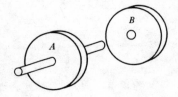

FIG. 7.7

4. Wheel A is turning with an angular velocity of 100 rpm, on a shaft of negligible mass (Fig. 7.7). Wheel B, initially at rest, whose moment of inertia is three times that of wheel A, is suddenly coupled to the same shaft. What is the final angular velocity of the combination?

Work and Energy

Terms

Define or describe briefly what is meant by the following terms. If you have difficulty, refer to the textbook section given in parentheses.

work (8.1)

joule (8.1)

kinetic energy (8.2, 8.3)

gravitational potential energy (8.2, 8.4)

elastic potential energy (8.2, 8.5)

thermal energy (8.2, 8.8)

system (8.6)

environment (8.6)

conservation of energy principle (8.6)

work-energy equation (8.6)

inelastic collision (8.9)

elastic collision (8.9)

power (8.10)

watt (8.10)

horsepower (8.10)

metabolic rate (8.10)

kcal (8.10)

Equation Review

For each equation, be able to state the situation to which it applies, what quantity each symbol represents, and the units for measuring each quantity in some consistent set of units.

$$W = F\Delta r \cos \theta \qquad\qquad (8.1)$$

$$\Delta KE = \tfrac{1}{2}mv^2 - \tfrac{1}{2}mv_0^2 \qquad\qquad (8.4)$$

$$\Delta KE_r = \tfrac{1}{2}I\omega^2 - \tfrac{1}{2}I\omega_0^2 \qquad\qquad (8.5)$$

$$\Delta PE_g = mg(y - y_0) \qquad\qquad (8.6)$$

$$\Delta PE_s = \tfrac{1}{2}kx^2 - \tfrac{1}{2}kx_0^2 \qquad\qquad (8.7)$$

$$F = kx$$

$$Q + W + \ldots = \Delta E_{\text{system}} \qquad\qquad (8.8)$$

$$W = \Delta KE + \Delta PE_g + \Delta PE_s + \Delta E_{th} + \ldots \qquad\qquad (8.10)$$

$$\Delta E_{th} = F_k \Delta x \qquad\qquad (8.12)$$

$$P = \frac{\Delta W}{\Delta t} \text{ or } \frac{\Delta E}{\Delta t} \qquad\qquad (\textbf{8.17 or 8.18})$$

Problems with Solutions and Discussion

PROBLEM: A 10-kg box is pulled 8 m up a 30° inclined plane with negligible friction by a 70-N force parallel to the plane. Determine the work done by each force acting on the box.

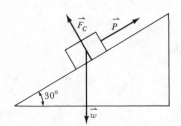

FIG. 8.1

Solution: The force diagram shows the three forces acting on the box (Fig. 8.1). The defining equation for the work done by a force $\vec{F}$ on an object moving a distance $\Delta \vec{r}$ is $W = F\Delta r \cos \theta$, where θ is the angle between $\vec{F}$ and $\Delta \vec{r}$. •

$$\text{Work done by } \vec{P}: \quad W = P\Delta r \cos 0° = (70 \text{ N})(8 \text{ m})(1)$$
$$W = \underline{560 \text{ J}}$$

Force $\vec{P}$ does 560 J of work on the box, where we have used the fact that the *mks* unit of work, the newton-meter, is called a joule, J.

$$\text{Work done by } \vec{w}: \quad W = w\Delta r \cos 120° = mg\Delta r \cos 120°$$
$$= (10 \text{ kg})(9.8 \text{ m/s}^2)(8 \text{ m})(-0.5)$$
$$W = \underline{-392 \text{ J}}$$

Here the negative sign indicates work has been done *against* the force of gravity in lifting the box upward.

$$\text{Work done by } \vec{F}_c: \quad W = F_c\Delta r \cos 90° = F_c(8 \text{ m})(0) = \underline{0}$$

No work is done by $\vec{F}_c$ because it is perpendicular to the displacement $\Delta \vec{r}$.

PROBLEM: If a car's speed is tripled, by what factor does its kinetic energy increase?

Solution: Initially, $KE_0 = \tfrac{1}{2}mv_0^2$. If v_0 increases to $3v_0$, then:

$$KE = \tfrac{1}{2}m(3v_0)^2 = \tfrac{1}{2}m(3^2 v_0^2) = 3^2(\tfrac{1}{2}mv_0^2) = 9KE_0$$

The kinetic energy increases by a factor of three squared, or <u>nine</u>.

PROBLEM: A rolling object has both translational and rotational motion. A disk of mass 3 kg rolls at 4 m/s along level ground. How much kinetic energy does it have?

Solution:

$$\text{Total } KE = KE + KE_r = \tfrac{1}{2}mv^2 + \tfrac{1}{2}I\omega^2$$

From Table 7.1, I for a disk is $\tfrac{1}{2}mr^2$. Also, $\omega = v/r$, where the tangential speed v of the edge of the rolling disk is in fact the linear speed v of the disk as a whole (see Discussion, page 44). So:

$$\text{Total } KE = \frac{1}{2} mv^2 + \frac{1}{2}\left(\frac{1}{2} mr^2\right)\left(\frac{v^2}{r^2}\right) = \frac{1}{2} mv^2 + \frac{1}{4} mv^2 = \frac{3}{4} mv^2$$

$$\text{Total } KE = \frac{3}{4}(3 \text{ kg})(4 \text{ m/s})^2 = 36 \; \frac{\text{kg} \cdot \text{m}^2}{\text{s}^2} = \underline{36 \text{ J}}$$

PROBLEM: A moving car's tires have rotational kinetic energy. Compare this energy to the translational kinetic energy for a car moving at 20 m/s. Take the car's mass as 1500 kg and let each tire be considered a disk of mass 10 kg and radius 25 cm.

Solution: For each tire, $KE_r = \tfrac{1}{2}I\omega^2$. For a disk, $I = \tfrac{1}{2}mr^2$ (from Table 7.1) and $\omega = v/r$, so

$$KE_r = \frac{1}{2}\left(\frac{1}{2} mr^2\right)\left(\frac{v^2}{r^2}\right) = \frac{1}{4} mv^2 = \frac{1}{4}(10 \text{ kg})(20 \text{ m/s})^2 = 1000 \text{ J}$$

Since there are 4 tires,

$$KE_r = 4(1000 \text{ J}) = 4000 \text{ J}$$

The translational kinetic energy of the car is

$$KE = \tfrac{1}{2}mv^2 = \tfrac{1}{2}(1500 \text{ kg})(20 \text{ m/s})^2 = 300{,}000 \text{ J}$$

The rotational kinetic energy is thus $\dfrac{4000}{300{,}000} = \underline{0.013}$, or 1.3 percent of the total kinetic energy. This is because the mass of the tires is such a small fraction of the total mass. We are quite justified in neglecting the rotational kinetic energy of moving cars.

PROBLEM: A 50-kg child is swinging on a swing 6 m long and of negligible mass (Fig. 8.2a). Her maximum angular displacement from the vertical is 37°. Find the change in her gravitational potential energy from the lowest to the highest point of the motion.

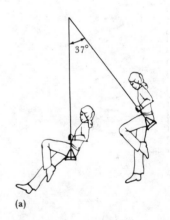

Solution: $\Delta PE_g = mg(y - y_0)$ where $(y - y_0)$ is the *vertical* change in height. A bit of geometry is needed. Taking $y_0 = 0$ at the bottommost position, the vertical displacement y is given by $y = l - a$, where a is one side of a right triangle and is given by $a = l \cos 37°$.

$$y = l - l \cos 37° = l(1 - \cos 37°)$$
$$y = 6 \text{ m}(1 - 0.8) = 6 \text{ m}(0.2) = 1.2 \text{ m}$$

Then

$$\Delta PE_g = mgy = (50 \text{ kg})(9.8 \text{ m/s}^2)(1.2 \text{ m}) = \underline{588 \text{ J}}$$

The change in gravitational potential energy is 588 J.

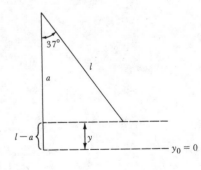

(b)

FIG. 8.2

PROBLEM: For the child in the previous problem, what is her speed as she passes through the lowest point of her swing?

Solution: As the girl moves from the highest to the lowest point, gravitational potential energy is being lost and kinetic energy is being gained. If no other energy changes occur and energy is conserved, then the loss in gravitational potential energy must equal the gain in kinetic energy.

51

$$\Delta KE \text{ gained} = \Delta PE_g \text{ lost}$$
$$\tfrac{1}{2}mv^2 - \tfrac{1}{2}mv_0^{\cancel{2}\,0} = 588 \text{ J (from above)}$$

where $v_0 = 0$ at the top of the swing.

$$\tfrac{1}{2}mv^2 = 588 \text{ J}$$
$$v = \sqrt{\frac{2(588 \text{ J})}{(50 \text{ kg})}} = \underline{4.8 \text{ m/s}}$$

Her speed at the lowest point is 4.8 m/s.

Discussion: It would be quite difficult to solve the above problem by dynamics and kinematics methods. Two forces act, the tension of the rope and the gravitational force (weight). The girl starts at the end point with $v_0 = 0$ and accelerates. The tangential acceleration to increase her speed comes from the component of $\vec{g}$ in the direction of her motion. (The component of $\vec{g}$ perpendicular to her motion combines with the tension to supply the centripetal force needed for the circular motion.) But the components of $\vec{g}$ depend on the angle θ which is constantly changing as she moves downward! So we cannot use a constant acceleration equation from kinematics because the acceleration components are not constant. The problem is simply not amenable to being solved by the dynamical and kinematic techniques we are familiar with. But the conservation of energy approach yields the required result in a quite straightforward way. Conservation of energy techniques will often work quickly and easily when other methods are very messy and difficult to apply. That is one reason for the popularity and widespread usage of the conservation of energy principle.

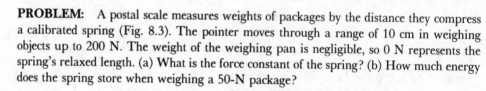

PROBLEM: A postal scale measures weights of packages by the distance they compress a calibrated spring (Fig. 8.3). The pointer moves through a range of 10 cm in weighing objects up to 200 N. The weight of the weighing pan is negligible, so 0 N represents the spring's relaxed length. (a) What is the force constant of the spring? (b) How much energy does the spring store when weighing a 50-N package?

FIG. 8.3

Solution: (a) We may use any force and its corresponding displacement to determine k from $F = kx$, the equation describing the behavior of a linear spring. For example, a 200-N weight extends the spring 10 cm, a 100-N weight extends it 5 cm, etc.

$$k = \frac{F}{x} = \frac{200 \text{ N}}{0.1 \text{ m}} = \underline{2000 \text{ N/m}}$$

The spring's force constant is 2000 N/m.

(b) When weighing a 50-N package, the spring's extension can be determined from $F = kx$.

$$x = \frac{F}{k} = \frac{50 \text{ N}}{2000 \text{ N/m}} = 0.025 \text{ m, or } 2.5 \text{ cm}$$

Or, x may be determined without knowing k, by taking advantage of the linearity of the spring and making a simple proportion:

$$\frac{50 \text{ N}}{200 \text{ N}} = \frac{x \text{ cm}}{10 \text{ cm}} \quad \text{or} \quad x = \frac{50}{200}(10 \text{ cm}) = 2.5 \text{ cm} = 0.025 \text{ m}$$

To get the elastic potential energy stored in the spring,

$$PE_s = \tfrac{1}{2}kx^2 = \tfrac{1}{2}(2000 \text{ N/m})(0.025 \text{ m})^2 = \underline{0.625 \text{ J}}$$

The spring stores 0.625 J of energy when weighing a 50-N package.

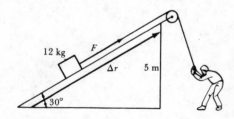

FIG. 8.4

PROBLEM: A man pulls a 12-kg crate up a frictionless 30° ramp 5 m high (Fig. 8.4). If the crate is moved from rest at the bottom to rest at the top (and the initial acceleration and final deceleration are negligible) what constant force did the man exert?

Solution: The work-energy equation for the box, $W = \Delta E_{box}$, says the work done on the box is equal to the change in energy of the box. Only its gravitational potential energy changes, since $v = v_0 = 0$.

$$W = \Delta PE_g = mg(y - y_0)$$

where

$$(y - y_0) = 5 \text{ m}$$

$$W = (12 \text{ kg})(9.8 \text{ m/s}^2)(5 \text{ m}) = 588 \text{ J}$$

Work is related to force by $W = F \Delta r \cos \theta$. Solving for F,

$$F = \frac{W}{\Delta r \cos \theta}$$

Now Δr is the displacement of the object. In this case, the plane is a right triangle:

$$\sin 30° = \frac{5 \text{ m}}{\Delta r} \quad \text{so} \quad \Delta r = \frac{5 \text{ m}}{\sin 30°} = 10 \text{ m}$$

θ, the angle between $\vec{F}$ and Δr, is $0°$.

$$F = \frac{588 \text{ J}}{(10 \text{ m})(1)} = \underline{58.8 \text{ N}}$$

The man pulls with a force of 59 N.

PROBLEM: A car is at the top of a 20-m hill on a frictionless track (Fig. 8.5). Farther along there is another hill 45 m high. What must the car's speed be at the top of the first hill if it can just make it to the top of the second hill?

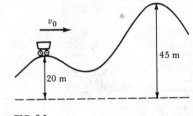

FIG. 8.5

Solution: Considering the earth, car, and track as our system, no *external* forces do any work, so the work-energy equation is:

$$W = \Delta E_{system} = \Delta PE_g + \Delta KE = 0$$

$$mg(y - y_0) + \tfrac{1}{2}m(v^2 - v_0^2) = 0$$

Here $y_0 = 20$ m, $y = 45$ m, and v at the top of the second hill is 0.

$$\cancel{m}g(45 - 20) + \tfrac{1}{2}\cancel{m}(-v_0^2) = 0$$

$$v_0^2 = 2g(45 - 20) = 2(9.8 \text{ m/s}^2)(25 \text{ m}) = 490 \text{ m}^2/\text{s}^2$$

$$v_0 = \underline{22 \text{ m/s}}$$

The car must have a speed of 22 m/s at the top of the first hill to be able to get to the top of the second hill.

Discussion: In cases where no external work is done, the work-energy equation can be rearranged to express energy conservation in the same form as the momentum conservation equation.

$$0 = \Delta PE_g + \Delta KE + \Delta PE_s + \cdots$$

$$0 = (mgy_2 - mgy_1) + (\tfrac{1}{2}mv_2^2 - \tfrac{1}{2}mv_1^2) + (\tfrac{1}{2}kx_2^2 - \tfrac{1}{2}kx_1^2) + \cdots$$

$$mgy_1 + \tfrac{1}{2}mv_1^2 + \tfrac{1}{2}kx_1^2 = mgy_2 + \tfrac{1}{2}mv_2^2 + \tfrac{1}{2}kx_2^2$$

On one side we have the types of energy of a system in a certain initial state and on the other side the types of energy when it has evolved to a final state, or

$$E_{initial} = E_{final}$$

The types of energy which do not change between the initial and final states would cancel if written on each side, therefore we don't bother to write them. In applying this equation it is traditional to assign the lowest position in the problem a y coordinate of zero. Then each potential energy term mgy will always be either positive or zero.

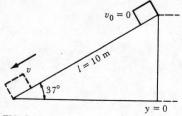

FIG. 8.6

PROBLEM: A 6-kg block slides from rest down a frictionless 37° inclined plane 10 m long (Fig. 8.6). What is its speed at the bottom?

Solution: Choosing the top and bottom as the initial and final states for the block, energies at the two positions must be equal, or

$$E_{top} = E_{bottom}$$
$$(PE_g + KE)_{top} = (PE_g + KE)_{bottom}$$

Taking $y = 0$ at the bottom of the plane,

$$\cancel{m}gy + \frac{1}{2}m\cancel{v_{top}^2}^0 = \cancel{m}g(0) + \frac{1}{2}\cancel{m}v_{bottom}^2$$
$$v_{bottom} = \sqrt{2gy}$$

Now

$$y = d \sin 37° = (10 \text{ m})(0.6) = 6 \text{ m, so}$$
$$v = \sqrt{2(9.8 \text{ m/s}^2)(6 \text{ m})} = \underline{11 \text{ m/s}}$$

Note that the body does not have the energies on both sides of the equation simultaneously. But the conservation of energy principle allows us to say that the energies at the two different positions are equal to each other.

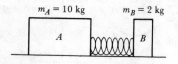

FIG. 8.7

PROBLEM: On a frictionless surface, two masses are pushed together to compress a massless spring of force constant 4800 N/m (Fig. 8.7). After the entire system is released from rest, block A is found to have a speed of 4 m/s. How much was the spring compressed by the blocks originally?

Solution: Since no net external forces act on the system after the moment of release, we may conserve both energy and momentum. Let the initial position be when the spring is compressed (the moment of release) and the final position be when the spring is relaxed and the masses are in motion.

Conservation of momentum: $\quad \vec{p}_{initial} = \vec{p}_{final}$

$$0 = m_A\vec{v}_A + m_B\vec{v}_B = (10 \text{ kg})(4 \text{ m/s}) + (2 \text{ kg})v_B$$

where the direction of v_A is taken as positive.

$$v_B = \frac{-40 \text{ kg} \cdot \text{m/s}}{2 \text{ kg}} = -20 \text{ m/s}$$

Here the negative sign means v_B is opposite in direction to v_A.

Conservation of energy: $\quad E_{initial} = E_{final}$

Elastic potential energy of the spring = Kinetic energy of the two blocks

$$\frac{1}{2}kx^2 = \frac{1}{2}m_Av_A^2 + \frac{1}{2}m_Bv_B^2$$
$$\frac{1}{2}(4800)x^2 = \frac{1}{2}(10)(4)^2 + \frac{1}{2}(2)(-20)^2$$
$$2400\,x^2 = 80 + 400$$
$$x^2 = \frac{480}{2400} = 0.2$$
$$x = \underline{0.45 \text{ m}}$$

The spring was compressed 45 cm.

PROBLEM: An 80-kg man starts from rest and runs up a flight of stairs 3 m high in 4 s, reaching the top with a speed of 3 m/s. What is his average power during this time interval?

Solution: From Eq. (8.18), average power is given by $\Delta E/\Delta t$. The man's energy changes ΔE are his increase in kinetic energy from zero to $\frac{1}{2}mv^2$ and his increase in gravitational potential energy.

$$P = \frac{\Delta E}{\Delta t} = \frac{\frac{1}{2}mv^2 + mg(y - y_0)}{t} = \frac{\frac{1}{2}(80)(3)^2 + (80)(9.8)(3)}{4}$$

$$P = \frac{360 \text{ J} + 2352 \text{ J}}{4 \text{ s}} = 678\,\frac{\text{J}}{\text{s}} = \underline{678 \text{ watts}}$$

His average power is about 680 watts.

Avoiding Pitfalls

1. The physicist's definition of work is $F\Delta r \cos\theta$. The work done by a force may be zero because the object doesn't move ($\Delta r = 0$) as in the case of holding a suitcase. Or it may be zero because F and Δr are perpendicular to one another ($\cos 90° = 0$). Two familiar forces which do no work are a contact force F_c between surfaces and a centripetal force, both of which act at right angles to the motion of the object.

2. The defining equation for work is valid only if the force is constant during the whole displacement Δr. If the force is variable, as in the case of the spring where the force required to stretch it increases as the displacement increases, a graphical method or a calculus method is required to determine the work done.

3. Work is a scalar quantity. It is added algebraically, not vectorially. The positive and negative signs do *not* indicate direction, but rather whether work is done *on* a system or *by* a system. Work done *on* a system is taken as positive because the system gains energy; work done *by* a system is negative because the system loses energy.

4. A change in gravitational potential energy, $\Delta PE_g = mg(y - y_0)$, does not depend in any way on *how* the body moved from y_0 to y. Even if it took a very long circuitous route, ΔPE_g depends only on the *vertical* separation of y and y_0. This energy change is termed *independent of path*. Contrast it to a change in thermal energy due to friction as an object moves on a surface. This energy change depends very much on the length of the path.

5. When gravitational potential energy is changing in a problem, choose $y = 0$ to be the lowest (closest to earth) position of any object in the problem. Then each y and mgy will be positive.

6. In calculating elastic potential energy stored in a spring, $PE_s = \frac{1}{2}kx^2$, x is the displacement of the spring *from its unstretched length*. When relaxed, $x = 0$, which is called the spring's equilibrium position.

7. Work terms and energy terms can be added to and subtracted from each other in the same equation, since they have the same units. If a system does external work, it loses that same amount of energy; if work is done *on* a system by an external force, the system gains that same amount of energy. So it is a matter of convenience whether the work term or the equivalent energy term is used in a particular equation.

Drill Problems
Answers in Appendix

1. A truck and an Audi have the same kinetic energy, but the truck's mass is 4 times the mass of the Audi. How do their velocities and momenta compare?

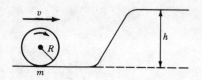

FIG. 8.8

2. A spherical ball of mass m and radius R is rolling toward a hill whose vertical height is h (Fig. 8.8). What must be its speed v at the bottom of the hill so that it will just get to the top as it comes to rest?

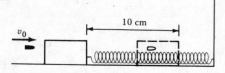

FIG. 8.9

3. A rifle bullet of mass 0.02 kg strikes and embeds itself in a block of mass 0.98 kg which rests on a horizontal surface and is attached to a relaxed spring as shown in Fig. 8.9. The spring's force constant is 400 N/m and the impact compresses the spring 10 cm. The coefficient of kinetic friction between block and surface is 0.41.

(a) Find the maximum potential energy stored in the spring.
(b) Find the velocity of the block just after the impact of the bullet.
(c) Find the initial velocity of the bullet.

56

Thermal Energy, Temperature, and Heat

9

Define or describe briefly what is meant by the following terms. If you have difficulty, refer to the textbook section given in parentheses.

thermal energy (9.1)

temperature (9.2)

thermometric property (9.2)

absolute temperature scale (9.2)

heat (9.3)

specific heat capacity (9.4)

basal metabolic rate (9.5)

change of state (9.6)

melting temperature (9.6)

freezing temperature (9.6)

boiling temperature (9.7)

condensation temperature (9.7)

heat of fusion (9.6)

heat of vaporization (9.7)

evaporation (9.7)

sublime (9.7)

heat of sublimation (9.7)

coefficient of linear expansion (9.8)

coefficient of volume expansion (9.8)

Equation Review

For each equation, be able to state the situation to which it applies, what quantity each symbol represents, and the units for measuring each quantity in some consistent set of units.

$$T_F = \tfrac{9}{5}T_C + 32 \qquad\qquad (9.1)$$

$$T_C = \tfrac{5}{9}(T_F - 32) \qquad\qquad (9.2)$$

$$W + Q = \Delta E_{\text{system}} \qquad\qquad (9.3)$$

$$Q = mc\Delta T \qquad\qquad (9.4)$$

melting $\qquad Q = mL_f \quad = -mL_f\ \text{freezing} \qquad\qquad (9.6)$

boiling $\qquad Q = mL_v \quad \approx -mL_v\ \text{condens} \qquad\qquad (9.9)$

$$\Delta L = \alpha L\Delta T \qquad\qquad (9.11)$$

$$\Delta V = \gamma \overset{\vee}{L}\Delta T \qquad\qquad (9.12)$$

Problems with Solutions and Discussion

PROBLEM: Mercury is the only metal which is a liquid at room temperature. Its freezing point is $-39°$C. Convert this to $°$F and K.

Solution: To get the Fahrenheit temperature, we use Equation (9.1):

$$T_F = \tfrac{9}{5}T_C + 32 = \tfrac{9}{5}(-39) + 32 = -70 + 32 = \underline{-38°F}$$

Mercury freezes at $-38°$F. To get the Kelvin temperature, we recall that Celsius and Kelvin degrees are the same size. Since the freezing point of mercury is 39 degrees below the freezing point of water on the Celsius scale, it will also be 39 degrees below the freezing point of water on the Kelvin scale, where water freezes at 273.15 K. Thus,

$$273 - 39 = \underline{234\ \text{K}}$$

Mercury freezes at 234 K.

Discussion: An alternate method of converting temperatures from one scale to another is based on the idea of simple proportions. It is useful if you have trouble remembering the equations for conversions between temperatures—if you can't remember whether the 32 gets added or subtracted, or whether the fraction is $\tfrac{9}{5}$ or $\tfrac{5}{9}$. It does require some memory— that of the boiling and freezing points of water (or any other two points), but those you probably already know.

The *proportions method* rests on the fact that even though the boiling and freezing points of water have different numbers on different scales, the intervals between them represent the same physical quantity. That is, interval B is equivalent to interval D in Fig. 9.1. Suppose you wanted to convert $62°$C to a Fahrenheit temperature. Note that interval A is to interval B as interval C is to interval D, or

$$\frac{A}{B} = \frac{C}{D}$$

Expressing the intervals in terms of end points,

$$\frac{62 - 0}{100 - 0} = \frac{T_F - 32}{212 - 32}$$

$$\frac{62}{100} = \frac{T_F - 32}{180}$$

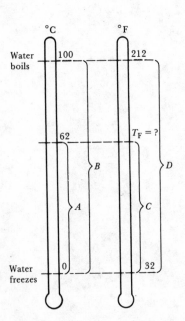

FIG. 9.1

Rearranging,
$$T_F - 32 = \frac{180}{100}(62)$$
$$T_F = \tfrac{9}{5}(62) + 32$$

Note that we have actually derived Eq. (9.1), $T_F = \tfrac{9}{5}T_C + 32$, but it is not necessary to remember it, or even to shake down the equation to that particular form. Continuing,
$$T_F = 112 + 32 = \underline{144°F}$$

62°C is the same temperature as 144°F.

PROBLEM: Oxygen changes from a gas to a liquid at −362°F. Convert this temperature to kelvins.

Solution: Making use of the two intervals indicated in Fig. 9.2,
$$\frac{A_F}{B_F} = \frac{A_K}{B_K}$$
$$\frac{212 - 32}{212 - (-362)} = \frac{373 - 273}{373 - T_K}$$
$$\frac{180}{574} = \frac{100}{373 - T_K}$$
$$373 - T_K = {}^{100}\!/_{180}(574) = \tfrac{5}{9}(574) = 319$$
$$T_K = 373 - 319 = \underline{54 \text{ K}}$$

Oxygen changes from a gas to a liquid at 54 K.

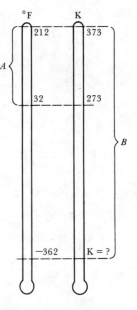

FIG. 9.2

PROBLEM: What is the result of adding 500 g of iron (specific heat capacity of iron = 450 J/kg·C°) at 500°C to 1 kg of water at 0°C?

Solution: Assuming no heat losses to the surroundings, the heat lost by the iron (negative Q) will equal the heat gained by the water (positive Q), so that
$$Q_i + Q_w = 0$$

Using $Q = mc\Delta T$ for each substance, the equation becomes
$$m_i c_i (T - T_0)_i + m_w c_w (T - T_0)_w = 0$$
$$\left(450 \, \frac{J}{kg \cdot C°}\right)(0.5 \text{ kg})(T - 500°C) + \left(4180 \, \frac{J}{kg \cdot C°}\right)(1 \text{ kg})(T - 0°C) = 0$$
$$(225 \text{ J/C°})T - 112{,}500 \text{ J} + (4180 \text{ J/C°})T = 0$$

Collecting terms in T_F,
$$(4405 \text{ J/C°})T = 112{,}500 \text{ J}$$
$$T = \frac{112{,}500 \text{ J}}{4405 \text{ J/C°}} = 26°C$$

The mixture of iron and water will come to an equilibrium temperature of 26°C.

PROBLEM: How much heat must be removed from 2 kg of steam at 110°C (called superheated steam) to change it into ice at −10°C?

Solution: This process can be thought of as occurring in five steps, two of them changes of state.

Q	=	$-mc\Delta T$	$-mL_v$	$-mc\Delta T$	$-mL_f$	$-mc\Delta T$
heat released		steam cools to 100°C	steam condenses at 100°C	water cools from 100°C to 0°C	water freezes at 0°C	ice cools to −10°C

59

The specific heat capacities of the substances may be obtained from Table 9.1 of the text, and the latent heats from Table 9.2. Substituting numerical values for symbols,

$$Q = -(2 \text{ kg})\left(1970 \frac{J}{\text{kg}\cdot\text{C}°}\right)(10 \text{ C}°) - (2 \text{ kg})\left(2.26 \times 10^6 \frac{J}{\text{kg}}\right)$$
$$- (2 \text{ kg})\left(4180 \frac{J}{\text{kg}\cdot\text{C}°}\right)(100 \text{ C}°)$$
$$- (2 \text{ kg})\left(3.35 \times 10^5 \frac{J}{\text{kg}}\right) - (2 \text{ kg})\left(2090 \frac{J}{\text{kg}\cdot\text{C}°}\right)(10 \text{ C}°)$$

Note that the specific heat capacities for ice, water, and steam are *not* equal; thus we cannot write just one term for the whole temperature change.

$$Q = -39,400 \text{ J} - 4,520,000 \text{ J} - 836,000 \text{ J} - 670,000 \text{ J} - 41,800 \text{ J}$$
$$Q = -6,107,200 \text{ J} = \underline{-6.1 \times 10^6 \text{ J}}$$

The heat removed is 6.1×10^6 J. The negative sign indicates it must flow out of the system.

PROBLEM: An electric coffeepot is advertised to bring one quart of water (about 1 kg) to a boil in 4 minutes. Assuming the water starts out at 10°C and that there are no heat losses to the surroundings, what power would be required to operate this pot?

Solution: To get the power P, we recall that power is work done per unit time. In this case, work is required to transfer heat to the water in the pot, so

$$P = \frac{Q}{t} = \frac{mc\Delta T}{t}$$

since the heat required to produce a temperature rise can be expressed as $mc\Delta T$. (Note that we want only to heat the water to a boil, not boil it all away, so the heat of vaporization is not required.) The heat is supplied in 4 minutes, or $4 \times 60 = 240$ s;

$$P = \frac{(4180 \text{ J/kg}\cdot\text{C}°)(1 \text{ kg})(100°\text{C} - 10°\text{C})}{240 \text{ s}} = \frac{376,200 \text{ J}}{240 \text{ s}} = \underline{1570 \text{ watts}}$$

The pot requires 1.57 kW of power for its operation.

PROBLEM: The Bridalveil waterfalls at Yosemite National Park is 620 ft high. Calculate the rise in temperature of the water due to its drop over the falls, assuming there are no heat losses to the surroundings.

Solution: The energy transformations occurring here could be diagrammed as follows:

$$\text{Gravitational potential energy} \Rightarrow \text{Kinetic energy} \Rightarrow \text{Thermal energy}$$

If there are no losses to surroundings, the entire change in gravitational potential energy ends up as thermal energy as the water dashes against the rocks at the bottom.

$$\Delta E_{th} = \Delta PE_g$$

Note that the same equation could be obtained from considering the conservation of energy equation,

$$W + Q = \Delta E_{system}$$
$$0 = -\Delta PE_g + \Delta E_{th}$$

since W and Q are zero. (No work is done and no heat flows in from another body.)

$$\Delta E_{th} = \Delta PE_g$$
$$mc\Delta T = mg\Delta y$$

60

$$\Delta T = \frac{mg\Delta y}{mc} = \frac{g\Delta y}{c}$$

To use c in *mks* units, y must be expressed in *mks* units; 620 ft $\left(\dfrac{1 \text{ m}}{3.28 \text{ ft}} \right) = 189$ m.

$$\Delta T = \frac{(9.8 \text{ m/s}^2)(189 \text{ m})}{4180 \dfrac{\text{kg} \cdot \text{m}^2}{\text{s}^2 \cdot \text{kg} \cdot \text{C}^\circ}} = \underline{0.44 \text{ C}^\circ}$$

Here the units of $c(\text{J/kg} \cdot \text{C}^\circ)$ have been explicitly spelled out to show that the units cancel out, leaving C° for the temperature rise. The water's temperature rises by about half a Celsius degree.

PROBLEM: A car of mass 1500 kg, traveling at 20 m/s (about 45 mi/hr), brakes sharply and comes to a stop. 75 percent of the kinetic energy is dissipated as thermal energy in the brake drums, whose temperature rises from 20°C to 35°C. The brake drums have a total mass of 20 kg. Find their average specific heat.

Solution: The energy transformation of interest is

$$75 \text{ percent Kinetic energy} \Rightarrow \text{Thermal energy}$$

which causes the temperature of the brake drums to rise. Stated as an equation,

$$0.75(\tfrac{1}{2}mv^2) = m_b c \Delta T$$

where m is the mass of the entire car and m_b is the mass of the brake drums. Substituting in the given values,

$$(0.75)(\tfrac{1}{2})(1500 \text{ kg})(20 \text{ m/s})^2 = (20 \text{ kg})c(35°C - 20°C)$$

Solving for c,

$$c = \frac{(0.75)(\tfrac{1}{2})(1500 \text{ kg})(400 \text{ m}^2/\text{s}^2)}{(20 \text{ kg})(15 \text{ C}^\circ)} = \underline{750 \text{ J/kg} \cdot \text{C}^\circ}$$

Brake drums are normally made from steel ($c = 460 \text{ J/kg} \cdot \text{C}^\circ$) lined with asbestos ($c = 815 \text{ J/kg} \cdot \text{C}^\circ$), with flanges to help dissipate the heat generated in them during braking.

PROBLEM: How much space must be left between abutting steel railroad tracks 20 m long if they are laid at 40°F and temperatures may go as high as 112°F?

Solution: If each rail is considered to exhibit *half* its expansion at each end, and each gap must accommodate the expansion of *two* abutting ends, the gap must equal the total expansion of *one* rail. Thus,

$$\Delta L = \alpha L \Delta T$$

where $\alpha = 12 \times 10^{-6}(\text{C}^{\circ -1})$ from Table 9.4, $L = 20$ m, and the temperature interval ΔT is $112°F - 40°F = 72F°$. However, α is expressed in units of $\text{C}^{\circ -1}$, so the Fahrenheit interval must be converted into a Celsius interval. Using the equivalence of intervals, $9 \text{ F}^\circ = 5 \text{ C}^\circ$,

$$72 \text{ F}^\circ \left(\frac{5 \text{ C}^\circ}{9 \text{ F}^\circ} \right) = 40 \text{ C}^\circ$$

(Alternatively, one may convert 40 F° and 112 F° both to Celsius temperatures and take their difference.)

$$\Delta L = (12 \times 10^{-6} \text{ C}^{\circ -1})(20 \text{ m})(40 \text{ C}^\circ) = \underline{9.6 \times 10^{-3} \text{ m}}$$

A gap of almost 10 mm, or 1 cm, should be left between abutting rails.

PROBLEM: The gas station attendant fills and tops off your 15-gallon tank and doesn' tighten the cap. Your car then sits in the sun from early morning (40°F) until late afternoon (85°F). If gas costs $1.50 per gallon, how much money have you lost from spillage due to expansion? The gas tank is made of sheet iron.

Solution: The tank expands, but the gasoline expands more, as can be seen by comparing coefficients of volume expansion from Table 9.4:

$$\text{Iron:} \quad \gamma = 36 \times 10^{-6} \, C^{\circ -1} \qquad \text{Gasoline:} \quad \gamma = 900 \times 10^{-6} \, C^{\circ -1}$$

The amount of gasoline spilling out will be the difference in the expansion of the gasoline and the expansion of the tank, or

$$\text{Spill} = \Delta V_{gas} - \Delta V_{tank}$$

Using Eq. (9.12), $\Delta V = \gamma V \Delta T$,

$$\text{Spill} = \gamma_{gas} V \Delta T - \gamma_{tank} V \Delta T$$

V and ΔT are the same for the gasoline and the inside tank volume, so

$$\text{Spill} = (\gamma_{gas} - \gamma_{iron}) V \Delta T$$

ΔT is $85°F - 40°F = 45F°$. The conversion factor for intervals is $9 \, F° = 5 \, C°$, so $45 \, F°(5 \, C°/9 \, F°) = 25 \, C°$. Putting in these values,

$$\text{Spill} = (900 \times 10^{-6} \, C^{\circ -1} - 36 \times 10^{-6} \, C^{\circ -1})(15 \text{ gal})(25 \, C°)$$
$$\text{Spill} = (864 \times 10^{-6})(15 \text{ gal})(25) = 0.32 \text{ gal}$$

At $1.50 per gallon, this spill represents a loss of

$$(0.32 \text{ gal})(\$1.50/\text{gal}) = \underline{\$0.48}$$

48¢ worth of gasoline spilled out.

Avoiding Pitfalls

1. Distinguish carefully between *points* and *intervals* when making conversions between temperature scales. The *point* 5°C (five degrees Celsius) is equal to the *point* 41°F, since $T_F = (\%)(5°C) + 32 = 41°F$, whereas an *interval* of 5 C° (five Celsius degrees) is equivalent to an *interval* of 9 F°.

2. The equation $Q = mc\Delta T$ is valid for heat calculations only if no changes of state occur within the temperature range ΔT. If a change of state occurs, $Q = mL$ is needed, and a different specific heat capacity will be required for continued heating or cooling. For example, water and steam have different specific heat capacities.

3. When ice is added to water and melts, be sure to add its mass to the original water mass in calculations involving any subsequent temperature rise.

4. Changes of state occur only at particular temperatures (for a given pressure, usually assumed to be one atmosphere). If there is insufficient heat gained or lost for all the mass to undergo a change of state, the final temperature of the mixture of states will be the temperature at which that change of state occurs.

5. In the first law of thermodynamics, $W + Q = \Delta E_{system}$, the signs of W and Q are important. W is positive if work is done *on* the system, negative if work is done *by* the system; Q is positive if heat flows *into* the system, negative if heat flows *out of* the system. The resulting sign of ΔE_{system} then indicates whether there has been a net gain or loss in the total energy of the system.

6. In using $\Delta L = \alpha L \Delta T$ for linear expansion, α and ΔT must be in a consistent set of units. Since α is ordinarily given in units of $C^{\circ -1}$, ΔT must be given in C°, so Fahrenheit intervals must be converted to Celsius intervals. (Alternatively, the units of α may be converted to $F^{\circ -1}$ by multiplying by the ratio 5 C°/9 F°.) L and ΔL may be measured in any convenient units. The same considerations apply to using the equation $\Delta V = \gamma V \Delta T$.

Drill Problems
Answers in Appendix

1. The blood temperature of the golden hamster sometimes falls as low as 38°F during hibernation. Convert this temperature to the Celsius and Kelvin scales.

2. What is the result of adding 100 grams of steam at 100°C to 500 grams of ice at 0°C?

3. A hiker, weighing 700 N with his equipment, is planning to climb a 1000-m mountain. How many hardboiled eggs (1 egg contains 80 kcal) must he eat if 80% of the food he consumes is required to maintain his metabolism and the rest goes into changing his potential energy? Recall that 1 kcal = 4180 J.

10

Heat Transfer

Terms

Define or describe briefly what is meant by the following terms. If you have difficulty, refer to the textbook section given in parentheses.

conduction (10.1)

heat transfer rate (10.1)

thermal conductivity (10.1)

insulator (10.1)

convection (10.2)

convection coefficient (10.2)

wind-chill temperature (10.2)

electromagnetic radiation (10.3)

heat transfer by radiation (10.3)

emissivity (10.3)

Stefan's law (10.3)

Stefan-Boltzmann constant (10.3)

thermogram (10.3)

evaporation (10.4)

first law of thermodynamics (10.5)

infiltration (10.5)

Equation Review

For each equation, be able to state the situation to which it applies, what quantity each symbol represents, and the units for measuring each quantity in some consistent set of units.

$$H = \frac{\Delta Q}{\Delta t} \tag{10.1}$$

$$H_{cd} = \frac{KA(T_2 - T_1)}{L} \tag{10.2}$$

$$H_{cv} = hA(T_2 - T_1) \tag{10.3}$$

$$R_e = e\sigma A T_1^4 \tag{10.4}$$

$$R_a = e\sigma A T_2^4 \tag{10.5}$$

$$H_r = R_a - R_e = e\sigma A(T_2^4 - T_1^4) \tag{10.6}$$

$$\text{evaporation rate} = \frac{\Delta m}{\Delta t} \tag{10.7}$$

$$H_e = -L_e\frac{\Delta m}{\Delta t} \tag{10.8}$$

$$\frac{\Delta Q}{\Delta t} + \frac{\Delta W}{\Delta t} = \frac{\Delta E_{\text{system}}}{\Delta t} \tag{10.10}$$

$$\frac{\Delta Q}{\Delta t} + \frac{\Delta W}{\Delta t} = \frac{\Delta E_{th}}{\Delta t} + \frac{\Delta E_{\text{chem}}}{\Delta t} \tag{10.11}$$

$$H_{cd} + H_{cv} + H_r + H_e + \frac{\Delta W}{\Delta t} = \frac{\Delta E_{th}}{\Delta t} + \frac{\Delta E_{\text{chem}}}{\Delta t} \tag{10.13}$$

Problems with Solutions and Discussion

PROBLEM: A rod 80 cm long with a radius of 8 cm conducts heat from a large container of boiling water to a large container of melting ice (Fig. 10.1). Assume the rod is well insulated and no heat escapes from its sides. Determine the heat transfer rate into the melting ice if the rod is made of (a) copper, and (b) brass.

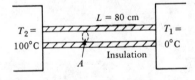

FIG. 10.1

Solution: (a) Choosing the melting ice as region 1, $T_1 = 0°C$ and $T_2 = 100°C$. Equation (10.2) gives the rate of heat transfer by conduction:

$$\frac{\Delta Q}{\Delta t} = \frac{KA(T_2 - T_1)}{L}$$

where

K for copper $= 385$ W/m·C° (from Table 10.1);
$L = 80$ cm $= 0.8$ m; $A = \pi r^2 = \pi(0.08$ m$)^2 = 0.02$ m²;
$T_2 - T_1 = 100°C - 0°C = 100$ C°

Putting in these values,

$$\frac{\Delta Q}{\Delta t} = \frac{(385 \text{ W/m·C°})(0.02 \text{ m}^2)(100 \text{ C°})}{0.8 \text{ m}} = \underline{963 \text{ W}}$$

The copper rod conducts 963 joules of heat into the cold container every second.

(b) For an identical brass rod, the same formula applies, except that for brass, $K = 109$ W/m·C°.

$$\frac{\Delta Q}{\Delta t} = \frac{(109 \text{ W/m·C°})(0.02 \text{ m}^2)(100 \text{ C°})}{(0.8 \text{ m})} = \underline{273 \text{ W}}$$

65

Or, since K_{brass} is smaller than K_{copper} by a factor of $\dfrac{K_{brass}}{K_{copper}} = \dfrac{109}{385} = 0.283$, the brass will transfer heat at 0.283 times the rate that the copper conducts.

$$963 \text{ W}(0.283) = \underline{273 \text{ W}}$$

By either method, the brass rod conducts 273 joules to the cold container each second.

QUESTION: A window can be fitted with a storm window, which consists of a second windowpane mounted outside the first with an air space between them. Discuss how you would expect the rate of heat transfer through a window to be affected by the addition of such a storm window to a single windowpane.

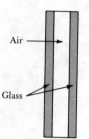
Air
Glass

FIG. 10.2

Answer: In putting in a storm window, the path the heat must pass through has increased from one thickness of glass to a glass-air-glass combination (Fig. 10.2). K_{glass} is 0.8 (from Table 10.1), but K_{air} is 0.024. So the rate of heat transfer through the air should be reduced by a factor of 0.8/0.024, or 33, compared to the rate through the same thickness of glass. And, with air and glass in combination, the heat transfer rate will be less than through either one alone, so the storm window will reduce the heat transfer rate by *more than a factor of 30*.

The air space between the two panes is variable; as it increases, heat loss by conduction decreases, but heat loss by *convection* may become significant. An air space of one-half to three-quarters of an inch is considered optimum.

PROBLEM: Members of the Polar Bear Club make a practice of going swimming in lakes in the middle of winter. A "polar bear" whose skin temperature is 34°C has 1.2 m² of skin area exposed at an outside temperature of 2°C. (a) What is the rate of heat loss due to convection if the wind is blowing at 3 m/s? (b) If a member jumps into a lake whose temperature near the surface is 2°C, what is the rate of heat loss? $h = 19 \text{ W}/\text{m}^2 \cdot \text{C}°$ for a body immersed in water.

Solution: (a)

$$\left(\frac{\Delta Q}{\Delta t}\right)_{cv} = hA(T_{air} - T_{skin}) = \left(26 \, \frac{\text{W}}{\text{m}^2 \cdot \text{C}°}\right)(1.2 \text{ m}^2)(2°\text{C} - 34°\text{C})$$

where the value for the convection coefficient h is obtained from Fig. 10.6 in the text (page 185), for a 3 m/s wind.

$$\frac{\Delta Q}{\Delta t} = (26)(1.2)(-32) \text{ W} = -998 \text{ W}$$

The "polar bear" loses almost 1000 joules of heat per second.

(b)

$$\left(\frac{\Delta Q}{\Delta t}\right)_{cv} = hA(T_{water} - T_{skin}) = \left(19 \, \frac{\text{W}}{\text{m}^2 \cdot \text{C}°}\right)(1.2 \text{ m}^2)(2°\text{C} - 34°\text{C}) = -730 \text{ W}$$

The calculation is similar to that in (a), except that we use the convection coefficient for water. In the water the swimmer loses 730 joules of heat each second—and is better off in cold water than in cold air with a 3 m/s wind. We can find from text Fig. 10.6 what wind speed gives a convection coefficient equal to h for water (19 W/m²·C°). It is a wind speed of about 1.5 m/s. For any wind speed exceeding this value, the swimmer will feel warmer in the water than in the air. And if the air temperature has dropped rather suddenly to 2°C, the lake might still be at a considerably higher temperature than the air. But when the swimmer emerges *wet* from the water into the cold wind, evaporative heat losses are added to convective ones, and he or she will feel *most* uncomfortable!

PROBLEM: A circular stove burner element has a surface area of 360 cm² and a net radiative heat transfer rate of 2000 watts when operating at a setting of "High." The tem-

66

perature of the room walls is 20°C and the emissivity of the burner is 0.95. What is the temperature of the burner?

Solution: Equation (10.6) relates the net radiative heat transfer rate to the temperature of a body and its surroundings:

$$H_r = e\sigma A(T_2^4 - T_1^4)$$

Here the temperatures *must* be absolute temperatures, so we immediately convert to kelvins.

T_1 = burner temperature   and T_2 = wall temperature = 20°C + 273 = 293 K

$A = 360 \text{ cm}^2(1 \text{ m}^2/10^4 \text{ cm}^2) = 0.036 \text{ m}^2$ and $\sigma = 5.67 \times 10^{-8} \text{ W/m}^2 \cdot \text{K}^4$

Rearranging the above equation to solve for T_1,

$$T_2^4 - T_1^4 = \frac{H_r}{e\sigma A}$$

$$(293)^4 - T_1^4 = \frac{-2000}{(0.95)(5.67 \times 10^{-8})(0.036 \text{ m}^2)}$$

where the heat transfer rate of 2000 watts is given a negative sign because it is a heat *loss*.

$$7.37 \times 10^9 - T_1^4 = -1.03 \times 10^{12}$$

$$T_1^4 = 1.04 \times 10^{12}$$

Extracting the square root twice, we find that $T_1 = \underline{1.01 \times 10^3 \text{ K}}$. The temperature of the burner is about 1000 K.

As you know, such a burner will glow red. When an object reaches about 800 K (about 1100°C), enough of the emitted electromagnetic radiation is in the region of visible light that it begins to glow. As temperature increases, more and more energy is radiated as visible light. Stars whose surface temperatures are relatively low (3000 K) have a reddish visual appearance; stars which look blue may have temperatures above 20,000 K. Our sun, an intermediate star, has a surface temperature of about 6000 K.

PROBLEM: The sun has a radius of 6.96×10^8 m. Its surface temperature is 5770 K and it radiates electromagnetic energy at the rate of 3.83×10^{26} joules per second. (These data come from C. W. Allen's *Astrophysical Quantities* [London: Athlone Press, 1973] and are given to three significant figures.) What is the sun's emissivity e to three significant figures?

Solution: The equation $R_e = e\sigma AT^4$ relates an object's temperature to its rate of emission of electromagnetic energy. The surface area A of the sun is needed. The surface area of a sphere is "four great circles," or $4\pi r^2$, where r is the sphere's radius. For the sun,

$$A = 4\pi r^2 = 4\pi(6.96 \times 10^8 \text{ m})^2 = 6.087 \times 10^{18} \text{ m}^2$$

Solving the above equation for e and putting in numerical values, we get

$$e = \frac{R_e}{\sigma AT^4} = \frac{3.83 \times 10^{26}}{(5.67 \times 10^{-8})(6.09 \times 10^{18})(5770)^4} = \underline{1.00}$$

To three significant figures, the sun has an emissivity of 1.

Discussion: An emissivity of 1 means that a body is a *perfect absorber,* absorbing all radiation which falls on it. The absorption of all this energy would cause the body to heat up indefinitely, were it not also a *perfect emitter,* reemitting all the energy it receives, when it is in equilibrium with its surroundings. When the reradiated energy is not visible, the body appears black because it absorbs all visible radiation and reflects none of it. For this reason it is called a *blackbody.* When the temperature of a blackbody is high enough, it radiates visible electromagnetic radiation and appears as a luminous body. Thus the statement that *the sun is a perfect blackbody,* while sounding a bit odd, is in fact true (to three significant figures). The sun is a perfect absorber and emitter of electromagnetic energy.

You may be surprised to realize that the human body is very nearly a perfect black-body, since its emissivity is 0.98, very close to 1. It is not a perfect absorber; it does reflect some visible light, but visible light is a very tiny percentage of all the electromagnetic radiation received by the body. And of course the body temperature is far too low for radiation of visible light; rather it emits infrared radiation.

PROBLEM: Lake Ontario has a surface area of 3560 square miles. If two grams of water evaporate from each square meter of the lake surface every minute, how much heat is lost by the lake in an hour?

Solution: Let us first calculate the number of grams evaporating each minute. This requires converting the lake's area in square miles to square meters:

$$3560 \text{ mi}^2 \left(\frac{1609 \text{ m}}{1 \text{ mi}} \right)^2 = 9.22 \times 10^9 \text{ m}^2$$

(Note that the conversion factor between meters and miles is squared because we are converting *square* miles, an area, rather than a length.) At a rate of 2 grams ($= 2 \times 10^{-3}$ kg) per minute per square meter of surface, the mass evaporated per minute is

$$\frac{\Delta m}{\Delta t} = \left(2 \times 10^{-3} \frac{\text{kg}}{\text{min} \cdot \text{m}^2} \right) (9.22 \times 10^9 \text{ m}^2) = 1.84 \times 10^7 \text{ kg/min}$$

The heat required to evaporate a liquid is given by $\Delta Q = \Delta m L_e$, where L_e for water is 2.4×10^6 J/kg. Therefore, the heat lost per minute due to evaporation of water is

$$\frac{\Delta Q}{\Delta t} = \frac{\Delta m}{\Delta t} L_e = \left(1.84 \times 10^7 \frac{\text{kg}}{\text{min}} \right) \left(2.4 \times 10^6 \frac{\text{J}}{\text{kg}} \right) = 4.4 \times 10^{13} \text{ J/min}$$

The heat lost in an hour will be 60 times greater, since there are 60 minutes in an hour.

$$\frac{\Delta Q}{\Delta t} = \left(4.4 \times 10^{13} \frac{\text{J}}{\text{min}} \right) \left(\frac{60 \text{ min}}{1 \text{ hr}} \right) = 2.6 \times 10^{15} \text{ J/hr}$$

The lake loses 2.6×10^{15} J of heat each hour due to evaporation of water from its surface. The heat loss is very large because the lake is very large. The lake temperature does not alter appreciably in an hour in spite of this loss because it is receiving energy from the sun and surrounding air.

PROBLEM: A woman is sunbathing on the deck of a sailboat moving at 4 m/s. Her exposed surface area is 0.8 m² and her skin temperature is 33°C. Her metabolic rate is 100 watts and she receives radiation from the sun at the rate of 480 watts. The surrounding air is at 28°C. Determine her evaporative heat transfer rate and her perspiration rate, $\Delta m / \Delta t$. Assume that no heat transfer by conduction occurs.

Solution: Equation (10.13)

$$\frac{\Delta Q}{\Delta t} = \frac{\Delta E_{\text{system}}}{\Delta t}$$

shows that the net heat flowing into the system in a given time equals the net energy change within the system during that time. In this case, the system is the woman's body, and her internal energy change is given by her metabolic rate, since we assume her core temperature does not change. The metabolic rate represents an expenditure of energy to keep bodily processes going, so $\Delta E_{\text{system}}/\Delta t = -100$ watts.

Heat enters and leaves the system by convection, radiation, and evaporation, so Eq. (10.13) becomes

$$H_{cv} + H_r + H_e = -100 \text{ W} \tag{1}$$

For convection,

$$H_{cv} = hA(T_{\text{air}} - T_{\text{skin}}) = \left(29 \frac{\text{W}}{\text{m}^2 \cdot \text{C}°} \right) (0.8\text{m}^2) (28°\text{C} - 33°\text{C})$$

Here h is obtained from text Fig. 10.6 for a wind speed of 4 m/s.

$$H_{cv} = -116 \text{ W}$$

For radiation, the woman absorbs solar energy at the rate R_a of 480 watts. She emits radiation at a rate governed by her own temperature:

$$R_e = e\sigma A T^4 = (0.98) \left(5.67 \times 10^{-8} \frac{\text{W}}{\text{m}^2 \cdot \text{K}^4} \right) (0.8 \text{ m}^2)(306 \text{ K})^4 = 390 \text{ W}$$

where e is taken as 0.98 for the human body, and T is converted to kelvins. Her net radiative heat transfer rate is given by Eq. (10.6):

$$H_r = R_a - R_e = 480 - 390 = 90 \text{ W}$$

This is a positive number; she absorbs more energy from the sun than she radiates in a given time. Putting all these heat transfer rate values into equation (1) above, we get

$$-116 \text{ watts} + 90 \text{ watts} + H_e = -100 \text{ W}$$

$$H_e = \underline{-74 \text{ W}}$$

The evaporative rate of heat transfer is 74 watts and it is a loss.
  Her perspiration rate, $\Delta m / \Delta t$, may be obtained from Eq. (10.8),

$$H_e = -\frac{\Delta m}{\Delta t} L_e$$

where L_e is taken as 2.4×10^6 J/kg for water. Solving for $\Delta m / \Delta t$,

$$\frac{\Delta m}{\Delta t} = \frac{-H_e}{L_e} = \frac{-(-74 \text{ J/s})}{2.4 \times 10^6 \text{ J/kg}} = 3.1 \times 10^{-5} \text{ kg/s} = \underline{0.031 \text{ g/s}}$$

Her perspiration rate is 0.031 grams per second, or 112 grams per hour.

PROBLEM: A 60-kg man is running at 2 m/s against a 2 m/s wind. His metabolic rate is 500 watts, and 20 percent of this chemical energy is used up in doing the work of running, while the rest must be released by his body as heat. The surroundings are at 20°C. His skin temperature is 34°C and the exposed surface area is 0.5 m². (a) Determine what percentage of the total heat loss is due to convection, to radiation, and to evaporation, assuming those to be the only heat loss mechanisms occurring. (b) If the body had no way of losing this heat, how much would the runner's temperature rise each hour? Assume that the specific heat capacity of body tissue is the same as for water.

Solution: (a) Since 20 percent of the chemical energy is required to do the running, 80 percent must be released as heat:

$$\text{Total heat loss rate} = 0.8(500 \text{ W}) = 400 \text{ W}$$

  Convective heat loss rate:

$$H_{cv} = hA(T_{\text{air}} - T_{\text{skin}})$$

where the convective coefficient, $h = 29 \text{ W/m}^2 \cdot \text{C}°$, is taken from text Fig. 10.6 for a wind speed of 4 m/s, the relative speed of the 2 m/s wind past the 2 m/s runner.

$$H_{cv} = \left(29 \frac{\text{W}}{\text{m}^2 \cdot \text{C}°} \right) (0.5 \text{ m}^2)(20°\text{C} - 34°\text{C}) = \underline{-203 \text{ W}}$$

203 watts out of 400, or <u>51 percent</u> of the total heat loss rate, is due to convection.
  Radiative heat loss rate:

$$H_r = e\sigma A(T_2^4 - T_1^4)$$

$$\text{where } T_2 = T_{\text{surroundings}} = 20°\text{C} + 273 = 293 \text{ K}$$

$$T_1 = T_{\text{skin}} = 34°\text{C} + 273 = 307 \text{ K}$$

and e is taken as 0.98 for the human body

$$H_r = (0.98)\left(5.67 \times 10^{-8}\,\frac{\text{W}}{\text{m}^2 \cdot \text{K}^4}\right)(0.5\ \text{m}^2)[(293\ \text{K})^4 - (307\ \text{K})^4] = \underline{-42\ \text{W}}$$

42 watts out of 400, or <u>10 percent</u> of the total heat loss rate is due to radiation.

Evaporative heat loss rate: Since this is the only other heat transfer mechanism occurring, it must account for the remainder of the heat loss rate.

$$H_e = \text{total heat loss rate} - H_{cv} - H_r$$
$$= -400\ \text{W} - (-203\ \text{W}) - (-42\ \text{W}) = \underline{-155\ \text{W}}$$

Evaporation therefore accounts for 155 out of 400 watts, or <u>39 percent</u> of the total heat loss rate.

(b) If the 400 J/s of heat could not be transferred out of the body, its temperature would rise according to $Q = cm\Delta T$. Dividing by Δt to get the rate of temperature rise,

$$\frac{\Delta Q}{\Delta t} = cm\,\frac{\Delta T}{\Delta t} \quad \text{or} \quad \frac{\Delta T}{\Delta t} = \frac{\Delta Q}{cm\Delta t}$$

Taking $c = 4180\ \text{J/kg} \cdot \text{C}°$, the specific heat capacity of water, we get

$$\frac{\Delta T}{\Delta t} = \frac{400\ \cancel{\text{J}}/\text{s}}{\left(4180\ \frac{\cancel{\text{J}}}{\cancel{\text{kg}} \cdot \text{C}°}\right)(60\ \cancel{\text{kg}})} = 0.00159\ \frac{\text{C}°}{\text{s}} \times \left(\frac{3600\ \text{s}}{1\ \text{hr}}\right) = \underline{5.7\ \text{C}°/\text{hr}}$$

The runner's temperature would rise at the rate of about 6 C° per hour if there were no heat transfer mechanisms.

The percentages of total heat loss due to convection, radiation, and evaporation obviously depend very much on a person's environment and level of activity. For a resting person, the heat loss rates would be in the neighborhood of 50 percent to radiation, 25 percent to convection, and 25 percent to evaporation of sweat from the skin and water from the lungs.

Avoiding Pitfalls

1. In solving problems involving conduction, convection, and radiation, the algebraic sign of the heat transfer rate will be determined by whether the hotter object or the cooler object is taken as region 1 with temperature T_1. The resulting positive or negative sign indicates whether heat is gained or lost, respectively, by region 1 with temperature T_1.

2. If you can't remember which object was assigned as region 1 with temperature T_1, just remember that in conduction, convection, and radiation, the net heat transfer occurs *from* the hotter region *to* the cooler region. If the region of concern is the hotter region, it will *lose* heat (the transfer terms will be negative). If the region of interest is the cooler region, it will *gain* heat (the transfer terms will be positive).

3. Evaporation is not really a form of heat transfer. It *always* has the effect of cooling a body; H_e always has a negative sign in a first law of thermodynamics equation.

4. The metabolic rate of a living creature is the rate at which chemical energy (derived from food) is made available for bodily processes and activities. In a first law of thermodynamics calculation, the body *expends* this energy, so it is a negative term representing a decrease in the body's internal energy.

5. In conduction and convection problems, temperatures need not be expressed in kelvins. ΔT is the quantity of importance, and since kelvins and Celsius degrees are the same size, ΔT will be the same number on either scale.

6. In radiation problems, temperatures *must* be expressed as absolute temperatures, usually kelvins, because the rate of radiation must increase with increasing temperature. (A temperature scale which passes through 0 and goes negative would not reflect this property of radiative emission. A body at 0°C does radiate, and its radiation rate at 0°C is greater than at -10°C.)

1. A picnic hamper made of styrofoam 1 cm thick has dimensions of 60 cm × 40 cm × 35 cm. If the inside temperature is 30° lower than the outside temperature,

    (a) how many joules of heat per second are entering due to convection through the styrofoam?

    (b) If this is the only heat transfer occurring, how much of the ice inside the hamper will melt in an hour?

2. One object is at 300°C and an identical object is at 900°C. What is the ratio of the energies radiated by the two objects?

3. A nude person with surface area of 1.4 m² is in a sauna where the temperature is 80°C. His skin temperature is 40°C and his metabolic rate is 100 watts. If the primary heat transfer mechanisms occurring are convection, radiation, and evaporation, determine his perspiration rate in grams per second. Take the convection coefficient h to be 5 W/ m²·C°.

11

Entropy and Thermodynamics

Terms

Define or describe briefly what is meant by the following terms. If you have difficulty, refer to the textbook section given in parentheses.

thermodynamics

irreversible process (11.1)

entropy (11.2, 11.4)

macrostate (11.2)

microstate (11.2)

state variable (11.2)

thermodynamic probability (11.2)

factorial (11.2)

Boltzmann constant (11.2)

second law of thermodynamics (11.2, 11.5)

equilibrium state (11.2)

heat engine (11.6)

efficiency (11.6)

heat pump (11.7)

refrigerator (11.7)

coefficient of performance (11.7)

recurrence paradox (11.8)

big bang (11.8)

heat death (11.8)

Equation Review

For each equation, be able to state the situation to which it applies, what quantity each symbol represents, and the units for measuring each quantity in some consistent set of units.

$$P_i = \frac{n!}{n_i! n_h!} \tag{11.1}$$

$$S_i = k \ln P_i \tag{11.2}$$

$$\Delta S = \frac{\Delta Q}{T} \quad \checkmark \quad {}^{J}\!/\!_{k} \tag{11.3}$$

$$\Delta S_{\text{system}} + \Delta S_{\text{environment}} \geq 0 \tag{11.4}$$

$$\Delta S_{\text{system}} \geq 0 \text{ (isolated system)} \tag{11.5}$$

$$e = \frac{\text{useful output}}{\text{input}} \tag{11.6}$$

For a heat engine:

$$e = \frac{W}{Q_{\text{hot}}} \tag{11.7}$$

$$e = 1 - \frac{Q_{\text{cold}}}{Q_{\text{hot}}} \tag{11.8}$$

$$\frac{Q_{\text{cold}}}{Q_{\text{hot}}} \geq \frac{T_{\text{cold}}}{T_{\text{hot}}} \tag{11.9}$$

$$e \leq 1 - \frac{T_{\text{cold}}}{T_{\text{hot}}} \tag{11.10}$$

For a heat pump:

$$\eta = \frac{Q_{\text{hot}}}{W} \tag{11.11}$$

$$\eta = \frac{Q_{\text{hot}}}{Q_{\text{hot}} - Q_{\text{cold}}} \tag{11.12}$$

$$\eta \leq \frac{T_{\text{hot}}}{T_{\text{hot}} - T_{\text{cold}}} \tag{11.14}$$

For a refrigerator:

$$\eta = \frac{Q_{\text{cold}}}{W} \leq \frac{T_{\text{cold}}}{T_{\text{hot}} - T_{\text{cold}}} \tag{11.16}$$

Problems with Solutions and Discussion

PROBLEM: Consider families with 6 children. Assume each child is equally likely to be a boy or a girl. Identify the different possible family makeups in terms of numbers of boys and girls. If these are the macrostates, calculate the number of microstates of each macrostate.

73

Solution: The number of microstates per macrostate is just the thermodynamic probability P_i of that macrostate. Let n = number of children = 6, n_b = number of boys, and n_g = number of girls. Then the number of microstates per macrostate is given by

$$P_i = \frac{n!}{n_b! n_g!}$$

Macrostate 1: All boys ($n_b = 6$, $n_g = 0$)

$$P_1 = \frac{6!}{6!\, 0!} = \underline{1} \quad \text{where } 0! = 1 \text{ by definition}$$

Macrostate 2: 5 boys, 1 girl ($n_b = 5$, $n_g = 1$)

$$P_2 = \frac{6!}{5!\, 1!} = \frac{6 \cdot \cancel{5} \cdot \cancel{4} \cdot \cancel{3} \cdot \cancel{2} \cdot \cancel{1}}{(\cancel{5} \cdot \cancel{4} \cdot \cancel{3} \cdot \cancel{2} \cdot \cancel{1})(1)} = \underline{6}$$

Macrostate 3: 4 boys, 2 girls ($n_b = 4$, $n_g = 2$)

$$P_3 = \frac{6!}{4!\, 2!} = \frac{6 \cdot 5 \cdot \cancel{4} \cdot \cancel{3} \cdot \cancel{2} \cdot \cancel{1}}{(\cancel{4} \cdot \cancel{3} \cdot \cancel{2} \cdot \cancel{1})(2 \cdot 1)} = \frac{30}{2} = \underline{15}$$

Macrostate 4: 3 boys, 3 girls ($n_b = 3$, $n_g = 3$)

$$P_4 = \frac{6!}{3!\, 3!} = \frac{6 \cdot 5 \cdot 4 \cdot \cancel{3} \cdot \cancel{2} \cdot \cancel{1}}{(3 \cdot 2 \cdot 1)(\cancel{3} \cdot \cancel{2} \cdot \cancel{1})} = \frac{120}{6} = \underline{20}$$

Macrostate 5: 2 boys, 4 girls. By symmetry with Macrostate 3, $P_5 = P_3 = \underline{15}.$

Macrostate 6: 1 boy, 5 girls. By symmetry with Macrostate 2, $P_6 = P_2 = \underline{6}.$

Macrostate 7: 0 boys, 6 girls. By symmetry with Macrostate 1, $P_7 = P_1 = \underline{1}.$

Thus, there are seven macrostates, with the numbers of microstates given above. As expected, 3 boys and 3 girls is the most probable distribution, although 4 boys and 2 girls or 2 boys and 4 girls is not far behind. Least probable is a family of all boys or all girls.

PROBLEM: (a) In 100 families with 6 children each, how many would you expect to have all boys? (b) How many would you expect to have two or more children of each sex?

Solution: (a) Adding all the microstates above, we find there are $1 + 6 + 15 + 20 + 15 + 6 + 1 = 64$ microstates within the 7 macrostates. A family of all boys is one such microstate, so the chance of its occurrence is 1 in 64. For 100 families, $\frac{1}{64}(100) = 1.56$. Thus, one or two families out of 100 would be expected to have all boys. (By symmetry, another one or two families would be expected to have all girls.)

(b) Macrostates 3, 4, and 5 have two or more children of each sex; their combined number of microstates is $15 + 20 + 15 = 50$. The chances of one of these microstates occurring is $\frac{50}{64}$, so $\frac{50}{64}(100) = 78$, or 78 families out of 100 would be expected to have two or more children of each sex.

PROBLEM: 500 g of water at 54°C is mixed with 500 g of ethyl alcohol at 46°C. Estimate the change in entropy of the system.

Solution: We need to know the final temperature of the mixture to determine the heat gains and losses occurring during the mixing process. If no heat escapes to the surroundings,

$$Q_{water} + Q_{alcohol} = 0$$
$$(cm\, \Delta T)_{water} + (cm\, \Delta T)_{alcohol} = 0$$

From Table 9.1, $c_{alcohol} = 2480$ J/kg·C° and $c_{water} = 4180$ J/kg·C°

$$\left(4180\, \frac{J}{kg \cdot C°} \right) (0.5\ kg)(T - 54°C) + \left(2480\, \frac{J}{kg \cdot C°} \right) (0.5\ kg)(T - 46°C) = 0$$

$$4180T - 225{,}720 + 2480T - 114{,}080 = 0$$
$$6660T = 339{,}800$$
$$T = 51.02°C$$

Notice that the final equilibrium temperature is *not* the midpoint between the two initial temperatures, even though the masses of the liquids are equal. (Their specific heat capacities are *not* equal.) To estimate ΔS, we will assume the heat gain or loss for each substance occurred at the average temperature during the mixing.

For water,

$$\overline{T} = \frac{54 + 51}{2} = 52.5°C = 52.5 + 273 = 325.5 \text{ K}$$

$$S = \frac{\Delta Q}{\overline{T}} = \frac{cm\,\Delta T}{\overline{T}} = \frac{(4180)(0.5)(51.02 - 54)}{325.5} = \frac{-6230 \text{ J}}{325.5 \text{ K}} = -19.14 \text{ J/K}$$

Here, four significant figures are carried for T to give ΔT to three significant figures.

For alcohol,

$$\overline{T} = \frac{46 + 51}{2} = 48.5°C = 48.5 + 273 = 321.5 \text{ K}$$

$$S = \frac{\Delta Q}{\overline{T}} = \frac{cm\,\Delta T}{\overline{T}} = \frac{(2480)(0.5)(51.02 - 46)}{321.5} = \frac{6230 \text{ J}}{321.5 \text{ K}} = 19.38 \text{ J/K}$$

The water lost heat, its entropy decreased; it became more orderly in cooling (less random motion of atoms and molecules). The alcohol gained heat, its entropy increased; it became less orderly (more random motion of atoms and molecules). The total entropy change for the system is

$$\Delta S_{\text{total}} = -19.14 \text{ J/K} + 19.38 \text{ J/K} = \underline{+0.24 \text{ J/K}}$$

The entropy of this system (assumed isolated) increased slightly during the process, as predicted by the second law of thermodynamics.

PROBLEM: Two grams of steam at 100°C condense into 108 grams of water at 90°C. What is the final change in entropy of the mixture?

Solution: In determining the final temperature of the mixture, recall that the steam will release 2.26×10^6 J of heat per kilogram in condensing to water at 100°C.

$$Q_{\text{steam}} + Q_{\text{water}} = 0$$

$$\underbrace{-mL_v}_{\substack{\text{steam releases}\\\text{heat in condensing}}} \quad \underbrace{+mc(T - 100°)}_{\substack{\text{steam cools}\\\text{to final } T}} \quad \underbrace{+mc(T - 90°)}_{\substack{\text{water warms}\\\text{to final } T}} = 0$$

$$-(0.002)(2.26 \times 10^6) + (0.002)(4180)(T - 100°) + (0.108)(4180)(T - 90°) = 0$$

$$-4520 + 8.36T - 836 + 451T - 40{,}630 = 0$$
$$459T = 45{,}990$$
$$T = 100°C$$

The steam gave up enough heat to raise the temperature of the 108 g of water to 100°C. Thus the steam, after condensing to water, did not cool further but remained at 100°C. Entropy change for the steam:

$$\Delta S = \frac{\Delta Q}{T} = \frac{-mL_v}{T} = \frac{-(0.002 \text{ kg})(2.26 \times 10^6 \text{ J/kg})}{(100 + 273)} = -12.12 \text{ J/K}$$

The steam lost heat and decreased in entropy.

Entropy change for the water: Here we must estimate ΔS, assuming the temperature rise to occur reversibly and the heat flow to occur at the average water temperature of 95°C.

$$\Delta S = \frac{\Delta Q}{\bar{T}} = \frac{cm\,\Delta T}{\bar{T}} = \frac{(4180)(0.108)(10)}{95 + 273} = 12.27 \text{ J/K}$$

The total change of entropy for the steam and water mixture is

$$\Delta S = -12.12 + 12.27 = \underline{0.15 \text{ J/K}}$$

A small amount of entropy, 0.15 J/K, was added to the system in coming to equilibrium.

PROBLEM: What is the maximum possible efficiency of a heat engine operating between 200°C and −100°C?

Solution: The maximum efficiency of a heat engine is given by Eq. (11.10):

$$e = 1 - \frac{T_{cold}}{T_{hot}}$$

Here the temperatures *must* be absolute temperatures for the results to be valid.

$$T_{hot} = 200 + 273 = 473 \text{ K}$$
$$T_{cold} = -100 + 273 = 173 \text{ K}$$

So

$$e = 1 - \frac{173 \text{ K}}{473 \text{ K}} = 1 - 0.37 = \underline{0.63}$$

The maximum possible efficiency for a heat engine operating between the above two temperatures is 63 percent. Notice what would happen if Celsius temperatures had been used.

$$e = 1 - \left(\frac{-100}{200}\right) = 1 + 0.5 = 1.5, \text{ an efficiency of 150 percent!}$$

This result would indicate that something had gone wrong with the calculation, but had both the Celsius temperatures been greater than zero, there would be no such indication. So one must just *remember* that the temperatures *must* be absolute (Kelvin or Rankine) for the results to be valid.

PROBLEM: A heat engine operating at a maximum efficiency of 20 percent does 100 J of work in each cycle. (a) How much heat is absorbed by this engine per cycle? (b) How much heat energy is discarded in each cycle? (c) If the heat is absorbed at 450 K, at what temperature is the heat discarded?

Solution: (a) From Eq. (11.7), $e = W/Q_{hot}$, where $W = 100$ J and $e = 20$ percent $= 0.2$. So

$$Q_{hot} = \frac{W}{e} = \frac{100 \text{ J}}{0.2} = \underline{500 \text{ J}}$$

The engine absorbs 500 J of heat per cycle.

(b) If the engine absorbs 500 J of heat per cycle and does 100 J of work, conservation of energy says it must discard 400 J of heat. This result can also be obtained more formally (but no more accurately) from Eq. (11.8):

$$e = 1 - \frac{Q_{cold}}{Q_{hot}}$$

$$0.2 = 1 - \frac{Q_{cold}}{500 \text{ J}}$$

$$\frac{Q_{cold}}{500 \text{ J}} = 0.8$$

$$Q_{cold} = 500 \text{ J}(0.8) = \underline{400 \text{ J}}$$

The engine discards 400 J of heat per cycle.

(c) Equation (11.10) relates T_{cold} and T_{hot}:

$$e = 1 - \frac{T_{cold}}{T_{hot}}$$

$$0.2 = 1 - \frac{T_{cold}}{450 \text{ K}}$$

$$T_{cold} = 450 \text{ K}(0.8) = \underline{360 \text{ K}}$$

The heat is discarded at a temperature of 360 K.

PROBLEM: A 100-megawatt coal-burning electrical power plant operates at 40 percent efficiency. If the burning of coal yields about 1.4×10^7 J of heat per pound of coal, how much coal must be burned per hour in this plant?

Solution: The plant's power is 100 MW $= 10^8$ W $= 10^8$ J/s, so it produces 10^8 J of electrical energy each second. Therefore in one hour it produces

$$10^8 \frac{\text{J}}{\text{s}} \left(\frac{3600 \text{ s}}{1 \text{ hr}} \right) = 3.6 \times 10^{11} \text{ J}$$

This is its useful work, W. At an efficiency of 40%, it requires a heat input Q_{hot} determined by Eq. (11.7)

$$e = \frac{W}{Q_{hot}} \text{ so } Q_{hot} = \frac{W}{e} = \frac{3.6 \times 10^{11} \text{ J}}{0.4} = 9 \times 10^{11} \text{ J}$$

Therefore each hour enough coal must be burned to produce 9×10^{11} J of heat. The coal yields 1.4×10^7 J of heat per pound, so the required coal each hour is

$$\frac{\text{Energy needed}}{\text{Energy/lb of coal}} = \frac{9 \times 10^{11} \text{ J}}{1.4 \times 10^7 \text{ J/lb}} = \underline{6.43 \times 10^4 \text{ lb}}$$

The plant must burn 6.43×10^4 lb, or about 32 tons of coal each hour.

PROBLEM: The earth's temperature increases toward its center at a rate of about 7 C° per 1000 ft below its surface. A mine 3000 ft beneath the earth's surface must have heat removed continuously for the comfort of the workers. A refrigerator operates between the mine and the earth's surface (at 15°C) at its maximum coefficient of performance. How much heat is removed when the refrigerator does 50,000 J of work?

Solution: The coefficient of performance of a refrigerator is defined as the ratio of the heat it removes to the work done in removing it, or

$$\eta = \frac{Q_{cold}}{W}$$

and is related to the hot and cold temperatures, for the ideal case, by

$$\eta \leq \frac{T_{cold}}{T_{hot} - T_{cold}}$$

Equating these two expressions for η and solving for Q_{out} gives

$$Q_{cold} = W \left(\frac{T_{cold}}{T_{hot} - T_{cold}} \right)$$

In this case the earth's surface is the cold reservoir, $T_{cold} = 15 + 273 = 288$ K. The mine's temperature is higher by 7 C° per thousand feet; at a 3000 ft depth, the increase in temperature is 21 C°. Since $15 + 21 = 36$°C, $T_{cold} = 36 + 273 = 309$ K. Putting in these values gives

$$Q_{cold} = (5 \times 10^4) \left(\frac{288}{309 - 288} \right) = \underline{6.86 \times 10^5 \text{ J}}$$

For each 50,000 J of work, this refrigerator removes 686,000 J of heat from the mine. It removes 13.7 times as many joules of heat as joules of work it performs; hence, its coefficient of performance is 13.7.

Avoiding Pitfalls

1. A system's entropy may increase, decrease, or remain constant during a particular process. The principle that during any real process entropy must increase or remain constant is true only when applied to the universe as a whole. If a system does undergo a process during which entropy decreases, the entropy of the environment must have increased by an equal amount or more.

2. In the equation defining entropy change, $\Delta S = \Delta Q/T$, the temperature T *must* be the *absolute* temperature at which the heat ΔQ enters or leaves the system. However, if ΔQ is given by $cm\,\Delta T$, ΔT may be in any units consistent with the temperature units of c. Recall also that Kelvin and Celsius intervals are equal because Kelvin and Celsius degrees are the same size.

3. In calculating entropy changes, ΔQ is positive if heat flows *into* the region of interest and negative if it flows *out of* the region of interest. The same sign convention for Q applies in equations describing heat engines, heat pumps, and refrigerators, except that the signs for the Q's and for work W have been incorporated into the equations so that the quantities are put in as positive numbers.

4. In the equations describing heat engines, heat pumps, and refrigerators, many ratios of heats and temperatures occur. Take care that the "hots" and the "colds" are not accidentally reversed in these equations. Recall also that the temperatures in these equations are *always* absolute temperatures.

5. A heat pump and a refrigerator operate identically; both transfer heat from a cold region to a warm region by means of doing work. Their names and coefficients of performance are different only because in the case of a heat pump we are interested in how much heat reaches the warm region, while in the case of the refrigerator we are interested in how much heat leaves the cold region.

Drill Problems
Answers in Appendix

1. A metal rod conducts heat from a very large container of boiling water to a very large container of melting ice. Determine the change of entropy in the system when 2000 J of heat flows from the hot container to the cold container. (The rod's entropy does not change.)

2. Two ice cubes, of mass 10 grams each, are put into a thermos containing 430 grams of water at 10°C. Find the approximate change of entropy of the system after it has reached equilibrium.

3. Consider a two-stage heat engine. The first stage takes in heat at 500 K and discards it at 350 K. The second stage takes in at 350 K the heat rejected by the first stage and discards it at 200 K. Both stages work at maximum efficiency.

   (a) What is the efficiency of the two-stage engine and how much work does it do per 1000 J of heat absorbed? (*Hint:* Consider each stage separately.)
   (b) Compare its efficiency to a single engine operating between 500 K and 200 K.

12

Gases

Define or describe briefly what is meant by the following terms. If you have difficulty, refer to the textbook section given in parentheses.

pressure (12.2)

manometer (12.2)

atmosphere (12.2)

bar (12.2)

pascal (12.2)

Torr (12.2)

gauge pressure (12.3)

state variable (12.4)

equation of state (12.4)

Boyle's law (12.4)

Charles' law (12.4)

Boltzmann's constant (12.5)

mole (12.5)

Avogadro's number (12.5)

universal gas constant (12.5)

ideal gas law (12.5)

kinetic theory of gases (12.6)

root-mean-square speed (12.6)

P-V diagram (12.8)

diffusion (12.9)

osmosis (12.10)

osmotic pressure (12.10)

molar concentration (12.10)

Equation Review

For each equation, be able to state the situation to which it applies, what quantity each symbol represents, and the units for measuring each quantity in some consistent set of units.

$$P = \frac{F}{A} \qquad\qquad\qquad (12.1)$$

$$P = P_{\text{atm}} + P_{\text{gauge}} \qquad\qquad\qquad (12.2)$$

$$PV = \text{constant (at constant } T) \qquad\qquad\qquad (12.3)$$

$$PV = NkT \qquad\qquad\qquad (12.5)$$

$$N = nN_A \qquad\qquad\qquad (12.7)$$

$$PV = nRT \qquad\qquad\qquad (12.8)$$

$$\tfrac{1}{2}m\overline{v^2} = \tfrac{3}{2}kT \qquad\qquad\qquad (12.12)$$

$$E_{\text{th}} = N(\tfrac{3}{2})kT \qquad\qquad\qquad (12.13)$$

$$v_{\text{rms}} = \sqrt{\overline{v^2}} = \sqrt{\frac{3kT}{m}} \qquad\qquad\qquad (12.14)$$

$$W = -P\,\Delta V \qquad\qquad\qquad (12.15)$$

$$P_{\text{os}} = \frac{nRT}{V} \qquad\qquad\qquad (12.18)$$

$$c = \frac{n}{V} \qquad\qquad\qquad (12.19)$$

$$P_{\text{os}} = cRT \qquad\qquad\qquad (12.20)$$

Problems with Solutions and Discussion

PROBLEM: The steel runner of an ice skate has a length of 8 inches and a width of $\frac{1}{10}$ inch. What pressure is exerted on the ice by a 160-lb skater when both skates are in contact with the ice? Express the result in lb/in^2 and in atmospheres.

Solution: Pressure is defined as perpendicular force per unit area. The area of contact of one skate is A_{skate} = length $\times$ width = (8 in)($\frac{1}{10}$ in) = 0.8 in^2. If both skates are in contact with the ice, the total area is twice the above, or 1.6 in^2. The force on this area is just the weight of the skater, 160 lb. So

$$P = \frac{F}{A} = \frac{160 \text{ lb}}{1.6 \text{ in}^2} = \underline{100 \text{ lb/in}^2}$$

The skater exerts a pressure of 100 lb/in^2. To express this pressure in atmospheres, we find the appropriate conversion factor in Table 12.1: 1 atm = 14.7 lb/in^2, so

$$P = 100 \text{ lb/in}^2 \left(\frac{1 \text{ atm}}{14.7 \text{ lb/in}^2} \right) = \underline{6.8 \text{ atm}}$$

The pressure exerted by the skater is almost 7 times the pressure exerted by the atmosphere at sea level.

PROBLEM: The atomic mass of gold is 197 u; therefore, 1 mole of gold weighs 197 grams. The volume of a mole of gold is about 10 cm³, or a cube about 2 cm on an edge. If one mole of gold were distributed equally among the four billion people on earth, how many atoms of gold would each person have?

Solution: One mole of gold contains Avogadro's number of atoms, or 6.02×10^{23} atoms. Dividing these among the world's population gives

$$\frac{6.02 \times 10^{23} \text{ atoms}}{4 \times 10^9 \text{ persons}} = 1.5 \times 10^{14} \frac{\text{atoms}}{\text{person}}$$

The number of atoms per person is still too large to comprehend easily. The real point of such calculations is that atoms are very very small.

PROBLEM: The best vacuum that can be produced in the laboratory is about 10^{-15} atmospheres. How many molecules still exist in one cubic centimeter of space at this pressure and at a temperature of 0°C?

Solution: Since the number of molecules N is asked for, we choose the form of the ideal gas law that expresses the amount of gas as a number of molecules rather than a number of moles.

$$PV = Nkt$$

where

$$P = 10^{-15} \text{ atm} \left(\frac{1.01 \times 10^5 \text{ N/m}^2}{1 \text{ atm}} \right) = 1.01 \times 10^{-10} \text{ N/m}^2$$

$$V = 1 \text{ cm}^3 \left(\frac{1 \text{ m}}{100 \text{ cm}} \right)^3 = 10^{-6} \text{ m}^3$$

$$k = 1.38 \times 10^{-23} \text{ J/K}$$

$$T = 0°C = 273 \text{ K}$$

Solving the above equation for N gives

$$N = \frac{PV}{kT} = \frac{(1.01 \times 10^{-10} \text{ N/m}^2)(10^{-6} \text{ m}^3)}{(1.38 \times 10^{-23} \text{ J/K})(273 \text{ K})} = 2.68 \times 10^4 \text{ molecules}$$

Even in this excellent vacuum, there are about 27,000 molecules per cubic centimeter. The vacuum of interstellar space contains about one atom per cubic centimeter.

PROBLEM: A mass of 110 grams of CO_2 gas at a pressure of 1 atm and 0°C is compressed to one third of its original volume and a pressure of 4 atm. What is its new temperature, in °C, and what was its original volume?

Solution: The ratio form of the general gas law can be used to obtain the new temperature, since the mass of gas remains unchanged during the compression.

$$\frac{P_1 V_1}{T_1} = \frac{P_2 V_2}{T_2} \quad \text{so} \quad T_2 = \left(\frac{P_2}{P_1} \right) \left(\frac{V_2}{V_1} \right) T_1$$

where $P_1 = 1$ atm, $P_2 = 4$ atm, $T_1 = 0°C = 273$ K, and $V_2 = \frac{1}{3}V_1$.

$$T_2 = \left(\frac{4 \text{ atm}}{1 \text{ atm}} \right) \left(\frac{\frac{1}{3} V_1}{V_1} \right) (273 \text{ K}) = \frac{4}{3} (273 \text{ K}) = 364 \text{ K}$$

In Celsius degrees, $T_2 = 364 - 273 = 91°C$. Note that in such ratios, any units of volume and absolute pressure may be used, since the units cancel out. However, temperatures *must* be absolute for the results to be valid, as can be seen by considering what would happen if a temperature of 0°C, a perfectly common gas temperature, were used in the calculation above.

The original volume can be obtained from the ideal gas law: $PV = nRT$. Solving for V gives

$$V = \frac{nRT}{P}$$

Now we must be careful to express all quantities in a consistent set of units.

$$R = 8.314 \text{ J/mole} \cdot \text{K}$$
$$T = 0°C + 273 = 273 \text{ K}$$
$$P = 1 \text{ atm} = 1.01 \times 10^5 \text{ N/m}^2 \text{ (from Table 12.1)}$$

To find n, the number of moles of CO_2, we first determine the number of grams in one mole. The molecular mass of CO_2 is 12 (carbon) $+ 2 \times 16$ (two oxygen) $= 12 + 32 = 44$; thus, one mole of CO_2 has a mass of 44 g. So 110 grams of CO_2 contains the number of moles given by

$$\frac{110 \text{ g}}{44 \text{ g/mole}} = 2.5 \text{ moles} = n$$

Therefore,

$$V = \frac{nRT}{P} = \frac{(2.5)(8.314)(273)}{1.01 \times 10^5} = \underline{5.62 \times 10^{-2} \text{ m}^3}$$

This volume of 0.056 m³ is the volume of a cube about 38 cm on an edge.

PROBLEM: Two constituents of air are oxygen (O_2) and water (H_2O) molecules. What is the ratio of the root-mean-square speeds of the water and the oxygen molecules in the same volume of air?

Solution: From Eq. (12.14)

$$v_{rms} = \sqrt{\frac{3kT}{m}}$$

So the desired ratio is

$$\frac{v_{rms}(H_2O)}{v_{rms}(O_2)} = \frac{\sqrt{3kT/m(H_2O)}}{\sqrt{3kT/m(O_2)}} = \sqrt{\frac{m(O_2)}{m(H_2O)}}$$

Note that while the root-mean-square speeds themselves depend on temperature, the *ratio* of the two speeds is independent of temperature, provided they have the same temperature. In this case they do, since they are in the same volume of air. We need the mass of each molecule.

For O_2, one mole has a mass of 32 grams and contains 6.02×10^{23} molecules, so

$$m(O_2) = \frac{32 \text{ g/mole}}{6.02 \times 10^{23} \text{ molecules/mole}} = 5.32 \times 10^{-23} \text{ g/molecule}$$

For H_2O, one mole has a mass of 18 grams and also contains 6.02×10^{23} molecules.

$$m(H_2O) = \frac{18 \text{ g/mole}}{6.02 \times 10^{23} \text{ molecules/mole}} = 2.99 \times 10^{-23} \text{ g/molecule}$$

Putting in the values for these masses,

$$\frac{v_{rms}(H_2O)}{v_{rms}(O_2)} = \sqrt{\frac{5.32 \times 10^{-23}}{2.99 \times 10^{-23}}} = \sqrt{1.78} = \underline{1.33}$$

The rms speed of the water molecules is greater than that of the oxygen molecules by a factor of 1.33. Since both types of molecules are at the same temperature, they have the same average kinetic energy. So the ones with the smaller mass (water) must have the larger speed, and vice versa.

PROBLEM: One mole of an ideal gas at a pressure of one atmosphere occupies a volume of 0.04 m³. It is compressed to half its volume at constant temperature. (a) Show this process on a P-V diagram. (b) What is the temperature at which this process occurs?

Solution: (a) We know that the new volume will be $\frac{1}{2}V_1 = \frac{1}{2}(0.04 \text{ m}^3) = 0.02 \text{ m}^3$. We can get the new pressure from Boyle's law:

$$P_1 V_1 = P_2 V_2$$

$$P_2 = P_1\left(\frac{V_1}{V_2}\right) = (1 \text{ atm})\left(\frac{0.04 \text{ m}^3}{0.02 \text{ m}^3}\right) = 2 \text{ atm}$$

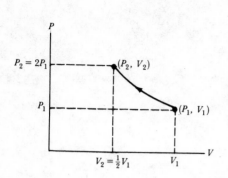

FIG. 12.1

The pressure has doubled. From Table 12.1, 1 atm = $1.01 \times 10^5 \text{ N/m}^2 = P_1$. Thus, $P_2 = 2P_1 = 2 (2 \times 10^5 \text{ N/m}^2)$. The P-V diagram is shown in Fig. 12.1. That the line is curved, not straight, can be seen by considering P as a function of V for constant temperature:

$$P = \frac{nRT}{V} = \frac{\text{constant}}{V}$$

Thus, P approaches infinity as V approaches zero. Were the curve a straight line, it would predict that at $3P_1$, the gas would have zero volume!

(b) The temperature can be determined from the ideal gas law for either the initial or the final state. From the initial state, $P_1 V_1 = nRT_1$, so

$$T_1 = \frac{P_1 V_1}{nR} = \frac{(1.01 \times 10^5 \text{ N/m}^2)(0.04 \text{ m}^3)}{(1 \text{ mole})(8.314 \text{ J/mole}\cdot\text{K})} = \underline{486 \text{ K}}$$

Verify for yourself that the same temperature results from the ideal gas law for the final state.

PROBLEM: Consider one mole of gas at a pressure of 1 atmosphere and a temperature of 27°C. It expands at constant pressure to three times its original volume. (a) Show this process on a P-V diagram. (b) How much work did the gas do?

Solution: (a) We are given that the pressure does not change; $P_1 = P_2 = P$; and that the volume triples; $V_2 = 3V_1$. Choosing an arbitrary point for (P_1,V_1) on the graph, we can locate the point $3V_1$ at this same P. The graph for this process is shown in Fig. 12.2.

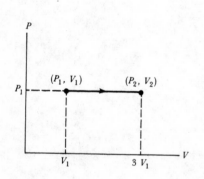

FIG. 12.2

(b) Work done by a gas is given by $W = -P \Delta V$. To get into the metric system of units, let us convert pressure in atmospheres to N/m². From Table 12.1, 1 atm = $1.01 \times 10^5 \text{ N/m}^2 = P$. We need $\Delta V = V_2 - V_1$, but V_1 was not given. It can, however, be obtained from the ideal gas law, $PV = nRT$, where $n = 1$ mole; $T = 27°C = 27° + 273 = 300$ K; and R is the universal gas constant.

$$V_1 = \frac{nRT_1}{P_1} = \frac{(1 \text{ mole})(8.314 \text{ J/mole}\cdot\text{K})(300 \text{ K})}{1.01 \times 10^5 \text{ N/m}^2}$$

$$V_1 = 2.47 \times 10^{-2} \text{ m}^3$$

Here the units have been analyzed as follows:

$$\frac{\text{J}}{\text{N/m}^2} = \frac{\text{J}\cdot\text{m}^2}{\text{N}} = \frac{(\text{N}\cdot\text{m})\text{m}^2}{\text{N}} = \text{m}^3$$

Now

$$V_2 = 3V_1 = 3(0.0247 \text{ m}^3) = 0.0741 \text{ m}^3, \text{ so } \Delta V = V_2 - V_1$$

$$= 0.0741 - 0.0247 = 0.0494 \text{ m}^3$$

Thus,

$$W = -P \Delta V = -(1.01 \times 10^5 \text{ N/m}^2)(0.0494 \text{ m}^3) = \underline{-4.99 \times 10^3 \text{ J}}$$

About 5000 J of work are done by the gas expanding. The negative sign indicates that the gas must lose internal energy in order to do the external work of expansion.

PROBLEM: For the process described in the previous problem, (a) what is the new temperature of the gas after expansion? (b) Did any heat flow into or out of the gas during expansion?

Solution: (a) The ideal gas law states that for a given quantity of gas,

$$\frac{P_1 V_1}{T_1} = \frac{P_2 V_2}{T_2}$$

In this case, $P_1 = P_2 = P$; $V_2 = 3V_1$; and $T_1 = 300$ K. Substituting these values and solving for T_2 we get

$$\frac{PV_1}{300 \text{ K}} = \frac{P(3V_1)}{T_2}$$

$$T_2 = 3(300 \text{ K}) = \underline{900 \text{ K}}$$

(b) The gas did some work, which would decrease its internal energy. However, its temperature *rose*, meaning, according to Eq. (12.13), that its internal energy increased in direct proportion to the temperature. Therefore, *heat must have flowed into the gas.* In fact, enough heat must enter the gas to do the external work and also to raise the temperature of the gas.

We can calculate the amount of heat that flowed in from the first law of thermodynamics, $Q + W = \Delta E_{internal}$, if we recall that for an ideal monatomic gas $E_{internal}$ is just the random kinetic energies of the molecules,

$$E_{internal} = N(\tfrac{1}{2}mv^2) = N(\tfrac{3}{2})kT = \tfrac{3}{2}nRT$$

Thus, $\Delta E_{internal} = \tfrac{3}{2}nR \, \Delta T$. T increased from 300 K to 900 K, so $\Delta T = 600$ K. So the change in internal energy is $\Delta E = \tfrac{3}{2}(1 \text{ mole})(8.314 \text{ J/mole} \cdot \text{K})(600 \text{ K}) = 7483 \text{ J} \approx 7500$ J. Solving the first law of thermodynamics for Q and substituting numbers gives

$$Q = \Delta E - W = 7500 \text{ J} - (-5000 \text{ J}) = \underline{12,500 \text{ J}}$$

12,500 joules of heat flowed into the gas during the expansion.

PROBLEM: Describe in words the processes AB, BC, and CA shown on the P-V diagram in Fig. 12.3 for a certain quantity of an ideal gas.

Solution: In process AB, the pressure decreases but the volume is unchanged. According to the ideal gas law, the only other state variable that can change is the temperature,

$$T = PV/nR$$

If P decreases, so does T, all other quantities remaining constant. Therefore, during the process AB, the gas was cooled at constant volume, decreasing its pressure.

In process BC, the volume increases; the gas is expanding. But the pressure remains constant. Ordinarily, expanding a gas decreases its pressure—molecules are now farther from each other and the walls than before. Therefore to keep the pressure constant during the expansion, the gas must be heated, raising its temperature. To see this from the gas law, again,

$$T = PV/nR$$

so if P is constant, T must increase as V increases. During the process BC, the gas was heated so that it expanded at constant pressure. With no scale on either axis of the graph, we do not have enough information to determine whether the rise in temperature during BC was equal to the drop in temperature during AB.

In process CA, the volume of the gas decreases and the pressure increases. The gas is being compressed. This requires external work to be done on the gas. (Gases do not compress themselves; they fill the entire volume available to them.) Since the gas returns to its original pressure and volume, it has also returned to its original temperature. During process CA, work is done on the gas, compressing it to its original volume and increasing its pressure.

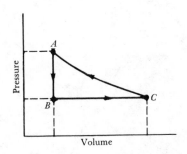

FIG. 12.3

85

Such a series of processes, in which a substance is returned to its initial state (original P, V, and T), is called a *cycle*. The net work done during the cycle is equal to the area enclosed by the paths on the P-V diagram.

PROBLEM: Seawater contains 450 moles/m³ of dissolved salt. Each dissolved salt molecule provides two ions, Na^+ and Cl^-. If seawater and pure water, both at 20°C, are separated by a membrane not permeable to the salt ions, what is the osmotic pressure and how high a mercury column would this pressure support?

Solution: Osmotic pressure is given by Eq. (12.18),

$$P_{os} = \frac{nRT}{V} = cRT$$

where c = concentration of ions in solution = $450 \frac{\text{moles}}{\text{m}^3}$ of salt $\times 2 \frac{\text{ions}}{\text{molecule}}$ = $900 \frac{\text{moles}}{\text{m}^3}$ of ions, and $T = 20 + 273 = 293$ K. So

$$P_{os} = (900)(8.314)(293) = \underline{21.9 \times 10^5 \text{ N/m}^2}, \text{ or } \underline{2190 \text{ kPa}}$$

where the correct *mks* units of pressure are assigned, since all quantities were in correct *mks* units. Dividing by 1.01×10^5 N/m² per atmosphere, we find that this is about 22 atm. From Table 12.1, 1 atm will support a column of mercury 76 cm high, so 22 atmospheres will support a mercury column $76 \times 22 = 1672$ cm, or about $\underline{16.7 \text{ meters}}$ high.

That the osmotic pressure should be correctly given (for low molar concentrations of solute) by the equation for an *ideal gas* is really quite remarkable, since (1) liquids, not gases, are usually involved, and (2) the number of moles of nonpenetrating particles are usually dissolved in a much larger number of solvent molecules. It is also of interest to note that the n in Eq. (12.18) refers to particles which will not pass through the membrane. Thus a dissolved substance which ionizes, like NaCl, producing two such particles per molecule, is more effective in producing osmotic pressure than an equal amount of a nonionizing substance like sugar.

Avoiding Pitfalls

1. The state of a given quantity of an ideal gas depends on only three properties: pressure, volume, and absolute temperature. Thus if any two are known, the third can be calculated from the ideal gas law.

2. If the quantity of gas does not change during a process, the ideal gas law can be used in the form: $P_1V_1/T_1 = P_2V_2/T_2$, and an unknown state variable can be determined without knowing the quantity of gas. In such cases, any convenient units of volume, absolute pressure, and absolute temperature can be used.

3. When using the ideal gas law in the form $PV = nRT$ or $PV = NkT$, *all* quantities must be in *one* consistent set of units.

4. Regardless of which form is being used, the ideal gas law and other equations of state for an ideal gas require *absolute* (not gauge) pressures and absolute temperatures.

5. A *mole* of any ideal gas contains the same number of molecules and occupies the same volume at any given temperature and pressure, however its *mass* depends on what kind of gas it is.

6. For an ideal gas, the average translational kinetic energy per molecule, $\frac{3}{2}kT$, depends on absolute temperature only, not on the species, or mass, or number of molecules, or volume, or pressure of the gas.

7. When a gas is compressed, work is done *on* it. A gas does not compress itself naturally; work must be done on the gas to achieve this.

8. The formula $W = P \Delta V$, for work done when a gas changes volume, is accurate only if pressure remains constant during the volume change. If it does not, the formula

gives only an approximate result when P is taken as the average pressure during the process. Graphical or calculus methods are required for accurate results.

9.   The sign convention for work W is such that if work is done *on* the gas (thereby adding energy to the gas), W is positive; if work is done *by* the gas (thereby causing it to lose energy), W is negative. The sign convention for heat is similar: Q is positive if heat flows *into* the gas, negative if it flows *out of* the gas.

10.   The ideal gas law can be used to relate certain state variables before and after a process, but it gives no direct indication of whether heat flowed in or out during the process. This information is usually obtained from the first law of thermodynamics.

11.   The equation describing osmotic pressure is identical in form to the ideal gas law. Recall, however, that in this case n is the number of moles of particles that do not pass through the membrane. Thus a solute which ionizes will produce twice as great a molar concentration of such particles as its own molar concentration indicates.

Drill Problems
Answers in Appendix

1.   A certain quantity of gas expands at constant temperature to three times its original volume. It is then heated at constant volume until it returns to its original pressure. By what factor has its temperature changed?

2.   Carbon has atomic mass number 12 and chlorine has atomic mass number 35.

(a) What is the mass of one mole of carbon tetrachloride, CCl_4?
(b) What is the volume of one mole of carbon tetrachloride at a pressure of one atmosphere and 0°C (often called *standard conditions,* or *standard temperature and pressure*)?

3.   One mole of an ideal gas at 300 K undergoes a process in which it decreases to half its original volume, while the pressure triples.

(a) How did the temperature of the gas change?
(b) Show the process on a P-V diagram.
(c) If it took 1000 J of work to compress the gas, how much heat, if any, flowed into or out of the gas?

13

Fluid Statics

Terms

Define or describe briefly what is meant by the following terms. If you have difficulty, refer to the textbook section given in parentheses.

fluid

density (13.2)

weight density (13.2)

specific gravity (13.2)

Pascal's principle (13.4)

Archimedes' principle (13.5)

buoyant force (13.5)

surface tension (13.6)

cohesion (13.7)

adhesion (13.7)

capillary action (13.7)

Equation Review

For each equation, be able to state the situation to which it applies, what quantity each symbol represents, and the units for measuring each quantity in some consistent set of units.

$$F = PA \tag{12.1}$$

$$\rho = \frac{m}{V} \tag{13.1}$$

$$P_1 = P_2 + \rho g(y_2 - y_1) \tag{13.3}$$

$$B = \rho_f g V_f \tag{13.6}$$

$$\gamma = \frac{F}{l} \tag{13.11}$$

$$h = \frac{2\gamma}{\rho g r} \tag{13.12}$$

Problems with Solutions and Discussion

PROBLEM: What volume would be occupied by 1 kg of cheese of density 1400 kg/m³? If this cheese were in the shape of a disk 3 cm thick, what would its radius be?

Solution: Density is defined as mass per unit volume, $\rho = m/V$. Solving for V gives

$$V = \frac{m}{\rho} = \frac{1 \text{ kg}}{1400 \text{ kg/m}^3} = 7.1 \times 10^{-4} \text{ m}^3$$

Converting to cubic centimeters, a more familiar volume unit,

$$7.1 \times 10^{-4} \text{ m}^3 \left(\frac{100 \text{ cm}}{1 \text{ m}} \right)^3 = 710 \text{ cm}^3$$

The volume of the cheese is 710 cm³. To determine its radius, recall that a disk is just a short cylinder, so its volume is given by the cross-sectional area times the thickness, $V = \pi r^2 t$. Substituting values,

$$710 \text{ cm}^3 = \pi r^2 (3 \text{ cm})$$
$$r^2 = \frac{710 \text{ cm}^3}{3.14(3 \text{ cm})} = 75.4 \text{ cm}^2$$
$$r = \underline{8.7 \text{ cm}}$$

This disk has a radius of about 9 cm.

PROBLEM: Now consider a 1 kg block of Swiss cheese of density 1400 kg/m³. If this block has dimensions 20 cm × 20 cm × 3 cm, what percentage of the total block volume is made up of holes?

Solution: The volume occupied by 1 kg of cheese of density 1400 kg/m³ is, as before, 710 cm³. But the block of Swiss cheese has a total volume of 20 × 20 × 3 = 1200 cm³. Therefore the cheese occupies a fractional volume of the block given by

$$\frac{V_{\text{cheese}}}{V_{\text{block}}} = \frac{710 \text{ cm}^3}{1200 \text{ cm}^3} = 0.59$$

Since the cheese occupies 59 percent of the total volume, the remaining 41 percent of the volume consists of holes.

PROBLEM: The standard queen-size waterbed has dimensions of 5 ft × 7 ft × 8 in. The weight density of water is 62.4 lb/ft³. Determine the total force, and the pressure, that such a waterbed exerts on the floor.

Solution: The force is the weight of the enclosed water, w. Since weight density D is defined as weight per unit volume, the water's weight may be expressed as

$$w = DV = (62.4 \text{ lb/ft}^3)(5 \times 7 \times \tfrac{2}{3})\text{ft}^3 = (62.4)(23.3)\text{lb} = \underline{1450 \text{ lb}}$$

The total force of the water on the floor is its weight, 1450 lb. You can understand the reluctance of landlords to allow waterbeds to be installed anywhere except on ground floors, even if you have insurance to cover damage due to leaks.

The pressure exerted on the floor by the waterbed is

$$P = \frac{F}{A} = \frac{\text{weight of water}}{\text{area of bed bottom}} = \frac{1450 \text{ lb}}{(5 \times 7)\text{ft}^2} = \underline{41.4 \text{ lb/ft}^2}$$

PROBLEM: Suppose you are getting a transfusion of blood plasma from a bottle 0.8 m above your arm (Fig. 13.1). What is the pressure difference between the blood entering your arm and that in the bottle?

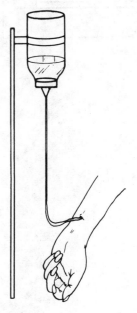

FIG. 13.1

89

Solution: Equation (13.3) relates pressure differences to differences of height in a fluid column:

$$P_1 = P_2 + \rho g(y_2 - y_1)$$

If y_1 is the lower position in the fluid column, its excess in pressure above the pressure at y_2 is given by

$$P_1 - P_2 = \Delta P = \rho g(y_2 - y_1)$$

In this case $(y_2 - y_1) = 0.8$ m, and ρ for blood plasma is 1030 kg/m³ from Table 13.1.

$$\Delta P = (1030 \text{ kg/m}^3)(9.8 \text{ m/s}^2)(0.8 \text{ m}) = \underline{8100 \text{ N/m}^2} = \underline{8.10 \text{ kPa}}$$

The pressure difference is 8100 N/m², or 8.10 kilopascals.

PROBLEM: A hydraulic "elephant lift" is used to lift elephants onto a circus train. The large circular platform which raises the elephant is 2 m in diameter. What is the diameter of the small piston if a 120-lb keeper can raise a 12,000-lb elephant by stepping onto it?

Solution: The pressure exerted by the keeper is transmitted undiminished through the hydraulic fluid, according to Pascal's principle, in particular to the piston supporting the elephant. Since pressure is defined as force per unit area,

$$\text{Pressure on small piston} = \frac{F_1}{A_1} = \frac{F_2}{A_2} = \text{pressure on large piston}$$

where

$$F_1 = \text{keeper's weight} = 120 \text{ lb, and } A_1 = \pi r_1^2$$
$$F_2 = \text{elephant's weight} = 12,000 \text{ lb}$$
$$A_2 = \pi r_2^2 = \pi(1 \text{ m})^2, \text{ where the piston's radius is half its diameter}$$

Solving for A_1 and substituting these values gives

$$A_1 = \left(\frac{F_1}{F_2}\right) A_2$$

$$\cancel{\pi} r_1^2 = \frac{120 \cancel{\text{ lb}}}{12,000 \cancel{\text{ lb}}} \times \cancel{\pi}(1 \text{ m})^2$$

$$r_1^2 = 0.01 \text{ m}^2; \quad \text{hence} \quad r_1 = 0.1 \text{ m}$$

The diameter of the small piston (twice its radius) need be only 0.2 m or <u>20 cm.</u>

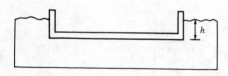

FIG. 13.2

PROBLEM: A raft is made in the form of a box 3 m × 4 m × 0.5 m deep. The raft has a mass of 1500 kg. Determine how far water comes up on the side, call it h, as the raft floats in fresh water (Fig. 13.2).

Solution: The raft will sink until it is in equilibrium with its downward weight exactly supported by the upward buoyant force.

$$\text{Weight of raft} \downarrow = \text{Buoyant force} \uparrow$$
$$mg_{\text{raft}} = B$$

According to Archimedes' principle, B is equal to the weight mg of the displaced water. The definition for density, $\rho = m/V$, allows us to replace m by ρV, so $B = \rho V g$, where ρ and V are properties of the displaced fluid.

$$mg_{\text{raft}} = \rho_f V_f g$$
$$1500 \text{ kg} = (1000 \text{ kg/m}^3) V_f$$
$$V_f = \frac{1500 \text{ kg}}{1000 \text{ kg/m}^3} = 1.5 \text{ m}^3$$

The raft must displace a volume of 1.5 m³ in order to float. The raft's submerged volume is given by $A \times h$, where A is its area.

$$V_f = Ah$$

$$h = \frac{V_f}{A} = \frac{1.5 \text{ m}^3}{(3 \times 4) \text{ m}^2} = 0.125 \text{ m} = \underline{12.5 \text{ cm}}$$

The raft floats with 12.5 cm submerged.

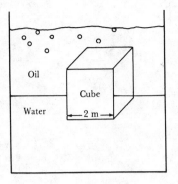

PROBLEM: A cube two meters on an edge floats as shown in Fig. 13.3 such that half the cube is above and half below the oil-water boundary. The oil's density is 600 kg/m³. (a) Use Archimedes' principle to determine the mass of the cube. (b) What is its density?

FIG. 13.3

Solution: (a) Archimedes' principle says that the upward buoyant force B is equal to the weight of the displaced fluid. In this case, some oil and some water are displaced.

$$B = m_{oil}g + m_{water}g$$

Since the cube is suspended in equilibrium, the downward weight is exactly balanced by the upward buoyant force, or

$$m_{cube}g = m_{oil}g + m_{water}g$$

Since $m = \rho V$, we may say that

$$m_{cube} = \rho_o V_o + \rho_w V_w$$

V_o and V_w are both equal to half the cube's volume, or ½(2 m $\times$ 2 m $\times$ 2 m) = 4 m³.

$$m_{cube} = (600 \text{ kg/m}^3)(4 \text{ m}^3) + (1000 \text{ kg/m}^3)(4 \text{ m}^3) = 2400 \text{ kg} + 4000 \text{ kg} = \underline{6400 \text{ kg}}$$

(b) The density of the cube is given by

$$\rho = \frac{m}{V} = \frac{6400 \text{ kg}}{(2 \times 2 \times 2)\text{m}^3} = \frac{6400 \text{ kg}}{8 \text{ m}^3} = \underline{800 \text{ kg/m}^3}$$

Its density is 800 kg/m³, halfway between the oil's density and the water's density.

What conclusion could you draw about its density if it floated such that ¾ of its volume were below the oil-water boundary? If you're not sure, work it out, following the pattern of the above solution.

QUESTION: A cubic meter of oak floats ⅝ submerged in a tank of water while a cubic meter of lead lies completely submerged at the bottom of the tank. Which experiences the greater upward buoyant force?

Answer: Archimedes' principle says that the upward buoyant force is equal to the weight of the displaced fluid. The lead and oak are of equal volume, but the oak is not completely submerged; it displaces only ⅝ as much water as the lead. Since more water is displaced by the lead, *the lead experiences a greater upward buoyant force than the oak.*

You may have been tempted to think the oak experienced a greater buoyant force than the lead because it floated. Whether or not an object floats depends on whether or not the buoyant force is equal to or smaller than the weight of the object. A small buoyant force is sufficient to support a very light object (of low density); even a very large buoyant force may be inadequate to support a very heavy (dense) object.

PROBLEM: A balloon is filled with helium ($\rho_{He} = 0.178$ kg/m³) until it has a radius of 60 cm. The mass of the balloon itself is 0.5 kg. Surrounded by the atmosphere of density $\rho_a = 1.25$ kg/m³, the balloon feels an upward buoyant force. (a) Use Archimedes' principle to determine the upward buoyant force on the balloon. (b) What will be its initial acceleration when released?

Solution: (a) By Archimedes' principle, the upward buoyant force equals the weight of the displaced air.

91

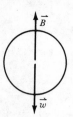

FIG. 13.4

$$B = m_{air}g = \rho_a V_a g = \rho_a(\tfrac{4}{3}\pi r^3)g$$

where $\tfrac{4}{3}\pi r^3$ is the volume of the balloon and also the volume of the displaced air.

$$B = (1.25 \text{ kg/m}^3)(\tfrac{4}{3}\pi)(0.6 \text{ m})^3(9.8 \text{ m/s}^2) = \underline{11.1 \text{ N}}$$

(b) The initial acceleration is determined by Newton's second law, $\Sigma\vec{F} = m\vec{a}$. Two forces act on the balloon, its downward weight and the upward buoyant force (Fig. 13.4). Taking *up* as positive,

$$\Sigma F = B - w(\text{balloon} + \text{helium})$$
$$\Sigma F = B - m_b g - m_{He}g$$

Solving Newton's second law for the acceleration,

$$a = \frac{\Sigma F}{m} = \frac{B - m_b g - m_{He}g}{m_b + m_{He}}$$

B has been determined in part (a), m_b is given, and m_{He} can be determined from the volume and density of helium:

$$m_{He} = \rho_{He}V = \rho_{He}(\tfrac{4}{3}\pi r^3) = (0.178 \text{ kg/m}^3)(\tfrac{4}{3}\pi)(0.6 \text{ m})^3 = 0.16 \text{ kg}$$

Substituting these values gives

$$a = \frac{11.1 \text{ N} - (0.5 \text{ kg})(9.8 \text{ m/s}^2) - (0.16 \text{ kg})(9.8 \text{ m/s}^2)}{(0.5 \text{ kg} + 0.16 \text{ kg})}$$
$$= \frac{11.1 \text{ N} - 4.9 \text{ N} - 1.57 \text{ N}}{0.66 \text{ kg}} = \frac{4.63 \text{ N}}{0.66 \text{ kg}} = \underline{7 \text{ m/s}^2, \text{ upward}}$$

The balloon's initial upward acceleration is 7 m/s². The balloon will not continue to accelerate at this rate, however, since air resistance cannot be ignored. Also, as the balloon rises, the density of the atmosphere decreases, reducing the upward buoyant force. Eventually the balloon will reach a height where the upward buoyant force is equal to the downward weight and the balloon will remain in equilibrium at this height.

PROBLEM: A manometer is being used to measure small changes in pressure (Fig. 13.5). The liquid in the manometer is water. (a) What is the excess height h in the open arm if the pressure P being measured is 5 percent greater than atmospheric pressure P_a? (b) Capillary action causes the water to rise higher in the tube. What is the maximum radius of the tube if this reading of h is to be accurate to 1 percent at 20°C?

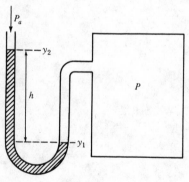

FIG. 13.5

Solution: (a) Equation (13.3) can be written as

$$P_1 - P_2 = \rho g(y_2 - y_1)$$

where, in this case, $(y_2 - y_1) = h$ and $P_1 - P_2 = \Delta P = P - P_a = 5$ percent P_a

$$\Delta P = \rho g h$$

Solving for h,

$$h = \frac{\Delta P}{\rho g} = \frac{(0.05)(1.01 \times 10^5 \text{ N/m}^2)}{(1000 \text{ kg/m}^3)(9.8 \text{ m/s}^2)} = \underline{0.52 \text{ m}}$$

The excess height in the open arm of the manometer is 52 cm.

(b) To be accurate to 1 percent, the additional height due to capillary action, say h_c, can be no greater than 1 percent of 0.52 m = $(0.01)(0.52) = 5.2 \times 10^{-3}$ m, or about 5 mm. h_c is related to the tube radius by Eq. (13.12):

$$h_c = \frac{2\gamma}{\rho g r}$$

where γ for water and air at 20°C is 0.0728 N/m. Solving for r gives

$$r = \frac{2\gamma}{\rho g h_c} = \frac{2(0.0728 \text{ N/m})}{(1000 \text{ kg/m}^3)(9.8 \text{ m/s}^2)(5.2 \times 10^{-3} \text{ m})} = 2.9 \times 10^{-3} \frac{\text{kg} \cdot \text{m/s}^2}{\text{kg/s}^2}$$

$$r = 2.9 \times 10^{-3} \text{ m} = \underline{2.9 \text{ mm}}$$

The tube radius should be no smaller than 3 mm (diameter = 6 mm) for the reading to be accurate to 1 percent.

Avoiding Pitfalls

1. Density $\left(\dfrac{m}{V}\right)$ and weight density $\left(\dfrac{mg}{V}\right)$ are easily confused. The units are a good indication of which is being given. Units of kg/m³ or g/cm³ indicate a *(mass)* density; units of lb/ft³ indicate a *weight* density. A mass density can be converted to a weight density by multiplying by g; similarly, dividing by g converts a weight density to a mass density.

2. If a solid has enclosed holes and/or airspaces, or if it is nonhomogeneous, its density is not constant throughout. The density given by $\dfrac{m}{V}$ will be an average density for the composite object.

3. The density of water in the *mks* system of units is 1000 kg/m³, not 1 kg/m³. In the *cgs* system, it is 1 g/cm³. (In the *fps* system it is 62.4 lb/ft³.)

4. The specific gravity of a substance has nothing to do with gravity; the meaning of this term is relative density (relative to the density of water).

5. In the equation $P_1 = P_2 + \rho g(y_2 - y_1)$, describing pressure variation with depth in a fluid, P_1 and P_2 may be gauge pressures, provided atmospheric pressure does not change noticeably between positions y_1 and y_2.

6. Pascal's principle says that it is the *pressure* produced by a force, not the force itself, that is transmitted undiminished throughout an enclosed fluid.

7. If Pascal's principle is expressed as ratios of forces to areas, $F_1/A_1 = F_2/A_2$, any convenient units of force or area may be used.

8. Archimedes' principle states that the buoyant force is equal to the *weight,* not the *volume,* of fluid displaced.

9. If an object is completely submerged in a fluid, it displaces a volume of fluid equal to its own volume. The equality of object volume to displaced water volume does *not,* of course, apply if the object is floating incompletely submerged.

10. If an object floats, the *weight* of the displaced fluid is equal to the *weight* of the object. (This is just the equilibrium condition.) A 10-ton floating ship displaces 10 tons of water. But the *volumes* of displaced water and ship are not the same, because the ship is not completely submerged.

11. An object that sinks still experiences an upward buoyant force as determined by Archimedes' principle; the buoyant force is just less than the object's weight.

12. Surface tension is defined as the force per unit length exerted by *one* surface; if two surfaces exert forces (for example, the two surfaces of a soap film), the surface tension force is twice as great as for one surface.

Drill Problems
Answers in Appendix

1. If blood represents about 7 percent of the body's mass, estimate the volume of blood in your body. What is the weight in pounds?

2. A hotel wishes to mark depths in the swimming pool without removing the water. A device which measures gauge pressure is lowered to the bottom of the pool at various depths.

(a) What pressure reading corresponds to 3 m depth?

(b) What depth does a pressure of 39,200 pascals correspond to?

3. A diver brings up a metal statue from a sunken pirate ship and asks you to determine whether or not it is a precious metal. You weigh the statue in air by hanging it on a spring balance, and again after lowering it into water. Its true weight is 82.3 N; when submerged in water, its apparent weight is 74.5 N. From this information, determine what the metal is.

Fluid Dynamics

14

Terms

Define or describe briefly what is meant by the following terms. If you have difficulty, refer to the textbook section given in parentheses.

laminar flow (14.1)

turbulent flow (14.1)

flow rate (14.2)

continuity equation (14.2)

Bernoulli's equation (14.3)

Torricelli's theorem (14.4)

viscosity (14.5)

coefficient of viscosity (14.5)

poise (14.5)

Poiseuille's law (14.5)

Reynolds number (14.6)

drag force (14.7)

Stokes' law (14.7)

terminal speed (14.7)

Equation Review

For each equation, be able to state the situation to which it applies, what quantity each symbol represents, and the units for measuring each quantity in some consistent set of units.

$$Q = \frac{V}{t} \tag{14.1}$$

$$Q = \bar{v} A \tag{14.2}$$

$$\bar{v}_1 A_1 = \bar{v}_2 A_2 \tag{14.3}$$

$$P_1 - P_2 = \tfrac{1}{2}\rho(v_2^2 - v_1^2) + \rho g(y_2 - y_1) \tag{14.4}$$

$$P_1 + \tfrac{1}{2}\rho v_1^2 + \rho g y_1 = P_2 + \tfrac{1}{2}\rho v_2^2 + \rho g y_2 \tag{14.5}$$

$$P_1 - P_2 = \frac{4\eta l}{r^2} v \tag{14.8}$$

$$Q = \frac{\pi r^4}{8\eta l}(P_1 - P_2) \tag{14.10}$$

$$Re = \frac{2\bar{v}r\rho}{\eta} \tag{14.11}$$

$$Re = \frac{vL\rho}{\eta} \tag{14.12}$$

$$F_D = 6\pi\eta r v \tag{14.13}$$

$$F_D \cong \tfrac{1}{2}C_D\rho A v^2 \tag{14.14}$$

Problems with Solutions and Discussion

PROBLEM: Water comes out of a faucet of 3-cm² cross-sectional area with a speed of 80 cm/s. (a) What is the flow rate of the water? (b) How long will it take to fill a 120-cm by 60-cm bathtub to a depth of 15 cm?

Solution: (a) Flow rate can be expressed, using Eq. (14.2), as $Q = \bar{v}A$, where $\bar{v} = 80$ cm/s and $A = 3$ cm².

$$Q = (80 \text{ cm/s})(3 \text{ cm}^2) = \underline{240 \text{ cm}^3/\text{s}}$$

The faucet delivers water at a flow rate of 240 cm³/s.

(b) By definition, the flow rate Q is the volume of liquid passing a certain position per unit time, $Q = V/t$. The volume of water under consideration is that in the bathtub:

$$V = 120 \text{ cm} \times 60 \text{ cm} \times 15 \text{ cm} = 108,000 \text{ cm}^3$$

Therefore,

$$t = \frac{V}{Q} = \frac{108,000 \text{ cm}^3}{240 \text{ cm}^3/\text{s}} = \underline{450 \text{ s}}$$

Note that part (b) can be solved without determining the flow rate, Q explicitly. Since $Q = \bar{v}A$ and also $Q = V/t$, it follows that

$$\bar{v}A = \frac{V}{t}$$

or

$$t = \frac{V}{\bar{v}A} = \frac{108,000 \text{ cm}^3}{(80 \text{ cm/s})(3 \text{ cm}^2)} = \underline{450 \text{ s}}$$

It will take 450 s, or 7½ minutes, to fill the bathtub.

PROBLEM: Water flows through three small pipes, each of radius 10 cm, at a speed of 3 m/s (Fig. 14.1). These three pipes come together into one larger pipe. What is the radius of this pipe if the water flows through it at a speed of 4 m/s?

Solution: Since water is virtually incompressible, it can't "pile up" anywhere. The volume of water into the junction from the 3 small pipes must equal the volume out via the

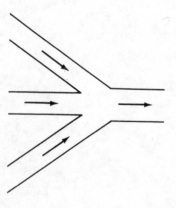

FIG 14.1

96

large pipe during any time period. This is what is meant by a constant flow rate, or continuity of flow. Equation (14.3) is one convenient way to express this continuity condition:

$$\bar{v}_1 A_1 = \bar{v}_2 A_2$$

Before the junction the water flows with speed $\bar{v}_1 = 3$ m/s through a certain total cross-sectional area A_1 given by

$$A_1 = (3 \text{ pipes})(\text{area of each pipe}) = 3(\pi r_1^2) = 3\pi(10 \text{ cm})^2 = 942 \text{ cm}^2$$

After the junction, the water speed is $\bar{v}_2 = 4$ m/s. Solving the continuity equation for A_2 gives

$$A_2 = \frac{\bar{v}_1}{\bar{v}_2} A_1 = \frac{3 \text{ m/s}}{4 \text{ m/s}}(942 \text{ cm}^2) = 706 \text{ cm}^2 = \pi r_2^2$$

$$r_2 = \sqrt{706 \text{ cm}^2/\pi} = \sqrt{225 \text{ cm}^2} = \underline{15 \text{ cm}}$$

The radius of the larger pipe is 15 cm.

PROBLEM: Water flows at a pressure of 3 atm through a pipe with a cross section of 60 cm². The pipe is gradually constricted until its cross-sectional area is 30 cm² (Fig. 14.2). After the constriction the pressure is 2 atm. What is the flow rate of the water?

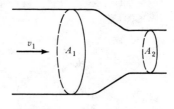

FIG. 14.2

Solution: The flow rate is given by Eq. (14.2) as $Q = \bar{v}A$. Thus Q is equal to either $\bar{v}_1 A_1$ (before the constriction) or $\bar{v}_2 A_2$ (after the constriction). Both areas are known, but neither velocity is given. We do, however, know that continuity requires the flow rates in both parts of the tube to be equal:

$$\bar{v}_1 A_1 = \bar{v}_2 A_2 \qquad\qquad (1)$$

We need a second equation relating the two velocities; Bernoulli's equation (Eq. 14.4) is such an equation:

$$P_1 - P_2 = \tfrac{1}{2}\rho(v_2^2 - v_1^2)$$

where the average vertical height of the two pipe sections is identical, $(y_1 = y_2)$, so the term involving change of elevation may be omitted. Solving equation (1) for v_2 and putting this into Bernoulli's equation gives

$$\bar{v}_2 = \frac{A_1}{A_2}\bar{v}_1$$

$$P_1 - P_2 = \tfrac{1}{2}\rho\left(\frac{A_1^2}{A_2^2}v_1^2 - v_1^2\right) = \tfrac{1}{2}\rho\left(\frac{A_1^2}{A_2^2} - 1\right)v_1^2$$

$$P_2 - P_1 = 3 \text{ atm} - 2 \text{ atm} = 1 \text{ atm} = 1.01 \times 10^5 \text{ N/m}^2$$

$$\frac{A_1^2}{A_2^2} = \frac{(60 \text{ cm}^2)^2}{(30 \text{ cm}^2)^2} = 4$$

Substituting these values gives

$$1.01 \times 10^5 \text{ N/m}^2 = \tfrac{1}{2}(1000 \text{ kg/m}^3)(4 - 1)v_1^2$$

$$v_1 = \sqrt{\frac{1.01 \times 10^5 \text{ N/m}^2}{1500 \text{ kg/m}^3}} = \sqrt{67.3} = 8.2 \text{ m/s}$$

The flow rate before the constriction, therefore, is

$$Q = \bar{v}_1 A_1 = (8.2 \text{ m/s})(60 \text{ cm}^2)\left(\frac{1 \text{ m}^2}{10^4 \text{ cm}^2}\right) = \underline{4.92 \times 10^{-2} \text{ m}^3/\text{s}}$$

The flow rate is about 0.049 m³/s. Square centimeters were converted to square meters to keep units consistent within the equation.

To check these results, let us calculate Q after the constriction:

$$\bar{v}_2 = \frac{A_1}{A_2}\,\bar{v}_1 = \frac{60\ \text{cm}^2}{30\ \text{cm}^2}\,(8.2\ \text{m/s}) = 16.4\ \text{m/s}$$

$$Q = \bar{v}_2 A_2 = (16.4\ \text{m/s})(30\ \text{cm}^2)\left(\frac{1\ \text{m}^2}{10^4\ \text{cm}^2}\right) = \underline{4.92 \times 10^{-2}\ \text{m}^3/\text{s}}$$

The flow rate after the constriction is 0.049 m³/s, identical to the flow rate before the constriction.

PROBLEM: A river flows with a speed of 0.5 m/s past a point where the riverbed is 400 m wide and 3 m deep. Later on, when it reaches a point 2 m lower in elevation, it is 150 m wide. How deep is the river now?

Solution: Continuity of flow rate requires that

$$\bar{v}_1 A_1 = \bar{v}_2 A_2 \tag{1}$$

Bernoulli's equation requires that

$$\tfrac{1}{2}\rho v_1^2 + \rho gy_1 = \tfrac{1}{2}\rho v_2^2 + \rho gy_2$$

where the pressure is considered to be the same at both points, the river being open to the atmosphere. Solving for v_2^2 gives

$$v_2^2 = v_1^2 + 2g(y_2 - y_1)$$

where $(y_2 - y_1) = 2$ m, and $v_1 = 0.5$ m/s.

$$v_2^2 = (0.5\ \text{m/s})^2 + 2(9.8\ \text{m/s}^2)(2\ \text{m}) = 39.45\ \text{m}^2/\text{s}^2$$

$$v_2 = 6.3\ \text{m/s}$$

Solving equation (1) above for A_2 gives

$$A_2 = \frac{\bar{v}_1 A_1}{\bar{v}_2}$$

where $A_1 = \text{width} \times \text{depth} = 400\ \text{m} \times 3\ \text{m} = 1200\ \text{m}^2$.

$$A_2 = \frac{(0.5\ \text{m/s})(1200\ \text{m}^2)}{6.3\ \text{m/s}} = 95.2\ \text{m}^2 = \text{width} \times \text{depth}$$

$$\text{depth} = \frac{95.2\ \text{m}^2}{150\ \text{m}} = \underline{0.63\ \text{m}}$$

Downstream the river is 63 cm deep.

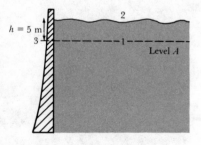

FIG. 14.3

PROBLEM: A dam is located at one end of a lake being used for a reservoir (Fig. 14.3). (a) Determine the pressure in the reservoir at the level labeled A, 5 m below the surface, using Bernoulli's equation. (b) If a hole is drilled through the dam at level A, what will be the speed of the water coming out?

Solution: (a) Let a point at level A be Position 1 and a point above the water be Position 2. Bernoulli's equation states that

$$P_1 + \rho gy_1 + \tfrac{1}{2}\rho v_1^2 = P_2 + \rho gy_2 + \tfrac{1}{2}\rho v_2^2$$

Note that no water is flowing yet, so $v_1 = v_2 = 0$. Rearranging,

$$P_1 = P_2 + \rho g(y_2 - y_1) \tag{1}$$

where $P_2 = P_a$, and $(y_2 - y_1) = h$, the height of the water above level $A = 5$ m.

$$P_1 = P_a + \rho gh = 1.01 \times 10^5\ \text{N/m}^2 + (1000\ \text{kg/m}^3)(9.8\ \text{m/s}^2)\,(5\ \text{m})$$

$$P_1 = 1.01 \times 10^5\ \text{N/m}^2 + 49{,}000\ \text{N/m}^2 = \underline{1.5 \times 10^5\ \text{N/m}^2}$$

The pressure at Position 1, and at all points at level A, is 49,000 N/m² above atmospheric pressure (a gauge pressure of 49,000 N/m²); the absolute pressure is 1.5×10^5 N/m².

Hopefully, you recognized equation (1) above as Eq. (13.3) in Chapter 13 on fluid statics. It relates pressure at two different depths in a stationary fluid. That relationship can be thought of as the "static case" ($v = 0$ throughout) of Bernoulli's equation.

(b) Let us choose as strategic points for the application of Bernoulli's equation the outer edge of the hole (Position 3) and the original Position 1. At Position 3, $P_3 = P_a$ because the edge of the hole is exposed to the atmosphere. Let $y_3 = y_1 = 0$ at level A. At position 1, $P_1 = P_a + \rho g h$ from part (a). Let us assume the reservoir is large enough that the speed of water at Position 1 is negligible compared to the speed at Position 3. Applying Bernoulli's equation to Positions 1 and 3 gives

$$P_a + \rho g h + 0 + 0 = P_a + 0 + \tfrac{1}{2}\rho v_3^2$$

Solving for v_3 and substituting numbers, we get

$$v_3 = \sqrt{2gh} \qquad\qquad (2)$$

$$v_3 = \sqrt{2(9.8 \text{ m/s}^2)(5 \text{ m})} = \sqrt{98 \text{ m}^2/\text{s}^2} = \underline{9.9 \text{ m/s}}$$

The initial speed of the water is 9.9 m/s. Note that equation (2) above is in fact Torricelli's theorem, since h is the total height of fluid above the point where the water emerges. You may wish to confirm that the same velocity is obtained using Positions 2 and 3 in Bernoulli's equation as we got above using Positions 1 and 3.

PROBLEM: Ice water is sucked through a straw of radius 2 mm at a rate of 10 cm³ per second. (a) Is the flow laminar or turbulent? (b) If the straw is 15 cm long, what pressure difference is needed to maintain this flow rate?

Solution: (a) The Reynolds number is useful in determining whether flow is laminar or turbulent.

$$Re = \frac{2\bar{v}r\rho}{\eta}$$

where $r = 2$ mm $= 2 \times 10^{-3}$ m; η (water at 0°C) $= 1.8 \times 10^{-3}$ N·s/m², from Table 14.1; and ρ for water $= 1000$ kg/m³.

We need the average speed $\bar{v}$. The flow rate Q is related to the average speed by $Q = \bar{v}A$. In this case, $A = \pi r^2 = \pi(2 \times 10^{-3} \text{ m})^2 = 1.26 \times 10^{-5}$ m², and $Q = 10$ cm³/s $= 10 \times 10^{-6}$ m³/s. Therefore:

$$\bar{v} = \frac{Q}{A} = \frac{10 \times 10^{-6} \text{ m}^3/\text{s}}{1.26 \times 10^{-5} \text{ m}^2} = 0.79 \text{ m/s}$$

Putting values into the expression for Re, we get

$$Re = \frac{2\bar{v}r\rho}{\eta} = \frac{2(0.79)(2 \times 10^{-3})(1000)}{1.8 \times 10^{-3}} = 1760$$

The Reynolds number is 1760. Assuming that a Reynolds number of 2000 marks the onset of turbulence, this flow is laminar.

(b) Solving Poiseuille's law, Eq. (14.10), for pressure difference, where $l = 0.15$ m,

$$P_1 - P_2 = \frac{8\eta l Q}{\pi r^4} = \frac{8(1.8 \times 10^{-3})(0.15)(10 \times 10^{-6})}{\pi(2 \times 10^{-3})^4} = \underline{4.3 \times 10^2 \text{ N/m}^2}$$

A pressure difference of 0.43 kilopascals is required to maintain the flow rate.

PROBLEM: A 5-kg salmon (about 12 lb) is swimming in still water at 0.4 m/s. It presents a cross-sectional area of 200 cm² to the water. If the drag coefficient is 1, (a) what drag force does the salmon experience? (b) At what rate does it do work against friction; that is, what power does it expend in swimming?

Solution: (a) Since a drag coefficient is given, we may assume that the flow of water past the salmon is turbulent and that Eq. (14.14) applies.

$$F_D = \tfrac{1}{2}C_D\rho Av^2 = \tfrac{1}{2}(1)(1000 \text{ kg/m}^3)(200 \times 10^{-4} \text{ m}^2)(0.4 \text{ m/s})^2 = \underline{1.6 \text{ N}}$$

Here the area A has been converted to m² to keep everything in *mks* units. The salmon experiences a drag force of 1.6 N.

(b) Power is defined as work done per unit time; the work done against the frictional drag force in traveling a distance x is $F_D x$.

$$P = \frac{W}{t} = \frac{F_D x}{t}$$

But $\dfrac{x}{t}$ is just the velocity of the salmon, so

$$P = F_D v = (1.6 \text{ N})(0.4 \text{ m/s}) = 0.64 \text{ J/s} = \underline{0.64 \text{ W}}$$

The salmon expends about 2/3 watt of power in swimming at 0.4 m/s.

PROBLEM: A lead BB of diameter 3 mm is shot from a gun at a speed of 185 m/s. (a) What is the drag force on the bullet? (b) What is the terminal speed of such a BB when falling vertically through the air? Take the density of air at 30°C to be 1.3 kg/m³, the density of lead to be 8500 kg/m³, and the drag coefficient to be 0.2.

Solution: (a) The drag force is given by Eq. (14.14), $F_D = \tfrac{1}{2}C_D\rho Av^2$ where

$$\rho = 1.3 \text{ kg/m}^3 \text{ for air}$$
$$v = 185 \text{ m/s}$$
$$r = \frac{1}{2}(3 \text{ mm})\left(\frac{1 \text{ m}}{1000 \text{ mm}}\right) = 1.5 \times 10^{-3} \text{ m}$$

so

$$A = \pi r^2 = \pi(1.5 \times 10^{-3} \text{ m})^2 = 7.1 \times 10^{-6} \text{ m}^2$$

Substituting these values, we get

$$F_D = \tfrac{1}{2}(0.2)(1.3 \text{ kg/m}^3)(7.1 \times 10^{-6} \text{ m}^2)(185 \text{ m/s})^2 = \underline{3.2 \times 10^{-2} \text{ N}}$$

The drag force on the BB moving at 185 m/s is 0.032 N. Note that in the expression for the drag force above (as well as the expression for Reynolds number) the density ρ is the density of the *fluid*, not of the object moving in the fluid.

(b) The falling BB has two forces acting on it, its downward weight $\vec{w}$ and the upward drag force $\vec{F}_D$ opposing the motion (Fig. 14.4). The weight of the BB is $mg = \rho Vg$, where $\rho = 8500 \text{ kg/m}^3$ (here we want the *object's* density) and the volume is

$$V = \tfrac{4}{3}\pi r^3 = \tfrac{4}{3}\pi(1.5 \times 10^{-3} \text{ m})^3 = 1.4 \times 10^{-8} \text{ m}^3$$

FIG. 14.4

Thus,

$$w = \rho Vg = (8500 \text{ kg/m}^3)(1.4 \times 10^{-8} \text{ m}^3)(9.8 \text{ m/s}^2) = 1.17 \times 10^{-3} \text{ N}$$

The terminal speed of the BB is the speed at which the downward weight and upward drag force are equal so that no further acceleration occurs:

$$w = F_D$$
$$w = \tfrac{1}{2}C_D\rho Av^2$$
$$1.17 \times 10^{-3} \text{ N} = \tfrac{1}{2}(0.2)(1.3 \text{ kg/m}^3)(7.1 \times 10^{-6} \text{ m}^2)v^2$$
$$v^2 = \frac{1.17 \times 10^{-3} \text{ N}}{9.23 \times 10^{-7} \text{ kg/m}} = 1.27 \times 10^3 \text{ m}^2/\text{s}^2$$
$$v = \underline{35.6 \text{ m/s}}$$

The terminal velocity of this BB is about 36 m/s.

Avoiding Pitfalls

1. The equation of continuity for steady-state fluid flow applies if the fluid is incompressible. This is a good assumption for liquids, but not for gases. The continuity equation makes no assumption about viscosity.

2. Bernoulli's equation has more restrictions than the continuity equation. In addition to being incompressible, the fluid is assumed to have no turbulence and no viscosity. It applies to the laminar, frictionless, steady-state flow of an incompressible fluid.

3. If a pipe has a constant cross section, the average velocity of steady-state fluid flow does not change. This is a direct outcome of the continuity equation, $\overline{v}_1 A_1 = \overline{v}_2 A_2$. Thus a change in elevation affects the pressure of the fluid, not the velocity.

4. Take care to use a consistent set of units in all terms of Bernoulli's equation. Do not mix pressures in atmospheres with speeds in meters per second and densities in grams per cubic centimeter.

5. Bernoulli's equation relates conditions at one point in the fluid to conditions at another point. The most strategic points to pick in applying Bernoulli's equation are points where pressure, speed, and elevation are either known or you wish to determine them.

6. If a pipe is open to the atmosphere, the pressure at the opening is atmospheric pressure.

7. If after writing Bernoulli's equation for two points in a fluid, you have more than one unknown, write the continuity equation for the two points. If that does not allow solution of the problem, choose a third strategic point and apply Bernoulli's equation again between that point and one of the other two points.

8. The two points chosen for applying Bernoulli's equation are not completely arbitrary. They must be such that fluid flows freely between them without crossing any artificial barriers. Another way of saying this is that they lie along the same *streamline*.

9. If you have trouble remembering the form of any one term in Bernoulli's equation, recall that its units must be identical to those of the terms you do remember. You may also wish to remember that the equation is a direct consequence of conservation of energy and each term has units of energy per unit volume (J/m^3).

Drill Problems
Answers in Appendix

1. (a) Show that each term in Bernoulli's equation has units of energy per unit volume.
(b) Show that the Reynolds number is dimensionless.

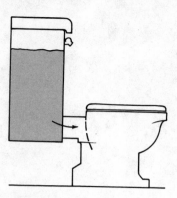

FIG. 14.5

2. The tank behind a toilet has a cross-sectional area of 600 cm² and the water stands 30 cm deep above the outlet valve, which has a diameter of 9 cm (Fig. 14.5). Determine the initial speed of water leaving the outlet valve when it is opened and the initial speed at which the water level in the tank drops. Assume laminar flow. (*Hint:* Since we are interested in how fast the water level drops, we are *not* assuming the velocity of water at the top surface is zero.)

3. Molasses has a viscosity of 20 N·s/m². What pressure difference is necessary to force it through a large hypodermic needle of radius 1 mm that is 3 cm long at a flow rate of 1 cubic centimeter every 10 seconds? Is the flow laminar or turbulent? The density of molasses is 1200 kg/m³.

Elastic Properties
of Solids

15

Terms

Define or describe briefly what is meant by the following terms. If you have difficulty, refer to the textbook section given in parentheses.

elastic deformation (15.1)

Hooke's law (15.1)

force constant (15.1)

stress (15.3)

strain (15.3)

modulus of elasticity (15.3)

tensile stress (15.4)

compressive stress (15.4)

Young's modulus (15.4)

ultimate tensile (or compressive) strength (15.4)

shear stress (15.5)

shear strain (15.5)

shear modulus (15.5)

volume stress (15.6)

volume strain (15.6)

bulk modulus (15.6)

compressibility (15.6)

Equation Review

For each equation, be able to state the situation to which it applies, what quantity each symbol represents, and the units for measuring each quantity in some consistent set of units.

$$F = kx \qquad (15.1)$$

$$\text{Stress} = F/A \tag{15.2}$$

$$\text{Strain} = \text{change in dimension/original dimension} \tag{15.3}$$

$$\text{Stress} = (\text{Modulus of elasticity}) (\text{Strain}) \tag{15.4}$$

$$Y = \frac{F_\perp/A}{\Delta L/L} = \frac{\text{stress}}{\text{strain}} \tag{15.5}$$

$$S = \frac{F_\parallel/A}{\Delta s/L} \tag{15.6}$$

$$B = -\frac{\Delta P}{\Delta V/V} \tag{15.7}$$

$$\text{Compressibility} = -\frac{1}{\Delta P}\frac{\Delta V}{V} \tag{15.8}$$

Problems with Solutions and Discussion

PROBLEM: A vertical spring stretches 10 cm when an unknown mass is hung from it. A 6-kg mass stretches the spring 3 cm further (Fig. 15.1). What is the unknown mass M?

Solution: If any displacement (from the equilibrium position) and corresponding force are known, k can be determined from Hooke's law, $F = kx$. Here, $x = 3$ cm when $F = mg = 6 \text{ kg}(9.8 \text{ m/s}^2) = 58.8$ N. Thus

$$k = \frac{F}{x} = \frac{58.8 \text{ N}}{0.03 \text{ m}} = 1960 \text{ N/m}$$

Knowing the force constant k allows us to determine any other force for a known displacement. Therefore, the original 10 cm displacement corresponds to a force of

$$F = kx = (1960 \text{ N/m})(0.1 \text{ m}) = 196 \text{ N}$$

and

$$M = \frac{196 \text{ N}}{g} = \frac{196 \text{ N}}{9.8 \text{ m/s}^2} = \underline{20 \text{ kg}}$$

Alternate Solution: The ratio of F to x is a constant (k), according to Hooke's law; therefore

$$\frac{F_1}{x_1} = \frac{F_2}{x_2} \quad \text{where } F = mg$$

$$\frac{(6 \text{ kg})(9.8 \text{ m/s}^2)}{3 \text{ cm}} = \frac{M(9.8 \text{ m/s}^2)}{10 \text{ cm}}$$

$$M = \text{\%}(10) = \underline{20 \text{ kg}}$$

The unknown mass is 20 kg by either method. The advantage of the first method is that we know the force constant of the spring also.

PROBLEM: A brass wire 13 m long and 1 mm in diameter is subjected to a tension of 300 N and its elongation is observed to be 5.51 cm. Find the strain and the stress for this wire and determine Young's modulus for brass.

Solution: Strain is defined by Eq. (15.3) as the ratio of the change in a dimension to the original dimension. In this case, the dimension is the length L, so

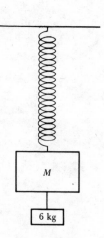

M

6 kg

FIG. 15.1

104

$$\text{Strain} = \frac{\Delta L}{L}$$

where $L = 13$ m and $\Delta L = 5.51$ cm $= 5.51 \times 10^{-2}$ m.

$$\text{Strain} = \frac{5.51 \times 10^{-2} \text{ m}}{13 \text{ m}} = \underline{4.24 \times 10^{-3}}$$

Strain is a unitless quantity; it is the *fractional* change in length.

Stress is defined by Eq. (15.2) as force per unit area. In this case, the wire is under tension, so we calculate the tensile stress, the perpendicular force applied per cross-sectional area of the wire. This area is $A = \pi r^2$, where $r = \frac{1}{2}(1 \times 10^{-3} \text{ m}) = 5 \times 10^{-4}$ m, so $A = \pi(5 \times 10^{-4} \text{ m})^2 = 7.85 \times 10^{-7}$ m^2. Therefore,

$$\text{Stress} = \frac{F}{A} = \frac{300 \text{ N}}{7.85 \times 10^{-7} \text{ m}^2} = \underline{3.82 \times 10^8 \text{ N/m}^2}$$

Young's modulus is the ratio of the stress to the strain and is a constant for a particular material within its elastic limits:

$$Y = \frac{\text{stress}}{\text{strain}} = \frac{3.82 \times 10^8 \text{ N/m}^2}{4.24 \times 10^{-3}} = \underline{9 \times 10^{10} \text{ N/m}^2}$$

Young's modulus for brass is 9×10^{10} N/m^2.

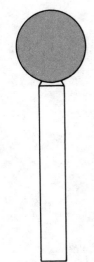

PROBLEM: A modern work of art consists of a solid brass sphere of mass 800 kg supported by a square pillar 1 m tall, made of solid glass (Fig. 15.2). The sphere causes the pillar to be compressed 5×10^{-3} cm. What is the stress on the pillar and what is its width?

Solution: Compressive stress and strain are related by Eq. (15.4):

$$\text{Stress} = Y(\text{Strain})$$

where strain $= \dfrac{\Delta L}{L} = \dfrac{5 \times 10^{-3} \text{ cm}}{100 \text{ cm}} = 5 \times 10^{-5}$

and Y for glass $= 7 \times 10^{10}$ N/m^2 from Table 15.1.

FIG. 15.2

$$\text{Stress} = (7 \times 10^{10} \text{ N/m}^2)(5 \times 10^{-5}) = \underline{3.5 \times 10^6 \text{ N/m}^2}$$

The glass pillar experiences a stress of 3.5×10^6 N/m^2. The defining equation for stress is Eq. (15.2):

$$\text{Stress} = \frac{F_\perp}{A}$$

Here the perpendicular force is the weight of the sphere, $w = mg = (800 \text{ kg})(9.8 \text{ m/s}^2) = 7840$ N.

Solving Eq. (15.2) for the cross-sectional A gives

$$A = \frac{F_\perp}{\text{stress}} = \frac{7840 \text{ N}}{3.5 \times 10^6 \text{ N/m}^2} = 2.24 \times 10^{-3} \text{ m}^2$$

Since the cross-sectional area is square, the width is just the square root of A,

$$\text{width} = \sqrt{A} = \sqrt{2.24 \times 10^{-3} \text{ m}^2} = 0.047 \text{ m} = \underline{4.7 \text{ cm}}$$

The width of the glass pillar is about 5 cm.

PROBLEM: Hooke's law, $F = kx$, relating applied force F to displacement x from equilibrium position, describes the behavior of an elastic spring. Assume that it also describes the behavior of an elastic solid; for example, a wire of length L, cross-sectional area A, and Young's modulus Y. Determine the value of the force constant k in terms of the properties of the wire.

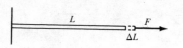

FIG. 15.3

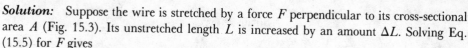

Solution: Suppose the wire is stretched by a force F perpendicular to its cross-sectional area A (Fig. 15.3). Its unstretched length L is increased by an amount ΔL. Solving Eq. (15.5) for F gives

$$F = \frac{AY \Delta L}{L} = \left(\frac{AY}{L}\right)\Delta L$$

where Y is Young's modulus for the material. This equation is of the form $F = kx$, since ΔL is in fact the wire's displacement from its equilbrium (unstretched) position. Therefore, by comparing the two equations, the force constant for the wire is

$$k = \frac{AY}{L}$$

Discussion: When we first studied energy, we learned that a stretched or compressed spring stores an amount of elastic potential energy

$$\Delta PE_s = \tfrac{1}{2}kx^2$$

The same is true for a stretched or compressed elastic solid. The equation for the energy stored by such a solid is also $\Delta PE_s = \tfrac{1}{2}kx^2$, where the force constant k is given by AY/L and the displacement is ΔL. For example, consider a steel wire 2 m long with cross-sectional area of one square millimeter $(= 1 \times 10^{-6} \text{ m}^2)$. When stretched 5 mm, it stores an elastic potential energy given by

$$\Delta PE_s = \frac{1}{2} kx^2 = \frac{1}{2}\left(\frac{AY}{L}\right)(\Delta L)^2$$

$$= \frac{1}{2}\frac{(1 \times 10^{-6} \text{ m}^2)(20 \times 10^{10} \text{ N/m}^2)}{2 \text{ m}}(0.005 \text{ m})^2 = \underline{1.25 \text{ J}}$$

The wire stores 1.25 J of energy. In fact, the equation $\Delta PE_s = \tfrac{1}{2}kx^2$ gives the elastic potential energy for any body whose displacement from its equilibrium position is directly proportional to the applied force. If a displacement x and the corresponding force F are known, k can be determined from $F = kx$ and the stored potential energy calculated without knowing any further details.

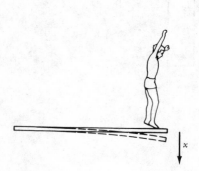

FIG. 15.4

PROBLEM: The displacement of a diving board from its equilibrium position is directly proportional to the applied force. What energy is stored in the board when a 60-kg diver stands on it, depressing it 8 cm (Fig. 15.4)?

Solution: Since we are told that F and x are directly proportional, $F = kx$, so

$$k = \frac{F}{x} = \frac{mg}{x} = \frac{(60)(9.8)}{0.08} = 7350 \text{ N/m}$$

Therefore

$$\Delta PE_s = \tfrac{1}{2}kx^2 = \tfrac{1}{2}(7350 \text{ N/m})(0.08 \text{ m})^2 = 24 \text{ J}$$

The diving board stores 24 J of elastic potential energy.

PROBLEM: A steel wire supports a uniform beam 2 m long of mass 30 kg. The wire has a cross-sectional area of 2×10^{-2} cm² and passes over a small frictionless pulley mounted on the ceiling (Fig. 15.5a). (a) What is the tensile strain of the wire? (b) What is the elongation of the wire?

Solution: (a) Tensile strain is defined as the fractional change in length $\Delta L/L$. Solving Eq. (15.5) for $\Delta L/L$ gives

$$\frac{\Delta L}{L} = \text{strain} = \frac{F}{AY}$$

106

For this wire,

$$A = \text{cross-sectional area} = 2 \times 10^{-2} \text{ cm}^2 = 2 \times 10^{-6} \text{ m}^2$$

$$Y \text{ for steel} = 20 \times 10^{10} \text{ N/m}^2 \text{ from Table 15.1}$$

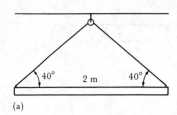

(a)

and we need to find the tension in the wire. It can be determined from the equilibrium equations for the beam. The force diagram is shown in Fig. 15.5b, with the wire tension F resolved into components. Recall that the one piece of wire has one and only one tension F. Because the two angles are equal, the two horizontal and the two vertical components are equal in magnitude. The equilibrium equation in the x direction simply confirms this symmetry:

$$F \cos 40° - F \cos 40° = 0$$

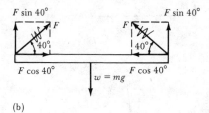

(b)

In the y direction,

$$F \sin 40° + F \sin 40° = w$$

$$2F \sin 40° = mg$$

$$F = \frac{mg}{2 \sin 40°} = \frac{(30 \text{ kg})(9.8 \text{ m/s}^2)}{2(0.643)} = 230 \text{ N}$$

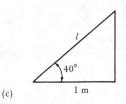

(c)

FIG. 15.5

Therefore,

$$\frac{\Delta L}{L} = \frac{F}{AY} = \frac{230 \text{ N}}{(2 \times 10^{-6} \text{ m}^2)(20 \times 10^{10} \text{ N/m}^2)} = \underline{5.75 \times 10^{-4}}$$

The strain, or fractional change in length, of the wire is 5.75×10^{-4}.

(b) Since $\Delta L/L = $ strain, the elongation ΔL is given by

$$\Delta L = (\text{strain})L = (5.75 \times 10^{-4})L$$

L is twice as long as l, the side of the triangle shown in Fig. 15.5c.

$$L = 2l \quad \text{where} \quad \cos 40° = \frac{1 \text{ m}}{l}$$

$$L = 2\left(\frac{1 \text{ m}}{\cos 40°}\right) = 2\left(\frac{1 \text{ m}}{0.766}\right) = 2(1.305 \text{ m}) = 2.61 \text{ m}$$

Therefore,

$$\Delta L = (2.61 \text{ m})(5.75 \times 10^{-4}) = \underline{1.5 \times 10^{-3} \text{ m}}$$

The elongation of the wire is 1.5 mm.

PROBLEM: A wooden dowel pin 4 cm long and half a centimeter in diameter is subjected to a shearing force of 10^4 N (Fig. 15.6). How far does one end move relative to the other? The shear modulus for the wood is 0.88×10^{10} N/m².

Solution: The shear modulus is the ratio of shear stress to shear strain, as given by Eq. (15.6):

$$S = \frac{F_{\parallel}/A}{\Delta s/L}$$

where

$$F = 10^4 \text{ N}$$
$$A = \pi r^2 = \pi(0.0025 \text{ m})^2 = 1.96 \times 10^{-5} \text{ m}^2$$
$$L = 4 \text{ cm} = 0.04 \text{ m}$$
$$S = 0.88 \times 10^{10} \text{ N/m}^2$$

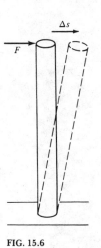

FIG. 15.6

107

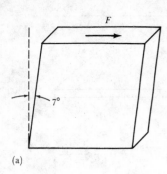

(a)

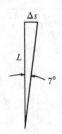

(b)

FIG. 15.7

Solving for Δs and putting in numerical values gives

$$\Delta s = \left(\frac{F_\parallel}{A}\right)\left(\frac{L}{S}\right) = \left(\frac{10^4 \text{ N}}{1.96 \times 10^{-5} \text{ m}^2}\right)\left(\frac{0.04 \text{ m}}{0.88 \times 10^{10} \text{ N/m}^2}\right) = \underline{2.3 \times 10^{-3} \text{ m}}$$

One end is displaced a distance of 2.3 mm.

PROBLEM: A rubber cube whose edge is 5 cm long is subjected to a shearing force which deforms it by 7° as shown in Fig. 15.7a. The shear modulus for rubber is 1.2×10^6 N/m². (a) How large is the shearing force? (b) How far did the top edge of the block move?

Solution: (a) Shear stress, $F_\parallel/A$, and shear strain, $\Delta s/L$, are related by Eq. (15.6)

$$S = \frac{F_\parallel/A}{\Delta s/L}$$

Solving for F gives

$$F = SA\left(\frac{\Delta s}{L}\right)$$

Δs is the side of a right triangle whose other side is L, the cube dimension (Fig. 15.7b); $\Delta s/L = \tan 7° = $ shear strain $= 0.123$; and $A = (0.05 \text{ m})^2 = 2.5 \times 10^{-3}$ m². Substituting these values gives

$$F_\parallel = (1.2 \times 10^6 \text{ N/m}^2)(2.5 \times 10^{-3} \text{ m}^2)(0.123) = \underline{369 \text{ N}}$$

The shearing force is about 370 N.

(b) Since $\dfrac{\Delta s}{L} = \tan 7°$, $\Delta s = $ L $\tan 7° = (0.05 \text{ m})(0.123) = \underline{6.15 \times 10^{-3} \text{ m}}$. The cube's edge moves about 6 mm.

PROBLEM: A cube of aluminum is 40 cm on an edge at a pressure of 1 atm. At what pressure in atmospheres would its density be increased by 2 percent?

Solution: Density is defined as mass per unit volume, $\rho = m/V$, or $m = \rho V$. The total mass of aluminum remains constant, so if ρ increases by 2 percent, V must decrease by 2 percent. Thus, for $\rho' = 1.02\rho$, $V' = 0.98$ V, and $\Delta V = V' - V = 0.98 V - V = -0.02 V$. Equation (15.7) states that the ratio of the pressure change ΔP to the fractional volume change is the bulk modulus, or

$$B = -\frac{\Delta P}{\Delta V/V}$$

Solving for ΔP gives

$$\Delta P = -B\left(\frac{\Delta V}{V}\right) = -\frac{(7.7 \times 10^{10} \text{ N/m}^2)(-0.02\,V)}{V} = 1.54 \times 10^9 \text{ N/m}^2$$

where B for aluminum is found in Table 15.3 on page 296 in the text.

$$\Delta P = 1.54 \times 10^9 \text{ N/m}^2 \left(\frac{1 \text{ atm}}{1.01 \times 10^5 \text{ N/m}^2}\right) = \underline{1.52 \times 10^4 \text{ atm}}$$

A pressure of over 15,000 atmospheres is required to decrease the volume, and thus increase the density, of aluminum by 2 percent. While there are very strong attractive forces holding atoms and molecules together, there are also very strong repulsive forces holding them apart. Aluminum and other solids resist compression very effectively. Tremendous pressures are required to alter their volume and density by even small amounts. Note that since ΔP depends on $\Delta V/V$, the *fractional* change in volume, we did not have to calculate the actual volume of the cube.

108

Avoiding Pitfalls

1. Young's modulus and the shear and bulk moduli are constants only within the region of elasticity of a material. When a material is deformed beyond its elastic limit, it does not return to its original configuration, and the constant moduli listed in Tables 15.1, 2, and 3 do not apply. Similar considerations apply to Hooke's law.

2. Strain $\left(\dfrac{\Delta L}{L} \text{ and } \dfrac{\Delta V}{V} \right)$ is a unitless ratio of two like quantities; they may be expressed in any convenient units as long as the two quantities forming the ratio have the same units.

3. Stress, Young's modulus, the shear modulus, and pressure all have the same dimensions. In the *mks* system, they are expressed in N/m^2.

Drill Problems
Answers in Appendix

1. A human hair 50 cm long is being stretched. What is its elongation just before it breaks, assuming that its behavior is elastic right up to the breaking point?

2. A steel cable 2 m long and 10^{-2} cm^2 in cross section is attached to a high ceiling. Connected to it is an aluminum cable of the same cross-sectional area but twice as long. A 100-kg mass is attached to its bottom (Fig. 15.8). What is the total elongation of the combination cable?

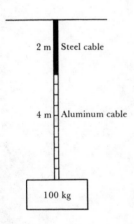

FIG. 15.8

3. A steel sphere of 20-cm radius at atmospheric pressure is subjected to a pressure which causes its radius to decrease by 2 mm. What is the pressure?

109

16   Vibration

Terms

Define or describe briefly what is meant by the following terms. If you have difficulty, refer to the textbook section given in parentheses.

restoring force (16.1)

vibration (16.1)

simple harmonic motion (16.1)

amplitude (16.1)

period (16.1)

frequency (16.1)

hertz (16.1)

simple pendulum (16.4)

small angle approximation (16.4)

damping (16.6)

critically damped oscillator (16.6)

resonant frequency (16.7)

Equation Review

For each equation, be able to state the situation to which it applies, what quantity each symbol represents, and the units for measuring each quantity in some consistent set of units.

$$F_r = -kx \tag{16.1}$$

$$f = \frac{1}{\tau} \tag{16.2}$$

$$E = \tfrac{1}{2}mv^2 + \tfrac{1}{2}kx^2 \tag{16.3}$$

$$\tau = 2\pi \sqrt{\frac{m}{k}} \tag{16.5}$$

$$f = \frac{1}{2\pi} \sqrt{\frac{k}{m}} \qquad\qquad (16.6)$$

$$f = \frac{1}{2\pi} \sqrt{\frac{g}{L}} \qquad\qquad (16.7)$$

$$x = A \sin 2\pi ft \qquad\qquad (16.9)$$

Problems with Solutions and Discussion

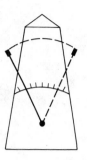

PROBLEM: A metronome is used by pianists to set the tempo of music. It resembles somewhat an inverted pendulum which ticks each time it reaches an end point of the vibration (Fig. 16.1). What is the period of a metronome set to tick 80 times per minute?

Solution: During one complete cycle, the metronome ticks twice, once at each end point. Therefore 80 ticks per minute corresponds to a frequency of 40 cycles per minute. Converting to seconds,

$$f = 40 \frac{\text{cycles}}{\text{min}} \left(\frac{1\ \text{min}}{60\ \text{s}} \right) = \frac{2}{3} \frac{\text{cycles}}{\text{s}} = \frac{2}{3}\ \text{Hz}$$

The period is the reciprocal of the frequency, or

$$\tau = \frac{1}{f} = \frac{1}{\tfrac{2}{3}\ \text{s}^{-1}} = \frac{3}{2}\ \text{s} = \underline{1.5\ \text{s}}$$

The time for one complete back-and-forth vibration is 1.5 s. The metronome ticks every 0.75 s.

PROBLEM: A 2-kg mass attached to a spring undergoes simple harmonic motion (Fig. 16.2a). In completing one vibration, it travels a total distance up and down of 40 cm in a time of 0.4 s. (a) What is the amplitude of the motion? (b) What is the frequency? (c) What is the spring constant of the spring?

Solution: (a) The amplitude is the maximum displacement *from the equilibrium position*, as shown in Fig. 16.2b. The distance traveled during one complete cycle is 4 times the amplitude, so

$$A = \frac{\text{total distance per cycle}}{4} = \frac{40\ \text{cm}}{4} = \underline{10\ \text{cm}}$$

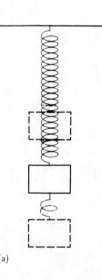

(a)

(b) The time for a complete cycle, 0.4 s, is the period of the motion. The frequency is the reciprocal of the period,

$$f = \frac{1}{\tau} = \frac{1}{0.4\ \text{s}} = \underline{2.5\ \text{Hz}}$$

(c) Since both mass and period are known, the period formula for simple harmonic motion can be used to obtain k, the force constant of the spring.

$$\tau = 2\pi \sqrt{\frac{m}{k}}, \text{ so } \tau^2 = 4\pi^2 \frac{m}{k}$$

$$k = \frac{4\pi^2\, m}{\tau^2} = \frac{4\pi^2 (2\ \text{kg})}{(0.4\ \text{s})^2} = 490\ \text{N/m}$$

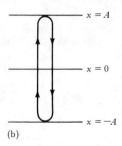

$x = A$

$x = 0$

$x = -A$

(b)

The harmonic motion has an amplitude of 10 cm, a frequency of 2.5 cycles per second, and a force constant of 490 N/m.

PROBLEM: A simple harmonic motion of amplitude 30 cm has a frequency of 2 Hz. What is the maximum acceleration?

Solution: Acceleration suggests Newton's second law, $\Sigma F = ma$. For simple harmonic motion, the net resultant force is called the restoring force $F_r = -kx$. Equating these two gives

$$ma = -kx \quad \text{or} \quad a = -\frac{k}{m}x$$

It appears that we need both k and m, neither of which is known. But actually we only need their ratio, and this can be obtained from the frequency equation for simple harmonic motion.

$$f = \frac{1}{2\pi}\sqrt{\frac{k}{m}}$$

Squaring and solving for $\dfrac{k}{m}$,

$$\frac{k}{m} = 4\pi^2 f^2 = 4\pi^2 (2 \text{ s}^{-1})^2 = 158 \text{ s}^{-2}$$

Since acceleration is largest when displacement is largest, the maximum acceleration occurs at the maximum displacement, $x = |A|$.

$$a_{\text{max}} = -\frac{k}{m}A = -(158 \text{ s}^{-2})(0.3 \text{ m}) = \underline{-47.4 \text{ m/s}^2}$$

The maximum acceleration, which occurs at the end points of the motion, is 47.4 m/s². The negative sign indicates that the direction of the acceleration opposes that of the displacement. An acceleration vector which is directly proportional to the displacement but in the opposite direction is the hallmark of simple harmonic motion.

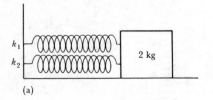

(a)

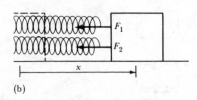

(b)

FIG. 16.3

PROBLEM: Two springs have equal lengths when unstretched but different force constants. The first spring has force constant $k_1 = 1200$ N/m; for the second spring, $k_2 = 2000$ N/m. They are attached to a 2-kg block at rest on a frictionless horizontal table. The block is pulled out and released (Fig. 16.3a). Will the subsequent motion be simple harmonic? If so, what is the frequency of the motion?

Solution: The motion will be simple harmonic motion if there is a restoring force which is directly proportional to the displacement. Consider the forces acting when the block is displaced a distance x to the right and released (b). Each spring exerts a force to the left:

$$F = -F_1 - F_2 \text{ (taking the direction of } x \text{ as positive)}$$

Using Hooke's law,

$$F = -k_1 x - k_2 x = -(k_1 + k_2)x \tag{1}$$

The net force is a *restoring* force (opposite in direction to the displacement from the equilibrium position) and is directly proportional to x. Therefore the subsequent motion is simple harmonic. Comparing Eq. (1) above with $F_r = -kx$, we see that the effective force constant k is $k_1 + k_2$.

$$k = k_1 + k_2 = 1200 \text{ N/m} + 2000 \text{ N/m} = 3200 \text{ N/m}$$

The frequency of any simple harmonic motion is related to the force constant by Eq. (16.5):

$$f = \frac{1}{2\pi}\sqrt{\frac{k}{m}} = \frac{1}{2\pi}\sqrt{\frac{3200 \text{ N/m}}{2 \text{ kg}}} = \frac{40 \text{ s}^{-1}}{2\pi} = \underline{6.37 \text{ s}^{-1}}$$

The mass oscillates 6.4 times per second.

PROBLEM: A mass hanging on a vertical spring is observed to vibrate with a period of 1.3 s. How much will the spring shorten when the mass is removed?

112

Solution: The spring will shorten by the same amount it stretched when the mass was attached. At that time, the spring lengthened by an amount x given by $F = kx$, where the force was the weight of the mass.

$$mg = kx \quad \text{or} \quad x = \frac{mg}{k}$$

The ratio $\frac{m}{k}$ can be obtained from the period equation, $\tau = 2\pi \sqrt{\frac{m}{k}}$. Squaring and solving for $\frac{m}{k}$ gives

$$\frac{m}{k} = \frac{\tau^2}{4\pi^2} = \frac{(1.3 \text{ s})^2}{4\pi^2} = 0.043 \text{ s}^2$$

Therefore

$$x = \frac{m}{k} g = (0.043 \text{ s}^2)(9.8 \text{ m/s}^2) = 0.42 \text{ m} = \underline{42 \text{ cm}}$$

The spring stretched 42 cm when the mass was attached; it will <u>shorten 42 cm</u> when the mass is removed.

PROBLEM: In a gasoline engine, the motion of a piston is approximately simple harmonic. The piston has a mass of 3 kg and a stroke of length 20 cm (Fig. 16.4). The total energy of the motion at any instant is 3 J. (a) What is the force constant k of the motion? (b) What is the speed of the piston when it is 6 cm from one end of the cylinder?

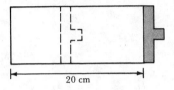

FIG. 16.4

Solution: (a) We know that $F = -kx$ for simple harmonic motion, but we are not given a force which produces a certain displacement. However, we are given the total energy of the motion and we do know it is constant. In general,

$$E = \tfrac{1}{2}mv^2 + \tfrac{1}{2}kx^2 = 3 \text{ J}$$

If we choose $x = A$, this will be the maximum displacement, where the piston momentarily comes to rest at the end point of the motion ($v = 0$).

$$\text{At } x = A, \quad E = 0 + \tfrac{1}{2}kA^2 = 3 \text{ J}$$

The amplitude of the motion is *not* 20 cm. That is the distance from one end point to the other, or twice the amplitude. The amplitude is the maximum displacement *from the equilibrium position*, which is at the center of the motion. Therefore, $A = 10 \text{ cm} = 0.1 \text{ m}$.

$$\tfrac{1}{2}k(0.1 \text{ m})^2 = 3 \text{ J}$$
$$k = \frac{2(3 \text{ J})}{0.01 \text{ m}^2} = \frac{6 \text{ N·m}}{0.01 \text{ m}^2} = \underline{600 \text{ N/m}}$$

The force constant is 600 N/m.

(b) At an arbitrary point, the total energy is shared between the kinetic and potential energy terms,

$$\tfrac{1}{2}mv^2 + \tfrac{1}{2}kx^2 = 3 \text{ J}$$

Here $x = 4 \text{ cm} = 0.04 \text{ m}$ (x is always measured from the equilibrium position; when the piston is 6 cm from one end, it is 4 cm from the equilibrium position).

$$\tfrac{1}{2}(3 \text{ kg})v^2 + \tfrac{1}{2}(600 \text{ N/m})(0.04 \text{ m})^2 = 3 \text{ J}$$
$$(1.5 \text{ kg})v^2 + 0.48 \text{ J} = 3 \text{ J}$$
$$v^2 = \frac{3 \text{ J} - 0.48 \text{ J}}{1.5 \text{ kg}} = 1.68 \frac{\text{m}^2}{\text{s}^2}$$
$$v = \underline{1.3 \text{ m/s}}$$

When the displacement of the piston is 4 cm, its speed is 1.3 m/s.

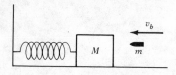

FIG 16.5

PROBLEM: A 0.49-kg block, attached to a spring of force constant 1500 N/m, rests on a frictionless table. A bullet of mass 0.01 kg traveling horizontally at 200 m/s strikes the block and embeds in it (Fig. 16.5). Determine the energy and the amplitude of the resulting simple harmonic motion.

Solution: The bullet imparts a speed V to the block at its equilibrium position, $x = 0$. This speed can be determined by conserving momentum during the collision.

$$p_{before} = p_{after}$$
$$mv_b = (M + m)V$$
$$(0.01 \text{ kg})(200 \text{ m/s}) = (0.49 \text{ kg} + 0.01 \text{ kg})V$$
$$V = \frac{(200 \text{ m/s})(0.01)}{0.5} = 4 \text{ m/s}$$

So at $x = 0$, the block (plus bullet) has a speed of 4 m/s. The energy at that position is

$$E = \tfrac{1}{2}mv^2 + \tfrac{1}{2}kx^2 = \tfrac{1}{2}(0.5 \text{ kg})(4 \text{ m/s})^2 + 0 = \underline{4 \text{ J}}$$

Since the energy during simple harmonic motion is constant, its total energy is 4 J at any point. At the end point of the motion, $x = A$ and $v = 0$, so

$$E = 0 + \tfrac{1}{2}kA^2$$
$$A^2 = \frac{2E}{k} = \frac{2(4 \text{ J})}{1500 \text{ N/m}} = 0.0053 \text{ m}^2$$
$$A = 0.073 \text{ m} = \underline{7.3 \text{ cm}}$$

The resulting simple harmonic motion has an energy of 4 J and an amplitude of 7.3 cm.

PROBLEM: A simple pendulum has a period of two seconds on earth. (It is sometimes called a "seconds pendulum." Why?) If it is taken to the moon, where $g = 1.6 \text{ m/s}^2$, what is its period there?

Solution: The frequency of a simple pendulum is given by

$$f = \frac{1}{2\pi} \sqrt{\frac{g}{L}}$$

Since period and frequency are reciprocals,

$$\tau = \frac{1}{f} = 2\pi \sqrt{\frac{L}{g}}$$

On earth, $\tau_E = 2\pi \sqrt{\dfrac{L}{g_E}}$, and on the moon, $\tau_M = 2\pi \sqrt{\dfrac{L}{g_M}}$. Taking the ratio of τ_M to τ_E gives

$$\frac{\tau_M}{\tau_E} = \frac{2\pi \sqrt{L/g_M}}{2\pi \sqrt{L/g_E}} = \sqrt{\frac{g_E}{g_M}} = \sqrt{\frac{9.8 \text{ m/s}^2}{1.6 \text{ m/s}^2}} = \sqrt{6.1}$$
$$\tau_M = 2.5\tau_E = 2.5(2 \text{ s}) = \underline{5 \text{ s}}$$

The pendulum swings two-and-a-half times more slowly on the moon. When the gravitational acceleration is reduced by a factor of 6.1, the period is increased by a factor of $\sqrt{6.1}$, or 2.5.

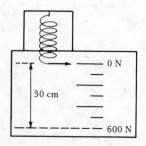

FIG. 16.6

PROBLEM: A spring used for weighing objects has a scale 30 cm long which reads from 0 to 600 N (Fig. 16.6). A package is hung vertically on the spring and is found to oscillate up and down 15 times in 10 seconds. (a) How much does the package weigh? (b) The package is then pulled down 8 cm and released. Write an equation which could be used to determine the displacement from the equilibrium position at any time t. (c) What is its displacement at $t = 1.1$ s?

114

Solution: (a) Since 600 N corresponds to a spring extension of 30 cm,

$$k = \frac{F}{x} = \frac{600 \text{ N}}{0.3 \text{ m}} = 2000 \text{ N/m}$$

The mass of the package can be obtained from the frequency equation.

$$f = \frac{1}{2\pi} \sqrt{\frac{k}{m}}, \text{ where } f = \frac{15 \text{ oscillations}}{10 \text{ s}} = 1.5 \text{ s}^{-1}$$

Squaring and solving for m gives

$$m = \frac{k}{4\pi^2 f^2} = \frac{2000 \text{ N/m}}{4\pi^2 (1.5 \text{ s}^{-1})^2} = 22.5 \text{ kg}$$

The weight of the package is

$$w = mg = 22.5 \text{ kg}(9.8 \text{ m/s}^2) = \underline{220 \text{ N}}$$

The package weighs 220 N, and when it comes to rest, the spring will extend by 220/600 (30 cm) or 11 cm of its full 30-cm scale.

(b) The equation giving displacement of a vibrating mass as a function of time is Eq. (16.9):

$$x = A \sin 2\pi f t$$

where A is the amplitude and f the frequency of the motion. Here the package is given an amplitude of 8 cm. The frequency of a simple harmonic motion is *independent of amplitude*. It depends only on m and k, neither of which has changed; f is thus still 1.5 Hz. Substituting these values gives

$$x = 8 \sin 2\pi (1.5)t$$

$$\underline{x = 8 \sin 9.42t} \ (x \text{ is in cm}, t \text{ is in s})$$

Note that when $t = 0$, $\sin 0 = 0$, so $x = 0$. Therefore to use this equation, we must start counting time at $x = 0$, when the mass is passing through the equilibrium position. And since the sine function is positive in the first quadrant, we would choose $t = 0$ when the mass is at $x = 0$ *going up* so that displacements *above* the equilibrium position would be positive.

(c) When $t = 1.1$ s,

$$x = 8 \sin 9.42(1.1) = 8 \sin 10.36 = 8(-0.805) = \underline{-6.44 \text{ cm}}$$

At $t = 1.1$ s, the mass is 6.44 cm *below* the equilibrium position. *Note:* Recall that the argument of the sine, $2\pi f t$, is in radians, not degrees, so the sine of 10.36 radians, not 10.36°, is needed above. The answer using 10.36° would be totally incorrect.

Avoiding Pitfalls

1. The force causing simple harmonic motion is *not* constant, but varies linearly with x, the displacement. Therefore, by Newton's second law, the acceleration is *not* constant either. The constant acceleration equations do *not* apply to simple harmonic motion.

2. Displacements are measured from the equilibrium position, as is the amplitude, since it is just the maximum displacement. The range of the simple harmonic motion is twice the amplitude; the total distance traveled during one cycle is four times the amplitude.

3. The equilibrium position for a body undergoing simple harmonic motion has the same meaning it has always had; it is the position where the resultant of all the forces acting on the body is zero. The displacement coordinate x is always equal to 0 at the equilibrium position.

4. The velocity, however, is *not* zero at $x = 0$; it has its maximum value there. When $x = A$ (at the end points of the motion), velocity is zero, as the object momentarily stops while reversing direction.

5. Newton's second law, $F = ma$, predicts that where F is largest, a is largest. In simple harmonic motion F is maximum at maximum x, since $F = -kx$. Thus acceleration is maximum at the end points of the motion, where velocity is zero!

6. The force constant k has now been generalized to mean whatever constant(s) of proportionality exist between force and displacement in an equation of the form $F = -(\text{constant})x$. The simple harmonic motion may or may not involve springs. The k which appears in $F = -kx$ is the same k which is in the frequency, period, and energy equations.

7. The term $\frac{1}{2}kx^2$ represents *all* the potential energy of a simple harmonic motion (regardless of whether or not any springs are involved), provided x is measured from the equilibrium position (where the resultant force on the object undergoing simple harmonic motion is zero).

8. The frequency of simple harmonic motion is independent of the amplitude of the motion. It depends on the ratio of k and m. It is not necessary to know either k or m if the value of their ratio is known.

9. The frequency and period formulas are very similar and thus easily confused. It is recommended that you memorize only one of them; the other is readily obtained if you know that period and frequency are reciprocals.

10. One can detect an accidental inversion of the ratio of m and k in the frequency and period formulas by carrying along units and making sure they produce the proper unit for the answer. Unfortunately, there is no such way to detect an omission or misplacement of the 2π factor; its position must simply be memorized.

11. Since x varies with time, so does the potential energy, $PE = \frac{1}{2}kx^2$. However, $E = \frac{1}{2}kA^2$ is the *total* energy of the oscillator, because when $x = A$, all its energy is potential energy. Do not confuse $\frac{1}{2}kx^2$ (which varies with time) with $\frac{1}{2}kA^2$ (which is the *constant total energy of the motion at any time*).

12. The same distinction exists between $KE = \frac{1}{2}mv^2$, which varies with time, and $\frac{1}{2}mv_{max}^2$, which is the constant total energy (evaluated at $x = 0$).

13. The equation $\frac{1}{2}kA^2 = \frac{1}{2}mv_{max}^2$ is another way of saying $E = E$, or the total energy of the motion is constant. They represent the energy of the oscillator at two different positions. It is conservation of energy that enables us to equate them to each other or to the energy at any other position during the motion:

$$\frac{1}{2}kA^2 = \frac{1}{2}mv^2 + \frac{1}{2}kx^2 = \frac{1}{2}mv_{max}^2$$

$$E_{\text{end point}} = E_{\text{intermediate } x \text{ and } v} = E_{\text{midpoint}}$$

14. In the equation giving displacement as a function of time, $x = A \sin 2\pi ft$, the angle $2\pi ft$ is measured in *radians*. The units of f and t cancel, leaving a unitless quantity, suggesting radians. Also, since 2π *radians* equals one cycle, the presence of the factor 2π may serve to remind you that this angle is in radians.

Drill Problems
Answers in Appendix

1. When a 2-kg mass is hung from a vertically suspended spring, the spring stretches 10 cm. When the mass is pulled down 15 cm further and released, simple harmonic motion occurs. Determine

  (a) the frequency of the motion,
  (b) the maximum speed of the mass,
  (c) the displacement of the mass from its equilibrium position when its speed is 1 m/s.

2. How does the period of a simple pendulum change if

   (a) the mass of the pendulum bob is doubled?
   (b) the length of the pendulum is doubled?

3. The displacement of a particle undergoing simple harmonic motion is given by

$$x = 4 \sin 5\pi t$$

where x is in centimeters and t in seconds. Determine the amplitude, the period, the maximum velocity, and the maximum acceleration of the motion.

17

Traveling Waves

Define or describe briefly what is meant by the following terms. If you have difficulty, refer to the textbook section given in parentheses.

wave (17.1)

impulsive wave (17.1)

periodic wave (17.1)

crest (17.1)

trough (17.1)

transverse wave (17.2)

longitudinal wave (17.2)

rarefaction (17.2)

waveform (17.2, 17.6)

amplitude (17.3)

speed (17.3)

frequency (17.3)

wavelength (17.3)

change in phase (17.4)

impedance (17.4)

sinusoidal wave (17.5)

superposition principle (17.5)

destructive interference (17.5)

constructive interference (17.5)

complex waves (17.6)

frequency spectrum (17.6)

beat (17.7)

beat frequency (17.7)

Doppler effect (17.8)

Equation Review

For each equation, be able to state the situation to which it applies, what quantity each symbol represents, and the units for measuring each quantity in some consistent set of units.

$$v \propto \sqrt{\frac{\text{elastic property}}{\text{inertial property}}} \qquad \text{(17.1)}$$

$$v = \sqrt{\frac{T}{\mu}} \qquad \text{(17.6)}$$

$$v = f\lambda \qquad \text{(17.8)}$$

$$f_{\text{beat}} = |f_1 - f_2| \qquad \text{(17.9)}$$

$$f' = f\left(\frac{v + v_0}{v - v_S}\right) \qquad \text{(17.12)}$$

Problems with Solutions and Discussion

PROBLEM: A surfer in the water sees a large ocean wave 40 m away. It is traveling at 5 m/s. How long does she have to get on her surfboard before the arrival of the wave?

Solution: Assuming the wave travels at constant speed, the time t taken to cover a distance x is

$$t = \frac{x}{v} = \frac{40 \text{ m}}{5 \text{ m/s}} = \underline{8 \text{ s}}$$

The wave will reach the surfer in 8 s.

PROBLEM: The speed of a television carrier wave is 3×10^8 m/s. If its frequency is 125 megahertz, determine the wavelength and period of this wave.

Solution: Wavelength, frequency, and speed of a periodic wave are related by the equation

$$v = f\lambda$$

Solving for λ and substituting numbers gives

$$\lambda = \frac{v}{f} = \frac{3 \times 10^8 \text{ m/s}}{125 \times 10^6 \text{ s}^{-1}} = \underline{2.4 \text{ m}}$$

The period of a wave is just the reciprocal of its frequency,

$$\tau = \frac{1}{f} = \frac{1}{125 \times 10^6 \text{ s}^{-1}} = \underline{8 \times 10^{-9} \text{ s}} = 8 \text{ ns}$$

The carrier wave has a wavelength of 2.4 meters and a period of 8 nanoseconds.

PROBLEM: An underwater swimmer hears an explosion and three seconds later, having surfaced to see where the sound came from, hears the explosion again. How far is the swimmer from the explosion?

Solution: The sound of the explosion travels much faster in water than in air, so the sound wave propagated through water reaches the swimmer sooner than the sound propagated through air. If x is the distance from the explosion to the swimmer, then

$$x = v_w t_w$$

where v_w is the speed of sound in water and t_w is the time taken for the sound wave to travel through the water. Similarly,

$$x = v_a t_a$$

for the sound wave traveling through air. Equating these two values for x gives

$$v_w t_w = v_a t_a$$

From Table 17.1, $v_w = 1500$ m/s and $v_a = 340$ m/s.

$$1000 t_w = 340 t_a$$

The sound took 3 s longer to traverse the distance through air, or $t_a = t_w + 3$.

$$1500 t_w = 340(t_w + 3)$$
$$1500\ t_w = 340\ t_w + 1020$$
$$t_w = \frac{1020}{1160} = 0.88\ \text{s}$$

Therefore,

$$x = v_w t_w = (1500\ \text{m/s})(0.88\ \text{s}) = \underline{1320\ \text{m}}$$

The explosion was 1320 m away from the swimmer. As a check of these results, the equation for the sound wave in air is $x = v_a t_a$, where $t_a = t_w + 3 = 3.88$ s

$$x = v_a t_a = (340\ \text{m/s})(3.88\ \text{s}) = \underline{1320\ \text{m}}$$

This result agrees with that obtained previously.

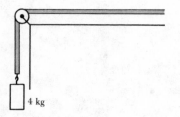

4 kg

FIG. 17.1

PROBLEM: Ten meters of a certain rope has a mass of 1 kg. The rope is stretched by hanging a 4-kg weight on its end as shown in Fig. 17.1. (a) What is the wavelength of a transverse wave of frequency 4 Hz traveling through this rope? (b) How much mass should be added to the end of the rope to double the speed of the transverse wave?

Solution: (a) Equation (17.8) relates speed, frequency, and wavelength of a wave,

$$v = f\lambda, \quad \text{so} \quad \lambda = \frac{v}{f}$$

The speed of a wave depends on the properties of the medium. For a rope under tension, Eq. (17.6) gives

$$v = \sqrt{\frac{T}{\mu}}$$

where μ = mass per unit length $= \dfrac{m}{l} = \dfrac{1\ \text{kg}}{10\ \text{m}} = 0.1\ \dfrac{\text{kg}}{\text{m}}$. The tension is supplied by the hanging mass. Since it is in equilibrium, $T - mg = 0$.

$$T = mg = (4\ \text{kg})(9.8\ \text{m/s}^2) = 39.2\ \text{N}$$

Therefore,

$$v = \sqrt{\frac{T}{\mu}} = \sqrt{\frac{39.2\text{ N}}{0.1\text{ kg/m}}} = \sqrt{392}\text{ m/s} = \underline{19.8\text{ m/s}}$$

So

$$\lambda = \frac{v}{f} = \frac{19.8\text{ m/s}}{4\text{ s}^{-1}} = \underline{4.95\text{ m}}$$

(b) Since μ is constant, we can change the speed only by altering the tension T. Multiplying Eq. (17.6) above by 2, since we want twice the speed,

$$2v = 2\sqrt{\frac{T}{\mu}} = \sqrt{\frac{4T}{\mu}}$$

Since v is proportional to the square root of T, to double v we must multiply T by 4. Now, $T = mg$, so increasing the mass by a factor of 4 will increase the tension the proper amount.

$$4m = 4(4\text{ kg}) = 16\text{ kg}$$

If we need 16 kg of mass, we must <u>add 12 kg</u> to the mass already present.

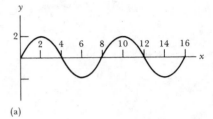
(a)

PROBLEM: A girl shouts across Echo Canyon and hears the echo of her voice 3 s later. How far is she from the opposite side of the canyon? Is the phase of the sound wave inverted upon undergoing reflection?

Solution: To return to the shouter, the sound traveled to the canyon wall and back again, a distance $x = 2D$, where D is the distance to the wall. Taking the speed of sound to be constant, $v = 340$ m/s,

$$2D = vt = 340\text{ m/s}(3\text{ s})$$
$$D = \frac{1020\text{ m}}{2} = \underline{510\text{ m}}$$

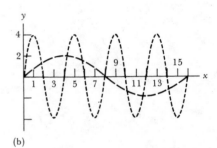
(b)

The canyon wall is 510 m from the shouter. Since the reflecting medium (the canyon wall) is more dense than the incident medium (air), the sound <u>will undergo a phase inversion</u> upon reflection.

PROBLEM: The graph (Fig. 17.2a) represents a wave of frequency 200 Hz. (a) Preserving the same scale, show a wave of frequency 400 Hz with twice the amplitude and a wave of frequency 100 Hz with the same amplitude. (b) Construct the complex waveform resulting in these three waves traveling simultaneously through the same medium. (c) Make a frequency spectrum for this complex wave.

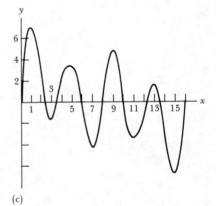
(c)

Solution: (a) The 400-Hz wave with twice the amplitude is shown by the dotted line in Fig. 17.2b; the 100-Hz wave with the same amplitude is shown by the dashed line.

(b) The superposition principle states that the net displacement is the sum of the individual displacements due to each wave. In this case three waves are superposed, so

$$y = y_{200} + y_{400} + y_{100}$$

We will take *up* as positive and *down* as negative and determine the net displacement at each tick (arbitrary units) on the diagram, starting at $x = 0$, by measuring the displacement of each wave and adding.

$x = 0$: $y_0 = 0 + 0 + 0 = 0$ units

$x = 1$: $y_1 = 1.4 + 4 + 0.8 = 6.2$ units

$x = 2$: $y_2 = 2 + 0 + 1.4 = 3.4$ units

$x = 3$: $y_3 = -4 + 0.8 + 1.4 = -1.8$ units

$x = 4$:   etc.

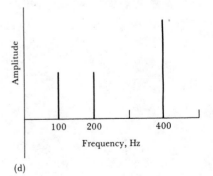
(d)

FIG. 17.2

121

The complex wave is shown in Fig. 17.2c. It repeats itself after $x = 16$, with a frequency of 100, the lowest frequency component present in the complex wave.

(c) The frequency spectrum is shown in (d). The lines mark the frequency components, and the height of each line indicates the relative amplitude of the component.

PROBLEM: When a tuning fork of unknown frequency is sounded simultaneously with a tuning fork of frequency 256 Hz, 3 beats per second are heard. (a) What are the possible frequencies of the unknown tuning fork? (b) If the unknown tuning fork is sounded with a 260-Hz fork, one beat per second is heard. What is the frequency of the unknown fork?

Solution: (a) The beat frequency is 3 Hz. Since the difference in frequencies is the frequency of the beats, the unknown fork has a frequency of either $256 + 3 = $ 259 Hz, or $256 - 3 = $ 253 Hz. Either of these frequencies, when sounded with 256-Hz fork, would produce 3 beats per second.

(b) Frequencies of 253 Hz and 260 Hz, heard together, produce a beat frequency of

$$f_{beat} = |f_1 - f_2| = |253 - 260| = 7 \text{ beats/s}$$

Frequencies of 259 Hz and 260 Hz, heard together, produce a beat frequency of

$$f_{beat} = |f_1 - f_2| = |259 - 260| = 1 \text{ beat/s}$$

Since one beat per second is heard, the unknown frequency must be 259 Hz.

PROBLEM: Compare the Doppler-shifted frequency received by (a) an observer moving at 20 m/s toward a stationary train which is blowing a whistle of frequency 800 Hz, and (b) a stationary observer, as the train approaches him at 20 m/s, blowing the same whistle.

Solution: When source and observer are in motion with respect to each other, the Doppler effect predicts the observer hears a frequency f' that differs from the frequency f emitted by the source, according to the equation

$$f' = f\left(\frac{v + v_O}{v - v_S}\right)$$

where v_O and v_S are the speeds of observer and source respectively, and v is the speed of the wave.

In case (a),

$v_S = 0$ (train is stationary)

$v_O = 20$ m/s (taken as positive if observer moves toward source)

$v = $ speed of sound in air $= 340$ m/s

$f = 800$ Hz

Substituting these values into the Doppler effect equation we get

$$f' = (800)\left(\frac{340 + (20)}{340 - (0)}\right) = 800\left(\frac{360}{340}\right) = 847 \text{ Hz}$$

The observer hears a higher frequency of 847 Hz, indicating a relative velocity of approach. In case (b),

$v_S = 20$ m/s (positive, since source approaches observer)

$v_O = 0$ (observer is stationary)

Therefore,

$$f' = (800)\left(\frac{340 + (0)}{340 - (20)}\right) = 800\left(\frac{340}{320}\right) = 850 \text{ Hz}$$

Again a higher frequency is heard, indicating a relative velocity of approach. But, interestingly enough, the frequency shift is not identical for both cases, even though in both cases

122

the relative velocity of approach between source and observer was the same. Apparently it makes a difference whether source moves toward observer or observer moves toward source. In fact, there is a very real physical difference between the two situations. The nonsymmetry arises from the presence of the medium propagating the waves, relative to which all speeds are measured. When the source moves, wave crests actually "pile up" in the medium ahead of the source. When the observer moves, the wavelength (distance between crests) of the wave in the medium is unaltered; the observer simply encounters crests more rapidly by virtue of his motion.

From this, we might boldly predict that if there were no medium of propagation, the asymmetry would disappear and only the relative velocity of source and observer would matter. This is in fact correct. Light and other electromagnetic waves require no medium. The Doppler effect for light is symmetric; only relative velocity is important. It doesn't matter if observer approaches source or source approaches observer. (Indeed, without a medium, there may be no way to tell which one is moving.)

PROBLEM: A car traveling at 30 m/s on a straight road is followed by a police car traveling at 30 m/s. The police car turns on its siren at a frequency of 1000 Hz. (a) What frequency is heard by the driver of the car? (b) The police car speeds up to 35 m/s. What frequency is heard by the driver, still moving at 30 m/s? (c) The driver slows down to 20 m/s. What frequency does she hear now? (d) The police car passes the driver (he was chasing someone else) and continues at 35 m/s. What frequency does the driver, still at 20 m/s, hear?

Solution: (a) In this case the police car is the source and the driver is the observer.

$v_O = -30$ m/s (taken as negative if observer moves away from the source)

$v_S = 30$ m/s (taken as positive if source moves toward observer)

$v =$ speed of sound in air $= 340$ m/s

$f = 1000$ Hz

According to the Doppler effect equation,

$$f' = f\left(\frac{v + v_O}{v - v_S}\right) = (1000)\left(\frac{340 + (-30)}{340 - (30)}\right) = (1000)\left(\frac{310}{310}\right) = \underline{1000 \text{ Hz}}$$

The driver hears the true frequency of the siren—there is no Doppler shift. Although both are moving, they are moving with identical velocities—therefore there is *no relative motion* between them. They are neither approaching nor separating from each other.

(b)

$v_O = -30$ m/s (negative if moving away from source)

$v_S = 35$ m/s (positive if approaching observer)

$$f' = f\left(\frac{v + v_O}{v - v_S}\right) = (1000)\left(\frac{340 + (-30)}{340 - (35)}\right) = (1000)\left(\frac{310}{305}\right) = \underline{1016 \text{ Hz}}$$

The driver hears a higher frequency of 1016 Hz, indicating that there is a relative velocity of approach (5 m/s) between the source and the observer. The former is approaching more quickly than the latter is receding.

(c)

$v_O = -20$ m/s (still going away from source)

$v_S = 35$ m/s (still approaching observer)

Before doing the calculation, we might notice that the relative velocity of approach has increased from 5 m/s in part (b) to 15 m/s; we therefore predict a Doppler shift to an even higher frequency than in (b).

$$f' = f\left(\frac{v + v_O}{v - v_S}\right) = (1000)\left(\frac{340 + (-20)}{340 - (35)}\right) = (1000)\left(\frac{320}{305}\right) = \underline{1049 \text{ Hz}}$$

As expected, the Doppler-shifted frequency of 1049 Hz is higher than in (b) because the relative velocity of approach increased.

(d)

$v_O = 20$ m/s (after police car passes, driver is now moving *toward* police car)

$v_S = -35$ m/s (after passing, police car moves *away* from driver)

There is now a relative velocity of separation between the two, even though neither have changed direction; we expect the frequency heard by the driver to be lower than the true frequency.

$$f' = f\left(\frac{v + v_O}{v - v_S}\right) = (1000)\left(\frac{340 + (20)}{340 - (-35)}\right) = (1000)\left(\frac{360}{375}\right) = \underline{960 \text{ Hz}}$$

The driver now hears a frequency of 960 Hz, lower in pitch than the true frequency of the siren.

Avoiding Pitfalls

1. Particles of a medium propagating a periodic wave vibrate back and forth about their equilibrium position. They do not move through the medium with the wave, their speed is not that of the wave, nor is the direction of their motion necessarily the direction of propagation of the wave.

2. The relationship $v = f\lambda$ is very general; it describes *any* periodic wave motion, regardless of whether it is transverse or longitudinal or a combination of both, and is independent of the nature of the medium of propagation.

3. A quick check for consistency of units will insure that the relationship $v = f\lambda$ is properly expressed and will help prevent such errors as writing $f = v\lambda$.

4. Despite the appearance of the equation $v = f\lambda$, v is *not* proportional to either f or λ; it depends only on the properties of the medium. Frequency f is determined by the source of the vibration, and λ depends on the independent variables v and f.

5. The amplitude of a transverse wave is the maximum displacement of a particle *from its equilibrium position,* not from trough to crest. Similarly the amplitude of a longitudinal wave is the distance from a point of maximum compression of the medium to the adjacent point of *equilibrium* pressure, not to the next rarefaction.

6. In the Doppler effect equation, the speed of the wave is always taken as positive. The speeds of the source (v_S) and the observer (v_O) are positive or negative depending on whether each is moving toward $(+)$ or away from $(-)$ the other.

7. The positive and negative signs in the Doppler effect equation are part of the equation. Signs for v_S and v_O are independently assigned according to the sign convention and put into the equation along with the magnitudes of the observer and source speeds.

8. It is extremely easy to make a sign error in using the Doppler equation. For this reason, many students find it helpful to first write out the equation with blanks as follows:

$$f' = f\left(\frac{v + v_O}{v - v_S}\right) = f\left(\frac{v + (\quad)}{v - (\quad)}\right)$$

and then substitute in values for v_S and v_O with their signs to indicate *toward* or *away from*.

9. Since the Doppler effect depends on ratios of speeds, the speeds of source, observer, and wave may be expressed in any convenient units as long as they all have the same units.

Drill Problems
Answers in Appendix

1. A screen displaying information stored in a computer is equipped to respond to touch. Compressional waves are sent both horizontally and vertically across the surface of a glass plate covering the screen and are reflected back from the opposite edge. When a person's finger touches the glass, waves are reflected from the finger back to receivers on the edges of the screen (Fig. 17.3).

(a) How long does it take a 4-MHz wave to cross the horizontal length of a 9 × 12-inch display screen?

(b) What is the wavelength of this wave?

(c) If a finger touches the screen in the center, what is the difference in time between a reflected wave's arrival at the nearest point on the horizontal edge and on the vertical edge?

For glass: $B = 3.7 \times 10^{10}$ N/m^2, $S = 2.3 \times 10^{10}$ N/m^2, $\rho = 2800$ kg/m^2

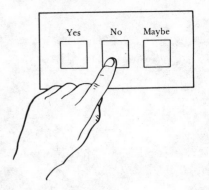

FIG. 17.3

2. Consider the wave motion in Fig. 17.4.

(a) On the diagram, show with a dashed line a wave with half the amplitude and twice the frequency.

(b) Use the idea of superposition to show with a solid line the approximate shape of wave motion that would result if these two waves moved simultaneously through the same medium.

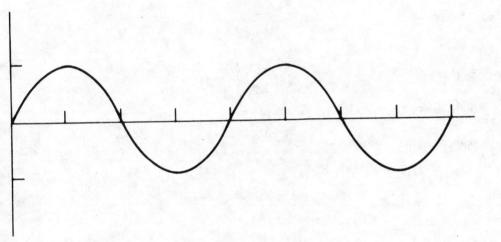

FIG. 17.4

3. An observer at a railroad crossing hears a frequency of 480 Hz as a train sounding its whistle approaches and a frequency of 420 Hz after the train has passed by. Determine the true frequency of the whistle and the speed of the train.

18

Standing Waves
and Sound

Terms

Define or describe briefly what is meant by the following terms. If you have difficulty, refer to the textbook section given in parentheses.

fundamental frequency (18.1)

harmonic (18.1)

overtone (18.1)

standing waves (18.1)

node (18.1)

antinode (18.1)

open pipe (18.2)

closed pipe (18.2)

pitch (18.3)

quality (18.4)

formant (18.4)

threshold of audibility (18.5)

intensity (18.5)

intensity level (18.5)

decibel (18.5)

Equation Review

For each equation, be able to state the situation to which it applies, what quantity each symbol represents, and the units for measuring each quantity in some consistent set of units.

$$f_1 = \frac{v}{2L} \tag{18.1}$$

$$f_n = n\left(\frac{v}{2L}\right), \ n = 1, 2, 3, 4, \ldots \tag{18.2}$$

$$\lambda_n = \frac{2L}{n}, \; n = 1, 2, 3, 4, \ldots \tag{18.3}$$

$$f_n = n\left(\frac{v}{4L}\right), \; n = 1, 3, 5, 7, \ldots \tag{18.4}$$

$$I = \frac{P}{A} = \frac{E}{tA} \tag{18.5}$$

$$\beta = 10 \log \frac{I}{I_0} \tag{18.6}$$

$$\beta_2 - \beta_1 = 10 \log \frac{I_2}{I_1} \tag{18.7}$$

Problems with Solutions and Discussion

PROBLEM: A violin string 0.33 m long has a fundamental frequency of 185 Hz. It needs to be tuned to 196 Hz. (a) Does the string require an increase or decrease in tension? (b) By what factor should the tension be changed?

Solution: (a) The fundamental frequency and string length of the violin are related by the equation

$$f = \frac{v}{2L}$$

The speed v of a transverse wave on a wire under tension is given by

$$v = \sqrt{\frac{T}{\mu}}$$

where T is the tension in the wire and μ is its mass per unit length. Thus,

$$f = \frac{1}{2L} \sqrt{\frac{T}{\mu}} \tag{1}$$

So f is directly proportional to the square root of the tension. To increase the frequency from 185 Hz to 196 Hz, one must increase the tension.

(b) To determine the amount of increase, let f' be the desired frequency of 196 Hz and T' the required tension to produce it. Then

$$f' = \frac{1}{2L} \sqrt{\frac{T'}{\mu}} \tag{2}$$

since the length of the string and the mass per unit length are constant. Dividing Eq. (2) by Eq. (1),

$$\frac{f'}{f} = \frac{\dfrac{1}{2L} \sqrt{\dfrac{T'}{\mu}}}{\dfrac{1}{2L} \sqrt{\dfrac{T}{\mu}}}$$

$$\sqrt{\frac{T'}{T}} = \frac{f'}{f} = \frac{196 \text{ Hz}}{185 \text{ Hz}} = 1.06$$

Squaring,

$$\frac{T'}{T} = (1.06)^2 = 1.12$$

$$T' = \underline{1.12 \, T}$$

The tension must be increased by a factor of 1.12, or 12 percent, to increase the frequency by 6 percent from 185 Hz to 196 Hz.

PROBLEM: A suspension bridge can be thought of as a loaded wire under tension, analogous to a stretched string. The speed of transverse waves through a certain bridge is 120 m/s, and the length of the central span between the two rigid support towers is 1000 m. What is the lowest natural frequency at which the central span can vibrate? What higher natural frequencies would cause large vibrations?

Solution: The central span is analogous to a stretched string fixed at both ends. The fundamental frequency is given by

$$f_1 = \frac{v}{2L} = \frac{120 \text{ m/s}}{2(1000 \text{ m})} = \underline{0.06 \text{ Hz}}$$

The other natural frequencies are integral multiples of the fundamental,

$$f_n = n\left(\frac{v}{2L}\right)$$

Thus, frequencies of 0.12, 0.18, 0.24, 0.3, 0.36 Hz, etc. will cause large vibrations. Note that the frequencies corresponding to $n = 20$ to 24 are in the range of the step frequency of marching groups. Armies and marching bands are instructed to break step when crossing bridges to avoid exciting a resonant frequency resulting in large amplitude vibrations.

Mechanical structures do have natural frequencies of vibration; when excited at these frequencies, vibrations of large amplitude can result. An example from recent history is the collapse of the Tacoma Narrows Bridge, following the growth of spectacular standing waves in the central span, caused by intermittent gusts of wind.

PROBLEM: A cello string is stretched between supports that are 0.75 m apart. Its fundamental frequency is 65.4 Hz. (a) What is the speed of a transverse wave in the string? (b) What is the wavelength of the transverse wave on the wire and the wavelength of the sound wave in air? (c) If the string is fingered one third of the way from one end while being played, what frequency will be heard?

Solution: (a) The fundamental frequency of a string fixed at both ends is given by

$$f = \frac{v}{2L}$$

Solving for v,

$$v = 2Lf = 2(0.75 \text{ m})(65.4 \text{ s}^{-1}) = \underline{98 \text{ m/s}}$$

The speed of this and other transverse waves in the string is 98 m/s.

(b) The wavelength of the transverse wave in the string λ_s is related to its frequency and speed by $v = f\lambda$. Thus,

$$\lambda_s = \frac{v}{f}$$

But

$$f = \frac{v}{2L},$$

so

$$\lambda_s = \frac{v}{v/2L} = 2L = 2(0.75 \text{ m}) = \underline{1.5 \text{ m}}$$

The wavelength of the sound wave in air is also given by $\lambda = v/f$, but although the frequency of the sound wave is identical to that of the vibrating string, the *speed* of sound in air is altogether different. Taking $v = 340$ m/s gives

128

$$\lambda_a = \frac{v}{f} = \frac{340 \text{ m/s}}{65.4 \text{ s}^{-1}} = \underline{5.2 \text{ m}}$$

The wavelength of a 65.4-Hz transverse wave in a string is 1.5 m; the wavelength of a 65.4-Hz compressional wave in air is 5.2 m. This situation occurs because λ depends on the speed of the wave in the medium, which in turn is determined by the properties of the medium.

(c) In the case of standing waves, the boundary conditions determine at which wavelengths (frequencies) large amplitude vibrations can be sustained. When fingered one-third of the way from the end, the cello string is effectively shortened to ⅔ of its original length by the introduction of a new boundary condition. The new length is $L' = ⅔L = $ ⅔(0.75 m) = 0.5 m. The corresponding new fundamental frequency is

$$f' = \frac{v}{2L'} = \frac{98 \text{ m/s}}{2(0.5 \text{ m})} = \underline{98 \text{ Hz}}$$

Shortening the string from 0.75 to 0.5 m increases the frequency from 65.4 Hz to 98 Hz. Playing a stringed instrument involves changing the fundamental frequency of the strings by using one's fingers to alter their effective lengths.

PROBLEM: Determine the length of a stretched string that can support a standing wave of a given wavelength λ.

Solution: Beginning with Eq. (18.2),

$$f_n = n\left(\frac{v}{2L}\right)$$

and solving for L gives

$$L = \frac{nv}{2f}$$

Substituting $v = f\lambda$ gives

$$L = \frac{nf\lambda}{2f} = n\left(\frac{\lambda}{2}\right)$$

This relationship gives a different way of thinking about the standing wave condition. It says that a standing wave can occur whenever the string length is equal to an integral number of half wavelengths. When vibrating at the fundamental frequency (n = 1, one loop), L is equal to half the fundamental wavelength. For $n = 2$ (two loops), the string length contains two half wavelengths (or one complete wave). For $n = 3$ (three loops), L contains three half wavelengths. Thus, counting loops is an easy way of determining the harmonic being observed (Fig. 18.1). For example, if a string is observed vibrating with 6 loops, we may immediately identify it as the *sixth* harmonic and write the relationship

$$L = 6\left(\frac{\lambda}{2}\right) \quad \text{or} \quad \lambda = \frac{L}{3}$$

The frequency is then

$$f = \frac{v}{\lambda} = \frac{3v}{L}$$

which, when written as

$$f = 6\left(\frac{v}{2L}\right)$$

is seen to correspond to the frequency formula

$$f_n = n\left(\frac{v}{2L}\right) \text{ for } n = 6$$

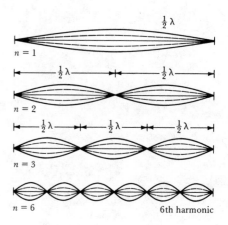

FIG. 18.1

PROBLEM: A certain pipe organ has open pipes with fundamental frequencies ranging from 50 Hz to 2500 Hz. What is the range in lengths of pipes in this organ?

Solution: The fundamental frequency of an open pipe is related to its length by the equation

$$f = \frac{v}{2L}$$

Solving for L,

$$L = \frac{v}{2f}$$

The length of the 50-Hz pipe is thus

$$L = \frac{340 \text{ m/s}}{2(50 \text{ s}^{-1})} = \underline{3.4 \text{ m}}$$

and the length of the 2500-Hz pipe is

$$L = \frac{340 \text{ m/s}}{2(2500 \text{ s}^{-1})} = 0.068 \text{ m} = \underline{6.8 \text{ cm}}$$

The pipes in this organ range in length from 3.4 m to 6.8 cm.

PROBLEM: Suppose you scream with a thousand times the intensity of your normal speaking voice. (a) What increase in decibels do you achieve? (b) How much louder does it sound to a listener?

Solution: (a) Let I_1 be the intensity of your normal voice and $I_2 = 1000\, I_1$ be the intensity of the scream. From Eq. (18.7),

$$\beta_2 - \beta_1 = 10 \log \frac{I_2}{I_1}$$

$$\Delta\beta = 10 \log \frac{1000\, I_1}{I_1}$$

$$\Delta\beta = 10(3) = \underline{30 \text{ dB}}$$

The scream is 30 dB louder than your normal speaking voice.

(b) Since a 10-dB increase has the sensation of doubling the loudness, the 30-dB increase will seem to double the loudness three times or $(2)^3 = 8$. The scream will sound eight times as loud.

PROBLEM: A rock fan wants to increase the sound level of the music he is listening to. His radio can play at an intensity level of 80 dB without too much distortion. If he turns on a second radio to the same level, what is the level of sound he now hears?

Solution: Turning on the second radio doubles the intensity of sound,

$$I_{\text{total}} = 2I_r$$

where I_r = intensity of one radio and can be determined from Eq. (18.6):

$$\beta = 10 \log \frac{I_r}{I_0}$$

where $\beta = 80$ dB and $I_0 = 10^{-12}$ W/m^2

$$\frac{\beta}{10} = \log \frac{I_r}{I_0}$$

130

Taking anti-logs gives

$$\frac{I_r}{I_0} = 10^{(\beta/10)}$$

$$I_r = I_0 10^{(\beta/10)} = (10^{-12} \text{ W/m}^2)10^{(80/10)} = (10^{-12} \text{ W/m}^2)(10^8) = 10^{-4} \text{ W/m}^2$$

Thus, $I_{total} = 2 \times 10^{-4}$ W/m^2. Using Eq. (18.6) again to get the intensity level in decibels corresponding to this total intensity,

$$\beta_t = 10 \log \frac{I_t}{I_0} = 10 \log \frac{2 \times 10^{-4}}{10^{-12}} = 10 \log(2 \times 10^8) = 10(8.301) = \underline{83 \text{ dB}}$$

The total intensity of two 80-dB radios is only 83 dB! Turning on the second radio only netted the rock fan an additional 3 decibels. (Recall that an increase of 10 dB is required for the sensation of doubling the loudness of a sound. Put another way, the intensity level would have to increase by a factor of ten before it sounded twice as loud.)

PROBLEM: A person speaking normally produces about 2×10^{-5} joules of sound energy per second. (a) If a person talked nonstop for one day, how much energy would she produce? (b) Another person 1 m away from the nonstop talker receives a sound intensity of 10^{-6} W/m^2. What total energy impinges on his eardrum, of area 60 mm^2, each minute?

Solution: (a) If the speaker produces 2×10^{-5} J/s of sound energy, her daily energy production is determined by multiplying by the number of seconds in a day.

$$E = (2 \times 10^{-5} \text{ J/s})(86,400 \text{ s/day}) = \underline{1.73 \text{ J}}$$

All that talking produces only 1.73 J, enough energy to raise the temperature of 1 g of water 0.4 C°.

(b) The listener receives an intensity I of 10^{-6} W/m^2, where

$$I = \frac{E}{tA}$$

In this case $t = 1$ min $= 60$ s and $A =$ area of eardrum $= 60$ mm$^2 \left(\dfrac{1 \text{ m}}{1000 \text{ mm}}\right)^2 =$ 6×10^{-5} m^2. Solving for E gives

$$E = ItA = (10^{-6} \text{ W/m}^2)(60 \text{ s})(6 \times 10^{-5} \text{ m}^2) = \underline{3.6 \times 10^{-9} \text{ J}}$$

The listener receives 3.6 nanojoules of sound energy each minute. The point of such calculations as these is that the amount of energy required to produce the sensation of sound is very small indeed. For this reason we do not include sound energy in conservation of energy problems involving car collisions, for example, even though the sounds produced seem quite loud. The sensitivity of the ear and its ability to detect the absorption of such tiny amounts of energy is truly incredible, as is the *range* over which the ear can respond, both in frequency and in intensity.

Avoiding Pitfalls

1. Standing waves can be sustained by a medium only for a discrete set of frequencies dependent on the boundary conditions of the medium.

2. A vibrating stretched string can produce sound waves at the frequency of its vibration. However, the wavelength of the transverse wave in the string is different from the wavelength of the longitudinal sound wave in air because the wave speed in the two mediums is different.

3. The harmonics for a stretched string, an open pipe, and a closed pipe, are integral multiples of the fundamental frequency. All multiples are possible in the case of the stretched string and the open pipe; only odd-integral multiples are possible for the closed pipe.

4. The expression of the standing wave condition is identical for stretched strings and open pipes, but not for closed pipes.

Drill Problems
Answers in Appendix

FIG. 18.2

1. A 60-cm string under tension has a fundamental frequency of 410 Hz. It is observed to be vibrating in a standing wave pattern with five loops (antinodes) as shown in Fig. 18.2. Determine the frequency of vibration and the speed of the wave in the string.

2. The typical fundamental frequency for a male voice is about 125 Hz; for a female voice it is about 250 Hz.

   (a) What length of open pipe has these two frequencies as consecutive harmonics?
   (b) What lengths of closed pipes have these fundamental frequencies?

3. One singer sings at an intensity level of 60 dB.

   (a) What is the intensity level when he is joined by 24 more people singing at the same intensity?
   (b) How many more singers are needed to again increase the intensity level by the same number of decibels?

Reflection and Refraction

19

Terms

Define or describe briefly what is meant by the following terms. If you have difficulty, refer to the textbook section given in parentheses.

corpuscular theory (19.1)

photon (19.1)

coherent wave (19.2)

incoherent wave (19.2)

wavefront (19.2)

ray (19.2)

diffuse reflection (19.3)

specular reflection (19.3)

law of reflection (19.3)

angle of incidence (19.3)

angle of reflection (19.3)

index of refraction (19.4)

refraction (19.5)

Snell's law (19.5)

angle of refraction (19.5)

critical angle (19.6)

total internal reflection (19.6)

dispersion (19.7)

Equation Review

For each equation, be able to state the situation to which it applies, what quantity each symbol represents, and the units for measuring each quantity in some consistent set of units.

$$\theta_i = \theta_r \tag{19.1}$$

$$n = \frac{c}{v} \tag{19.2}$$

$$\frac{\sin \theta_1}{v_1} = \frac{\sin \theta_2}{v_2} \tag{19.3}$$

$$n_1 \sin \theta_1 = n_2 \sin \theta_2 \tag{19.4}$$

$$\sin \theta_c = \frac{n_2}{n_1} \tag{19.5}$$

Problems with Solutions and Discussion

PROBLEM: Light strikes a mirror at an angle of incidence of 35° (Fig. 19.1). (a) What is the angle of reflection? (b) How does this result change if the medium surrounding the mirror is water instead of air?

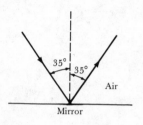

FIG. 19.1

Solution: (a) The law of reflection states that the angle of incidence equals the angle of reflection, $\theta_i = \theta_r$. Therefore the angle of reflection is 35°.

Both angles are traditionally measured from the normal to the reflecting surface (as shown), not from the surface itself. However, the law of reflection would still be true if both angles were measured with respect to the surface.

(b) The result is unchanged if the mirror is surrounded with water. The law of reflection is independent of the nature of the transparent medium surrounding the mirror.

PROBLEM: The index of refraction of turpentine is 1.472. What is the speed of light in turpentine?

Solution: The index of refraction n of a medium is the ratio of the vacuum speed of light to its speed in the medium:

$$n = \frac{c}{v}$$

Therefore,

$$v = \frac{c}{n} = \frac{3.00 \times 10^8 \text{ m/s}}{1.472} = 2.04 \times 10^8 \text{ m/s}$$

The speed of light in turpentine is 2.04×10^8 m/s, considerably slower than in a vacuum. Light travels fastest in a vacuum; the presence of any material medium reduces its speed.

PROBLEM: The index of refraction of Lucite is 1.49. Light in Lucite has a wavelength of 390 nm. (a) What is its frequency? (b) What is its wavelength in air? (c) What color is this?

Solution: (a) The general wave relationship $v = f\lambda$ applies to light. The speed v of light in Lucite can be determined from its index of refraction, defined as $n = \frac{c}{v}$.

$$v = \frac{c}{n} = \frac{3.00 \times 10^8 \text{ m/s}}{1.49} = 2.01 \times 10^8 \text{ m/s}$$

134

Solving $v = f\lambda$ for f gives

$$f = \frac{v}{\lambda} = \frac{2.01 \times 10^8 \text{ m/s}}{390 \times 10^{-9} \text{ m}} = \underline{5.15 \times 10^{14} \text{ Hz}}$$

The frequency of the light is 5.15×10^{14} Hz.

(b) In passing from one optical medium to another, the frequency of a wave does *not* change. Therefore, if the speed does change, the wavelength must change accordingly ($\lambda = v/f$). Since light travels with greater speed in air than in Lucite, its wavelength in air will be greater than in Lucite.

$$v_{\text{air}} \approx c = 3.00 \times 10^8 \text{ m/s}$$

$$\lambda_{\text{air}} = \frac{v}{f} = \frac{3.00 \times 10^8 \text{ m/s}}{5.15 \times 10^{14} \text{ Hz}} = 5.83 \times 10^{-7} \text{ m} = 583 \times 10^{-9} \text{ m} = \underline{583 \text{ nm}}$$

(c) Comparison of colors and wavelengths in Table 19.2 shows this wavelength to be yellow light.

PROBLEM: A light ray is incident on a water-glass interface at an angle of incidence of 30° as shown in Fig. 19.2. Determine the angle of refraction.

Solution: Snell's law relates the two angles to the indices of refraction of the two media:

$$n_1 \sin \theta_1 = n_2 \sin \theta_2$$

In this case medium 1 is water and medium 2 is the glass.

$$(1.33) \sin 30° = (1.50) \sin \theta_2$$

$$\sin \theta_2 = \frac{(1.33)(0.5)}{1.50} = 0.443$$

$$\theta_2 = \underline{26.3°}$$

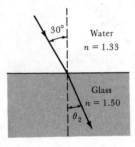

FIG. 19.2

The angle of refraction in the glass is 26.3°. Since the glass is more optically dense (has a greater index of refraction), the bending is toward the normal.

PROBLEM: Suppose the path of the light ray in the previous problem is reversed. It is incident in the glass at an angle of 26.3° (Fig. 19.3). Determine the angle of refraction in the water.

Solution: Again applying Snell's law, we find that

$$n_1 \sin \theta_1 = n_2 \sin \theta_2$$

where now the glass is medium 1 and the water is medium 2.

$$(1.50) \sin 26.3° = (1.33) \sin \theta_2$$

$$\sin \theta_2 = \frac{(1.50)(0.443)}{1.33} = 0.500$$

$$\theta_2 = \underline{30°}$$

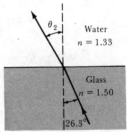

FIG. 19.3

The angle in the air is 30°; the bending is away from the normal as expected. Comparison of this problem and the previous one reveals that the path of light is reversible; that is, when we have determined the path of light in one direction across a boundary, the path in the opposite direction may be obtained by just reversing the direction of the arrows. The calculation need not be repeated. This property is sometimes referred to as the *reversibility of light rays*. Because of this, the labeling of angles as angle of incidence and angle of refraction is somewhat arbitrary. We could simply talk of angle θ_1 in medium 1 and angle θ_2 in medium 2.

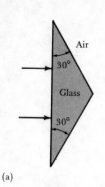

(a)

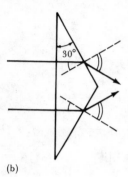

(b)

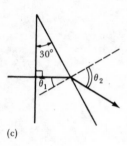

(c)

FIG. 19.4

PROBLEM: Consider two light rays incident on a prism as shown in Fig. 19.4a. (a) Sketch the path of the rays through the prism and into the air again, and (b) determine the angle made by each emerging ray with the prism if the small prism angles are 30° and its index of refraction is 1.5.

Solution: (a) At the first surface, each ray is incident normally, or $\theta_1 = 0°$. Since $\sin \theta_1 = \sin 0° = 0$, Snell's law gives

$$n_1 \sin \theta_1 = n_2 \sin \theta_2$$

$$0 = n_2 \sin \theta_2$$

$$\theta_2 = 0$$

The angle of refraction must be zero, since n_2 is not zero. The ray continues with its direction unchanged. (Its *speed* is still reduced in glass, but the change in speed does not result in a direction change in the special case of $\theta_1 = 0$.) Both rays continue undeviated across the first boundary.

At the second boundary, the normal is constructed for each ray [shown by dashed lines in (b)]. Since the light travels from a more optically dense medium to a less optically dense one, the bending will be away from the normal. The angle of refraction will be larger than the angle of incidence. Notice that the rays which were parallel before entering the prism have undergone changes in direction such that they are now going to intersect in front of the prism. We may say that the rays are focused by the prism. If the edges of the prism are smoothed down, its shape could become similar to the shape of a lens $\mathbb{D}$. A lens (or any piece of refracting material) that is thicker in the center than at the edges has the effect of converging or focusing parallel rays. Two such prisms back to back Φ double the converging effect of the single prism.

(b) To determine the final angle of refraction, we apply Snell's law to one of the rays at the second boundary. A little geometry reveals that θ_1 is 30° (Fig. 19.4c). (The missing angle of the top triangle is 60°; since the normal makes a right angle with the boundary, $60° + \theta_1 = 90°$; $\theta_1 = 30°$.) Snell's law states that

$$n_1 \sin \theta_1 = n_2 \sin \theta_2$$

where the glass is medium 1 and medium 2 is air.

$$(1.5) \sin 30° = (1) \sin \theta_2$$

$$\sin \theta_2 = 0.75$$

$$\theta_2 = 48.6°$$

The final angle of refraction for both rays is 48.6°; therefore the angle between each ray and the prism surface is 41.4°.

PROBLEM: A light ray is incident on a glass-alcohol boundary as shown in Fig. 19.5a. Sketch the reflected and refracted rays and determine the angles of reflection and refraction if the angle of incidence is (a) 50°; (b) 60°.

Solution: (a) For $\theta_1 = 50°$, the reflected angle is also 50°, in accordance with the law of reflection, $\theta_i = \theta_r$. The refracted angle is determined by Snell's law:

$$n_1 \sin \theta_1 = n_2 \sin \theta_2$$

$$(1.65) \sin 50° = (1.33) \sin \theta_2$$

$$\sin \theta_2 = \frac{(1.65)(0.766)}{1.33} = 0.95$$

$$\theta_2 = 72°$$

The sketch is shown in Fig. 19.5b.

(b) For $\theta_1 = 60°$, the reflected angle is 60°, since angle of incidence equals angle of reflection. Using Snell's law to determine the refracted ray gives

136

$$(1.65) \sin 60° = (1.33) \sin \theta_2$$
$$\sin \theta_2 = \frac{(1.65)(0.866)}{1.33} = 1.07!!$$

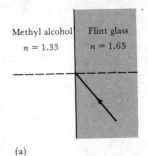

Methyl alcohol $n = 1.33$ Flint glass $n = 1.65$

(a)

The calculation of a sine greater than 1 indicates the angle of incidence is greater than the critical angle; hence there is <u>no refracted ray;</u> the entire ray is reflected. This phenomenon of total internal reflection, shown in (c), can occur when light travels from a more optically dense to a less optically dense medium. To confirm this result, let us calculate the critical angle for this glass-alcohol boundary. The critical angle is the angle of incidence for which the angle of refraction is 90°. Thus

$$(1.65) \sin \theta_c = (1.33) \sin 90°$$
$$\sin \theta_c = \frac{(1.33)(1)}{1.65} = 0.806; \quad \text{thus } \theta_c = 53.7°$$

The critical angle is about 54°; the 60° angle of incidence *is* larger than the critical angle; therefore total internal reflection occurs. No light penetrates the methyl alcohol if the angle of incidence exceeds the critical angle.

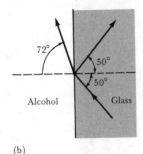

72° 50° 50°
Alcohol   Glass

(b)

Discussion: With no more sophisticated tools than the law of reflection, $\theta_i = \theta_r$, and Snell's law, along with the idea of total internal reflection, we can now trace the path of any ray through any combination of optical materials, provided we know their dimensions and indices of refraction, and the initial angle of incidence. A sample problem follows, another is found in the drill problems, and you are encouraged to make up problems of your own for practice.

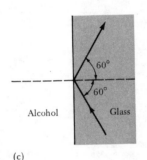

60° 60°
Alcohol   Glass

(c)

FIG. 19.5

PROBLEM: A piece of zircon ($n = 1.92$) is submerged in water ($n = 1.33$). A ray of light is incident normally on the zircon, as shown in Fig. 19.6a. (a) Determine the critical angle for a zircon-water boundary. (b) Sketch the path of the light ray until it emerges into the water again, and determine the last angle of refraction.

Solution: (a) The critical angle is determined by letting the refracted angle equal 90° in Snell's law:

$$n_z \sin \theta_c = n_w \sin 90°$$
$$(1.92) \sin \theta_c = (1.33)(1)$$
$$\sin \theta_c = \frac{1.33}{1.92} = 0.693; \quad \theta_c = \underline{43.9°}$$

The critical angle for a zircon-water boundary is about 44°; any larger angle of incidence will result in total internal reflection.

(b) The sketch is shown in Fig. 19.6b.

First boundary: $\theta_a = 0$ (normal incidence); no change in direction occurs.

Second boundary: The angle of incidence, 50°, is determined geometrically using the triangle at the bottom left and the fact that the normal makes a 90° angle with the first boundary. The angle of incidence is larger than the critical angle of 43.9° determined in part (a); therefore *total internal reflection* occurs, and the ray is reflected with an angle of reflection equal to the angle of incidence.

Third boundary: The angle of incidence is determined geometrically using the lower right triangle and the fact that the normal is always perpendicular to the boundary. It is 40°; this is less than the critical angle. Therefore refraction governed by Snell's law occurs.

$$n_z \sin \theta_z = n_w \sin \theta_w$$
$$(1.92) \sin 40° = (1.33) \sin \theta_w$$
$$\sin \theta_w = \frac{(1.92)(0.643)}{1.33} = 0.928$$
$$\theta_w = \underline{68.1°}$$

The final angle of refraction as the ray emerges from the zircon into the water is 68.1°.

Incident light
90°   Zircon $n = 1.92$
Water $n = 1.33$
50°

(a)

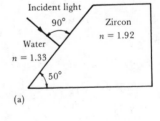

θ_a θ_w
40°
50° 50° 50°
50° 40° 40° 40°

(b)

FIG. 19.6

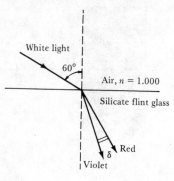

White light

60°

Air, $n = 1.000$

Silicate flint glass

δ Red

Violet

FIG. 19.7

PROBLEM: Determine the difference in angle of refraction (labeled δ in Fig. 19.7) for red and violet light when white light is incident at 60° to a boundary between air and silicate flint glass.

Solution: Because the index of refraction of a medium (and therefore the speed of light in that medium) depends on wavelength, different wavelengths are refracted by different angles for the same angle of incidence at a boundary. Let θ_R be the angle of refraction for red light. From Snell's law,

$$n_1 \sin 60° = n_R \sin \theta_R$$

where n_R = index of refraction of silicate flint glass for red light = 1.613 from Table 19.2.

$$(1.000) \sin 60° = (1.613) \sin \theta_R$$

$$\sin \theta_R = \frac{(1.000)(0.8660)}{1.613} = 0.5369$$

$$\theta_R = 32.5°$$

Here we have carried 4 significant figures throughout the calculation and rounded the answer to 3 significant figures.

For the violet light, n_V = index of refraction of silicate flint glass for violet light = 1.661 from Table 19.2. Snell's law gives

$$(1.000) \sin 60° = (1.661) \sin \theta_V$$

$$\sin \theta_V = \frac{(1.000)(0.8660)}{1.661} = 0.5214$$

$$\theta_V = 31.4°$$

The difference in these angles is

$$\delta = \theta_R - \theta_V = 32.5 - 31.4 = \underline{1.1°}$$

The angle of 1.1° between the red and violet light is a simple measure of the *dispersion* (variable refraction as a function of wavelength) of the light by the silicate flint glass for this angle of incidence.

Avoiding Pitfalls

1. Unlike mechanical waves, light requires no material medium for its propagation. In fact, the presence of any medium reduces the speed of light.

2. The speed of light in a vacuum is a constant, independent of wavelength. In a material medium, however, the speed of light, and thus the index of refraction of the medium, varies with wavelength. The index of refraction given in tables is usually for yellow light, an intermediate wavelength.

3. When light travels from one medium to another of different index of refraction, its frequency *does not change*. Therefore its change in speed is accompanied by a change in wavelength such that $v = f\lambda$ is satisfied.

4. The index of refraction of a medium is defined as $n = c/v$, the ratio of the vacuum speed of light and its speed in the medium. The erroneous inversion of this ratio can be avoided by recalling that for any material medium, n is greater than 1.

5. When light is incident on a boundary between two media of different indices of refraction n, the bending is *toward* the normal if light travels into a medium of higher n, *away from* the normal if into a medium of lower n.

6. No change in direction occurs for light incident normally on a boundary between media of different indices of refraction, even though the speed of light changes in accordance with the change in index of refraction. This is a direct outcome of Snell's law with $\theta_1 = 0$.

7. In general a light beam incident on a boundary between media of different indices of refraction is partially transmitted (at an angle given by Snell's law) and partially reflected (at an angle equal to the angle of incidence).

8. Total internal reflection (no transmitted beam) can occur only if light is incident on a medium of lower n and only if the angle of incidence is greater than the critical angle (the angle of incidence for which the refracted angle is 90°).

9. When light enters a dispersive medium, the longest wavelengths (red) are bent least; the shortest wavelengths (violet) are bent most.

Drill Problems
Answers in Appendix

1. The speed of light in ice is 2.292×10^8 m/s.

   (a) What is the index of refraction of ice?
   (b) When light in air strikes ice at an angle of incidence of 60°, what is the angle of refraction?
   (c) What is the critical angle for an ice-air boundary?

2. A light ray enters a glass slab with parallel sides and index of refraction n_2 at an angle of incidence θ_1 as shown in Fig. 19.8. Sketch its path through the slab and determine the change in direction caused by the slab.

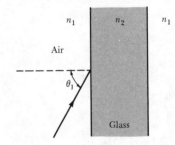

FIG. 19.8

3. A ray of light in air is incident on a piece of glass ($n = 1.5$) at an angle of incidence of 45° (Fig. 19.9). Determine by calculation and sketch the path of the ray as it is transmitted through the glass until it reemerges into the air.

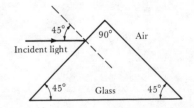

FIG. 19.9

4. The index of refraction of crown glass is 1.514 for red light and 1.532 for violet light. What is the difference in the speed of red and violet light in crown glass? Which travels faster?

20

Optical Instruments

Terms

Define or describe briefly what is meant by the following terms. If you have difficulty, refer to the textbook section given in parentheses.

lens (20.1)

converging lens (20.1)

diverging lens (20.1)

focal point (20.1)

focal length (20.1)

ray diagram (20.2)

principal axis (20.2)

Rays 1, 2, and 3 (20.2)

real image (20.2)

virtual image (20.2)

image distance (20.2)

object distance (20.2)

thin-lens equation (20.3)

linear magnification (20.3)

accommodation (20.4)

far point (20.4)

near point (20.4)

nearsightedness (20.4)

farsightedness (20.4)

angular magnification (20.5)

spherical aberration (20.7)

circle of least confusion (20.7)

chromatic aberration (20.7)

astigmatism (20.7)

converging (concave) mirror (20.8)

diverging (convex) mirror (20.8)

Equation Review

For each equation, be able to state the situation to which it applies, what quantity each symbol represents, and the units for measuring each quantity in some consistent set of units.

$$\frac{1}{f} = \frac{1}{s} + \frac{1}{s'} \tag{20.1}$$

$$m = \frac{h'}{h} = -\frac{s'}{s} \tag{20.2}$$

$$M = \frac{\theta'}{\theta} \tag{20.6}$$

$$M = \frac{\text{near point}}{s} \tag{20.8}$$

Problems with Solutions and Discussion

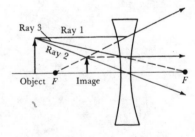

FIG. 20.1

PROBLEM: A diverging lens has a focal length of -30 cm. An object is placed 40 cm from this lens. Show by means of a ray diagram, approximately to scale, the location of the image formed by the lens. From your diagram, state whether the image is real or virtual, upright or inverted, enlarged or reduced.

Solution: Three rays are shown (Fig. 20.1). Ray 1, parallel to the principal axis, is refracted such that it appears to come from the left focal point. Ray 2 passes undeviated through the center of the lens. Ray 3, heading for the right focal point, emerges parallel to the axis. These rays do *not* converge after refraction. If we extend them backward (shown by dashed lines), we locate the virtual image, from which the refracted light appears to have come. The image may be described as <u>virtual, upright, and reduced</u> in size with respect to the object.

PROBLEM: An object is moved on the principal axis of a converging lens from a location that is 4 focal lengths from the lens to a location that is 1.5 focal lengths from the lens. By constructing ray diagrams, determine what happens to the size of the image formed and its distance from the lens.

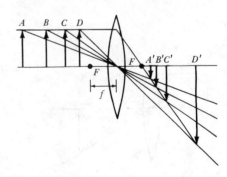

FIG. 20.2

Solution: The lens is indicated with its focal points labeled F. As the object approaches the lens, it is shown in four positions, at distances $4f$, $3f$, $2f$, and $1.5f$ from the lens center, labeled A, B, C, and D respectively (Fig. 20.2). The corresponding images are labeled A', B', C' and D'. The rays used for the construction are Ray 1 (parallel ray which passes

141

through the focal point after refraction) and Ray 2 (which passes through the center of the lens with no change in direction). The <u>image distance increases</u> from A' to D' and the <u>image gets larger</u>. The image size is proportional to the image distance, but neither is proportional to the object distance.

Notice that all objects located from A to infinitely far away must form images between image A' and the focal point of the lens, a very short distance. An object at infinity forms an image at the focal point (actually in a plane through the focal point called the *focal plane*), because rays from such an object are essentially parallel to each other when they reach the lens.

There is a condition for which the object and image distances are equal, as are the object and image sizes. This occurs when the object distance is twice the focal length. When $s > 2f$, the image is reduced; when $s < 2f$, the image is enlarged.

Notice also that when the object moves between $2f$ and f, the image size and distance are increasing quite rapidly. The increase becomes more and more dramatic the closer the object gets to F, until the object reaches F, at which location the image is infinitely large and infinitely distant. (When the object moves inside the focal point, light from it is so divergent that the lens cannot converge it enough to bring it to a focus. In this region, with the object between F and the lens, a *virtual, erect, enlarged* image is formed on the same side of the lens as the object.)

These conclusions about the behavior of the image as a function of object distance have been reached by studying the ray diagrams. You might be interested in confirming each of them mathematically, using the thin-lens equation and the linear magnification equation.

PROBLEM: Lenses come in many different shapes. A converging lens is thicker at the center than at the edges, while a diverging lens is thinner at the center than at the edges. Using this criterion, classify the common lenses in Fig. 20.3a–f as diverging or converging.

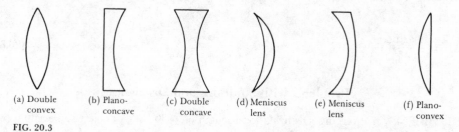

(a) Double convex   (b) Plano-concave   (c) Double concave   (d) Meniscus lens   (e) Meniscus lens   (f) Plano-convex

FIG. 20.3

Solution: (a) Converging; (b) diverging; (c) diverging; (d) converging; (e) diverging; (f) converging. Converging lenses are often called *positive* lenses because of the convention that their focal lengths are taken as positive numbers; similarly, diverging lenses are sometimes called *negative* lenses.

PROBLEM: (a) An object is 10 cm to the left of a lens of focal length 6 cm. Locate the image by means of a ray diagram and describe it. (b) Confirm your results by calculations.

Solution: (a) A lens with a positive focal length is a converging lens. Rays 1 and 2 in Fig. 20.4 are used to locate the image. It lies to the right of the lens and can be described as <u>real</u> (rays of light actually intersect there), <u>inverted</u>, and <u>enlarged</u> with respect to the original object.

(b) The thin-lens equation can be used to determine the image distance.

$$\frac{1}{f} = \frac{1}{s} + \frac{1}{s'}$$

$$\frac{1}{s'} = \frac{1}{f} - \frac{1}{s} = \frac{1}{6 \text{ cm}} - \frac{1}{10 \text{ cm}} = \frac{10 - 6}{60} = \frac{4}{60} = \frac{1}{15} \text{ cm}^{-1}$$

$$s' = \underline{15 \text{ cm}}$$

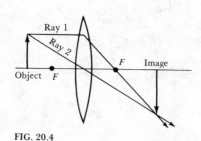

FIG. 20.4

A positive sign on s' means the image is *real* and formed to the right of the lens. The magnification is given by

$$m = -\frac{s'}{s} = -\frac{15 \text{ cm}}{10 \text{ cm}} = \underline{-1.5}$$

The image is 1.5 times *larger* than the object. The negative sign means the image is *inverted*. The image is located 15 cm to the right of the lens and is <u>real, inverted and enlarged by a factor of 1.5.</u>

PROBLEM: A thin lens forms a virtual image 60 cm from a lens when a real object is 12 cm from the lens. What is the focal length of this lens? Is it a converging or diverging lens?

Solution: We are given $s = 12$ cm (positive because the object is real)

$\qquad\qquad s' = -60$ cm (negative because the image is virtual)

Using the thin-lens equation to determine f gives

$$\frac{1}{f} = \frac{1}{s} + \frac{1}{s'} = \frac{1}{12 \text{ cm}} + \frac{1}{-60 \text{ cm}} = \frac{5-1}{60} = \frac{4}{60} = \frac{1}{15} \text{ cm}^{-1}$$
$$f = \underline{15 \text{ cm}}$$

The focal length is 15 cm; since it came out positive, it is a <u>converging</u> lens. The object distance is consistent with this result, since a converging lens does form a virtual image of an object closer than its focal point.

PROBLEM: (a) Design a pair of glasses for yourself assuming you are nearsighted and cannot see anything well beyond 3 m. (b) What is the power of these lenses? (c) Where do they form the image of an object that is 1 m from the lens?

Solution: Your eyes converge incoming light too rapidly forming an image in front of the retina. They need a diverging lens to increase the divergence of incoming rays from a distant object.

(a) You wish the lens to form an image at 3 m, where you can see it comfortably, of any distant object ($s = \infty$). Therefore, $s = \infty$ and $s' = -3$ m (negative because the image is to be upright and on the same side of the lens as the object, hence a virtual image). The desired focal length is found from the thin-lens equation:

$$\frac{1}{f} = \frac{1}{s} + \frac{1}{s'} = \frac{1}{\infty} + \frac{1}{-3 \text{ m}}; \quad \text{so} \quad f = \underline{-3 \text{ m}}$$

The glasses should have diverging lenses (indicated by the negative sign) of 3-m focal length.

(b) The power in diopters is the reciprocal of the focal length in meters,

$$P = \frac{1}{f} = \frac{1}{-3 \text{ m}} = \underline{-0.33 \text{ diopters}}$$

(c) For $s = 1$ m,

$$\frac{1}{s'} = \frac{1}{f} - \frac{1}{s} = \frac{1}{-3 \text{ m}} - \frac{1}{1 \text{ m}} = \frac{-1-3}{3} = \frac{-4}{3} \text{ m}^{-1}$$
$$s' = -\frac{3}{4} \text{ m} = \underline{-75 \text{ cm}}$$

This object distance of 100 cm is within the range of comfortable focus without glasses, but so is the virtual image, at 75 cm. Therefore you don't have to take off your glasses when looking at objects already within your range of comfortable vision.

PROBLEM: An electronics technician whose near point is 25 cm uses a magnifying glass of focal length 4 cm. (a) Where should she place her circuit board to achieve an angular magnification of 7. (b) Will she be able to see this magnified image clearly?

Solution: (a) Angular magnification is given by

$$M = \frac{\text{near point}}{s}$$

Solving for object distance s gives

$$s = \frac{\text{near point}}{M} = \frac{25 \text{ cm}}{7} = \underline{3.6 \text{ cm}}$$

If the circuit board is placed 3.6 cm beneath the magnifying glass, an angular magnification of 7 will result.

(b) She will be able to see the magnified image clearly if it is formed beyond her near point. The image distance can be obtained from the thin-lens equation:

$$\frac{1}{s'} = \frac{1}{f} - \frac{1}{s} = \frac{1}{4} - \frac{1}{3.6} = 0.25 - 0.278 = -0.028 \text{ cm}^{-1}$$

$$s' = -36 \text{ cm}$$

Since the technician's near point is 25 cm, she <u>will be able to see</u> clearly the virtual image (negative s') formed 36 cm behind the lens. The large angular magnification results because, with a near point of 25 cm, the technician would have to move the circuit board back to 25 cm to see it clearly. By using the glass, she can bring it in much closer so that it (and its image) subtend a much larger angle at her eye.

PROBLEM: Two lenses of focal lengths -20 cm and 10 cm respectively have their centers 3 cm apart. An object is placed 30 cm to the left of the diverging lens. Determine the position and magnification of the final image formed by this lens system.

Solution: The technique for analyzing more than one lens is to locate the image formed by the first lens and let it be the object for the second lens, and so on down the line. The location of the first image is determined from the thin-lens equation, where $s = 30$ cm and $f = -20$ cm.

$$\frac{1}{s'} = \frac{1}{f} - \frac{1}{s} = \frac{1}{-20 \text{ cm}} - \frac{1}{30 \text{ cm}} = \frac{-3 - 2}{60} = \frac{-5}{60} = \frac{-1}{12} \text{ cm}^{-1}$$

$$s' = \underline{-12 \text{ cm}}$$

The first image is virtual (s' is negative) and is located 12 cm to the left of the first lens. This makes it 15 cm from the second lens, so $s_2 = 15$ cm. (Or, $s_2 = d - s_1 = 3 - (-12) = 15$ cm.) This image becomes the object for the second lens. The object distance is considered positive because the object is on the left of the second lens, the side from which the light enters the lens. For the second application of the thin-lens equation, $f = 10$ cm, and $s_2 = 15$ cm.

$$\frac{1}{s_2'} = \frac{1}{f} - \frac{1}{s_2} = \frac{1}{10} - \frac{1}{15} = \frac{3 - 2}{30} = \frac{1}{30} \text{ cm}^{-1}$$

$$s_2' = \underline{30 \text{ cm}}$$

The image distance is positive, so a <u>real</u> image is formed 30 cm to the <u>right</u> of the second lens. The ray diagram is shown in Fig. 20.5. The first lens diverged the incoming light but not enough to prevent the second lens from converging it to form a real image. The magnification of the final image is the product of the magnifications of the individual lenses. For the diverging lens,

$$m_1 = -\frac{s_1'}{s_1} = -\frac{-12 \text{ cm}}{30 \text{ cm}} = 0.4$$

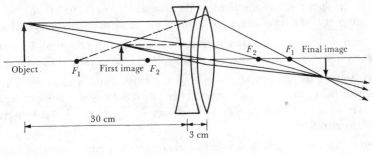

Object F_1 First image F_2 F_2 F_1 Final image

30 cm

3 cm

FIG. 20.5

For the converging lens,

$$m_2 = -\frac{s_2'}{s_2} = -\frac{30 \text{ cm}}{15 \text{ cm}} = -2$$

$$\text{Total } m = m_1 m_2 = (0.4)(-2) = -0.8$$

The final image is <u>reduced</u>, 0.8 times the size of the original object, and <u>inverted</u> (negative m). This is verified by the ray diagram.

PROBLEM: An employee leaving a clothing store at night observes that when he looks into the window from 3 m away, his reflected image is located very near where a mannikin of his height is standing. However he appears to be 10 percent larger than the mannikin. (a) What is the focal length of the window when acting as a mirror? (b) How far is the mannikin from the window?

Solution: Since the image appears 10 percent larger, the magnification is 1.1, positive because the image is upright.

$$m = 1.1 = -\frac{s'}{s}; \quad \text{therefore} \quad s' = -1.1 \, s$$

The object distance is the employee's distance from the window, $s = 3$ m.

$$s' = -1.1(3 \text{ m}) = -3.3 \text{ m}$$

The negative sign corresponds to the fact that the image is virtual. The mirror equation can now be used to obtain f.

$$\frac{1}{f} = \frac{1}{s} + \frac{1}{s'} = \frac{1}{3 \text{ m}} + \frac{1}{-3.3 \text{ m}} = 0.333 - 0.303 = 0.03 \text{ m}^{-1}$$

$$f = \underline{33 \text{ m}}$$

The ray diagram is shown in Fig. 20.6. The focal length of the glass window is 33 m. It is positive, indicating a concave (converging) mirror. This we could have deduced earlier because the image is enlarged; diverging mirrors and lenses never produce enlarged images of real objects.

(b) The image distance is -3.3 m. If the mannikin is very near the virtual image, it is also about <u>3.3 m</u> behind the window.

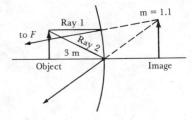

FIG. 20.6

Avoiding Pitfalls

1. Converging lenses and mirrors always have positive focal lengths, although they can form both real and virtual images. Diverging lenses and mirrors always have negative focal lengths and can form only virtual, upright images of real objects.

2. When values of s, s' and f are put into the thin-lens equation and the linear magnification equation, they must be accompanied by the appropriate algebraic sign, if known,

according to the sign conventions given in the chapter. These signs are distinct from the signs that are part of the equations or result from algebraic manipulation of the equations.

3. The sign conventions for s and s' presume that light traveling from object to lens or mirror travels from left to right, that is, that the object is to the left of the lens or mirror.

4. When solving for the image distance s', a positive number indicates a real image (formed where light rays actually intersect after reflection or refraction); a negative sign indicates a virtual image (light rays do *not* intersect at its position, they are diverging as if they had originated there).

5. When an image is formed in a region where no reflected or refracted rays actually travel (for example, behind a mirror, or to the left of a lens), that is a sure indication that the image is virtual.

6. When solving for the linear magnification m, a positive sign indicates an upright image and a negative sign an inverted image, with respect to the object.

7. The thin-lens equation is stated in terms of reciprocals of distances. When using it, do not forget the final inversion to obtain the distance of interest from its reciprocal.

8. Since each term in the thin-lens equation has the same dimensions, the distances s, s', and f may be measured in any convenient unit as long as the same unit is used for all three quantities.

9. The technique for handling a combination of lenses is to let the image formed by the first lens become the object for the second lens. The image formed by the second lens is then the object for the third and so on. This technique also works for combinations of mirrors or even for lenses and mirrors. The magnification of the combination is the product of the magnifications of the individual components.

10. The design of a lens (or lens system) to extend the capabilities of the human eye must always take into account the properties of the eye. There is no point in forming a huge image if it is too close for the eye to be able to focus on it.

Drill Problems
Answers in Appendix

1. A lens whose focal length has a magnitude of 20 cm forms an image with an image distance of -10 cm.

   (a) What is the object distance and what is the magnification?
   (b) Construct a ray diagram. (There are two solutions, one for a converging lens and one for a diverging lens.)

2. Two lenses, of focal lengths 15 cm and −20 cm respectively, are 10 cm apart. An object is placed 60 cm to the left of the converging lens as shown in Fig. 20.7. Determine the position and magnification of the final image.

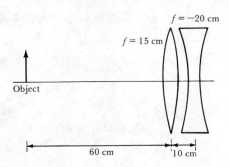

FIG. 20.7

3. A concave mirror has a focal length of 10 cm.

(a) Determine where an object must be placed to form a real inverted image twice the size of the object. Draw a ray diagram, approximately to scale, for this situation.

(b) Do the same for a virtual upright image twice the size of the object.

21

Wave Interference

Terms

Define or describe briefly what is meant by the following terms. If you have difficulty, refer to the textbook section given in parentheses.

Huygens' principle (21.1)

wavelet (21.1)

Young's double-slit experiment (21.2)

zeroth-order interference maximum (21.2)

n-th order interference maximum (21.2)

grating (21.3)

spectrograph (21.3)

thin-film interference (21.4)

diffraction (21.5)

holography (21.6)

Equation Review

For each equation, be able to state the situation to which it applies, what quantity each symbol represents, and the units for measuring each quantity in some consistent set of units.

$$\sin \theta = \frac{n\lambda}{d} \text{ (bright bands)} \tag{21.1}$$

$$\tan \theta_n = \frac{y_n}{D} \tag{21.2}$$

$$\frac{n\lambda}{d} \simeq \frac{y_n}{D} \tag{21.3}$$

$$\lambda_n = \frac{\lambda_{\text{air}}}{n}$$

$$2t = \frac{1}{2}\frac{\lambda_{air}}{n}, \frac{3}{2}\frac{\lambda_{air}}{n}, \frac{5}{2}\frac{\lambda_{air}}{n}, \ldots$$
(21.4, 21.5)

$$\sin\theta = \frac{n\lambda}{w} \text{ (dark bands)}$$
(21.6)

$$\frac{n\lambda}{w} \simeq \frac{y_n}{D}$$
(21.6)

Problems with Solutions and Discussion

PROBLEM: In a Young's double-slit experiment, light of wavelength 550 nm forms an interference pattern on a screen 2 m away. The third bright fringe is observed to be 1 cm from the central maximum. Determine the separation of the two slits.

Solution: Equation (21.3) relates the slit separation d to the displacement y_n of the n-th fringe from the central maximum:

$$\frac{n\lambda}{d} \simeq \frac{y_n}{D}$$

where D = distance to screen = 2 m, and λ = wavelength of light = 550 nm = 550 $\times 10^{-9}$ m. Solving for d gives

$$d \simeq \frac{n\lambda D}{y_n}$$

For the third bright fringe, $n = 3$, and $y_n = 1$ cm $= 10^{-2}$ m.

$$d \simeq \frac{(3)(550 \times 10^{-9} \text{ m})(2 \text{ m})}{10^{-2} \text{ m}} = 3.3 \times 10^{-4} \text{ m} = 0.33 \text{ mm}$$

The slit separation is 0.33 mm.

PROBLEM: (a) Find an expression for the linear separation of two adjacent bright fringes on the screen in a Young's double-slit experiment. (b) What is the separation between fringes when light of wavelength 500 nm is incident on two slits separated by 0.2 mm, when the screen is 3 m away?

Solution: (a) Solving Eq. 21.3 for y_n gives the linear displacement from the central maximum:

$$y_n \simeq \frac{n\lambda D}{d}$$

The next bright fringe would have order $n + 1$ and be located a distance from the central maximum given by

$$y_{n+1} \simeq \frac{(n + 1)\lambda D}{d}$$

The separation of the two fringes, y, is given by

$$\Delta y = y_{n+1} - y_n \simeq \frac{(n + 1)\lambda D}{d} - \frac{n\lambda D}{d} = \frac{n\lambda D}{d} + \frac{\lambda D}{d} - \frac{n\lambda D}{d} = \underline{\frac{\lambda D}{d}}$$

The expression for Δy is a constant for a given double-slit experiment, so the spacing of the fringes is uniform. The tenth fringe will be twice as far from the central maximum as the fifth fringe. The distance between any fringe and the one that is three fringes beyond will be $\frac{3\lambda D}{d}$.

(b) In this case:

$$\lambda = 500 \text{ nm} = 500 \times 10^{-9} \text{ m}$$
$$d = 0.2 \text{ mm} = 2 \times 10^{-4} \text{ m}$$
$$D = 3 \text{ m}$$
$$\Delta y = \frac{\lambda D}{d} = \frac{(500 \times 10^{-9} \text{ m})(3 \text{ m})}{2 \times 10^{-4} \text{ m}} = \underline{7.5 \times 10^{-3} \text{ m}}$$

The bright fringes of this pattern are separated by 7.5 mm.

PROBLEM: In a Young's double-slit experiment, light of wavelength 600 nm is incident on two slits separated by two-tenths of a millimeter. (a) What is the angular displacement of the fourth order maximum? (b) If the whole experiment is submerged in water ($n = 1.33$), what will be the angular displacement of the fourth order maximum?

Solution: (a) Angular displacement of maxima is given by

$$\sin \theta = \frac{n\lambda}{d}$$

In this case, $n = 4$, $\lambda = 600 \text{ nm} = 600 \times 10^{-9} \text{ m}$, and $d = 0.2 \text{ mm} = 2 \times 10^{-4} \text{ m}$. Putting in these values gives

$$\sin \theta = \frac{(4)(600 \times 10^{-9} \text{ m})}{2 \times 10^{-4} \text{ m}} = 1.2 \times 10^{-2}$$
$$\theta = \underline{0.69°}$$

(b) The wavelength of the light in water is reduced in direct proportion to the reduction of light's speed in water, or:

$$\lambda_{\text{water}} = \frac{\lambda_{\text{air}}}{1.33} = \frac{600 \times 10^{-9} \text{ m}}{1.33} = 450 \times 10^{-9} \text{ m}$$

The integer n, representing the order, and d, the slit separation, are unchanged, so

$$\sin \theta = \frac{4(450 \times 10^{-9} \text{ m})}{2 \times 10^{-4} \text{ m}} = 9 \times 10^{-3}$$
$$\theta = \underline{0.52°}$$

Increasing the index of refraction of the medium in which the light travels decreases its wavelength, causing the interference pattern to be compressed.

PROBLEM: A grating has 5000 lines/cm. The grating is illuminated by light of wavelength 450 nm. (a) Determine the angles of all the orders that can be seen. (b) What is the linear displacement from the central maximum of the third order on a screen 2 m away?

Solution: (a) The grating spacing d is the reciprocal of the number of lines per cm, or,

$$d = \tfrac{1}{5000} \text{ cm} = 2 \times 10^{-4} \text{ cm} = 2 \times 10^{-6} \text{ m}$$

Equation (21.1) is used to locate the maxima:

For $n = 1$: $\sin \theta = \dfrac{n\lambda}{d} = \dfrac{(1)(450 \times 10^{-9} \text{ m})}{2 \times 10^{-6} \text{ m}} = 0.225;$ $\theta = \underline{13.0°}$

For $n = 2$: $\sin \theta = 2(0.225) = 0.45;$ $\theta = \underline{26.7°}$

For $n = 3$: $\sin \theta = 3(0.225) = 0.675;$ $\theta = \underline{42.5°}$

For $n = 4$: $\sin \theta = 4(0.225) = 0.900;$ $\theta = \underline{64.2°}$

For $n = 5$, $5(0.225)$ produces a sine greater than one. Thus *four orders* are visible. Since $\theta = 90°$ is the limiting angle (at which diffracted light hits the grating itself), the fifth order is not seen. Note that the maxima are *not* equally spaced. Sin θ increases linearly with n, but the angle θ does not.

150

(b) For the third order, $\theta_3 = 42.5°$. The linear displacement y_3 is seen to be

$$y_3 = D \tan \theta_3 \quad \text{where } D = 2 \text{ m}$$

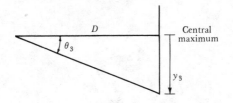

FIG. 21.1

Thus,

$$y_3 = (2 \text{ m})(0.916) = \underline{1.83 \text{ m}}$$

Note that Eq. (21.3),

$$y_n \simeq \frac{n\lambda D}{d}$$

is valid only when θ is small enough that $\sin \theta \simeq \tan \theta$. In this case Eq. 21.3 would give

$$y_3 = \frac{(3)(450 \times 10^{-9} \text{ m})(2 \text{ m})}{2 \times 10^{-6} \text{ m}} = 1.35 \text{ m}$$

A very large error results because θ is not a small angle.

PROBLEM: A grating is illuminated simultaneously with green light of wavelength 500 nm and red light whose wavelength we wish to determine. It is noted that on the screen the fourth order green fringe coincides with the third order red fringe. (a) What is the wavelength of the red light? (b) If the angular position at which these maxima coincide is 53.1° from the central maximum, how many lines per centimeter does the grating have?

Solution: (a) Writing Eq. (21.1) for the fourth order maximum of green light,

$$\sin \theta_g = \frac{4\lambda_{\text{green}}}{d} \tag{1}$$

Similarly, for the third-order red maximum,

$$\sin \theta_r = \frac{3\lambda_{\text{red}}}{d} \tag{2}$$

Since these two maxima coincide, $\sin \theta_g = \sin \theta_r$.

$$\frac{4\lambda_{\text{green}}}{d} = \frac{3\lambda_{\text{red}}}{d}$$

$$\lambda_{\text{red}} = \tfrac{4}{3}\lambda_{\text{green}} = \tfrac{4}{3}(500 \text{ nm}) = \underline{667 \text{ nm}}$$

The wavelength of the red light is 667 nm.

(b) If the angle where these maxima coincide is 53.1°, the grating spacing d can be determined from either of equations (1) and (2) above. Using equation (1) gives

$$\sin \theta_g = \frac{4\lambda_{\text{green}}}{d}$$

$$d = \frac{4\lambda_{\text{green}}}{\sin 53.1°} = \frac{4(500 \times 10^{-9} \text{ m})}{0.8} = 2.5 \times 10^{-6} \text{ m} = 2.5 \times 10^{-4} \text{ cm}$$

The number of lines per centimeter is the reciprocal of the grating spacing in centimeters:

$$\frac{1}{d} = \frac{1}{2.5 \times 10^{-4} \text{ cm}} = \underline{4 \times 10^3 \text{ cm}^{-1}}$$

The grating has 4000 lines/cm.

PROBLEM: An oil film ($n = 1.5$) floating on water is 750 nm thick. When white light is incident normally on it, what wavelengths of visible light will be intensified in the reflected beam?

Solution: Equation (21.5) tells us that constructive interference occurs with the condition

$$2t = \frac{1}{2}\frac{\lambda}{n}, \frac{3}{2}\frac{\lambda}{n}, \frac{5}{2}\frac{\lambda}{n}, \ldots$$

In this case n is the index of refraction of the oil film, t is its thickness, and λ is the wavelength of the light in air. Another way of stating this condition is

$$2t = \left(m + \frac{1}{2} \right) \frac{\lambda}{n} \quad \text{where } m = 0, 1, 2, 3, \ldots$$

Here m is used to represent an integer, since n is being used for the index of refraction of the film. For a given thickness, each value of m will enable us to solve for a correspoonding wavelength that will undergo constructive interference in the reflected beam.

For $m = 0$:

$$2t = \frac{1}{2} \frac{\lambda}{n} \quad \text{where } t = 750 \text{ nm and } n = 1.5$$

$$\lambda = 4tn = 4(750 \text{ nm})(1.5) = 4500 \text{ nm (infrared)}$$

For $m = 1$:

$$2t = \frac{3}{2} \frac{\lambda}{n}$$

$$\lambda = \frac{4tn}{3} = \frac{4500 \text{ nm}}{3} = 1500 \text{ nm (infrared)}$$

For $m = 2$:

$$2t = \frac{5}{2} \frac{\lambda}{n}$$

$$\lambda = \frac{4tn}{5} = \frac{4500 \text{ nm}}{5} = 900 \text{ nm (near infrared)}$$

For $m = 3$:

$$2t = \frac{7}{2} \frac{\lambda}{n}$$

$$\lambda = \frac{4tn}{7} = \frac{4500 \text{ nm}}{7} = \underline{643 \text{ nm (red)}}$$

For $m = 4$:

$$2t = \frac{9}{2} \frac{\lambda}{n}$$

$$\lambda = \frac{4tn}{9} = \frac{4500 \text{ nm}}{9} = \underline{500 \text{ nm} \text{ (green)}}$$

For $m = 5$:

$$2t = \frac{11}{2} \frac{\lambda}{n}$$

$$\lambda = \frac{4tn}{11} = \frac{4500 \text{ nm}}{11} = \underline{409 \text{ nm} \text{ (violet)}}$$

For $m = 6$:

$$2t = \frac{13}{2} \frac{\lambda}{n}$$

$$\lambda = \frac{4tn}{13} = \frac{4500 \text{ nm}}{13} = 346 \text{ nm (ultraviolet)}$$

There are three wavelengths in the visible range, corresponding to $m = 3$, 4, and 5, which will undergo constructive interference when reflected from this oil film.

Note that for constant t, as m increases, λ decreases. For the smaller λ's, more wavelengths fit into the path difference $2t$ traveled by the wave reflected from the bottom of the film. When m increases by one, an additional wavelength fits into the path difference $2t$. For example, when $m = 3$, 3½ wavelengths fit into the distance $2t$; when $m = 4$, 4½ wavelengths fit into the same distance. Both conditions, when occurring in conjunction with one abrupt phase change at a boundary, result in constructive interference of that particular wavelength.

PROBLEM: An optical system is being designed for use with a ruby laser which emits light of wavelength 693.4 nm. The lenses are coated with magnesium fluoride ($n = 1.38$) to make them "nonreflecting" at this wavelength. (a) What is the minimum thickness of the coating that will achieve this result? (b) What color will the lenses appear to be when coated in this way?

Solution: (a) For destructive interference of particular wavelengths, the thickness of a "nonreflective" coating on a glass lens must satisfy the condition

$$2t = \left(m + \frac{1}{2} \right) \frac{\lambda}{n}$$

The minimum thickness will result for $m = 0$. Solving for t gives

$$t = \frac{\lambda}{4n} = \frac{693.4 \text{ nm}}{4(1.38)} = \underline{126 \text{ nm}}$$

The minimum thickness of coating that will cause 693-nm light to interfere destructively in the reflected beam is 126 nm.

(b) The lens will appear to have the color of the visible wavelength which undergoes constructive interference upon reflection. The condition for constructive interference will differ by half a wavelength $\frac{1}{2}\lambda/n$ from the condition for destructive interference. Thus for constructive interference, $2t = m\lambda/n$. Solving for λ, with $m = 1$, gives

$$\lambda = \frac{2tn}{m} = \frac{2(126 \text{ nm})(1.38)}{1} = 348 \text{ nm}$$

This wavelength is in the ultraviolet; higher values of m produce even shorter wavelengths. No visible wavelength undergoes *complete* constructive interference in the reflected beam. However, those visible wavelengths nearest to 348 nm, in the violet region of the spectrum, interfere more constructively than the longer wavelengths which are approaching the wavelength of maximum destructive interference (in the red). Therefore the lens will appear violet.

PROBLEM: What is the angular width of the central maximum of the diffraction pattern of a single slit whose width is 0.002 mm when illuminated by light of wavelength 520 nm?

Solution: Since the diffraction pattern is symmetric about the central maximum at $\theta = 0$, the width of the central maximum is twice the angular spread to the first minimum on either side. The first minimum is located by setting $n = 1$ in Eq. (21.6).

$$\sin \theta_1 = \frac{n\lambda}{w} = \frac{(1)(520 \times 10^{-9} \text{ m})}{2 \times 10^{-6} \text{ m}} = 0.26$$

where w is the width of the slit, in this case 0.002×10^{-3} m $= 2 \times 10^{-6}$ m.

$$\theta_1 = 15.1°$$

This angle is measured from the center of the diffraction pattern to the first minimum. Therefore, the total width of the central maximum will be twice as great:

$$\text{Width} = 2\theta_1 = \underline{30.2°}$$

PROBLEM: A prison guard signals the end of the recreation period by blowing a whistle of frequency 650 Hz from inside an open window of width 0.75 m. The recreation area extends 100 m in front of the window and 50 m on either side. All the prisoners return except one who says he didn't hear the whistle. Is that possible? Take the speed of sound to be 340 m/s.

Solution: The sound wave is diffracted by the window in the same manner light is diffracted by a slit. Let us determine where the first minimum of the pattern occurs.

$$\sin \theta = \frac{n\lambda}{w}$$

where w = width of the window = 0.75 m

n = 1, for the first minimum

and $\lambda = \dfrac{v}{f} = \dfrac{340 \text{ m/s}}{650 \text{ s}^{-1}} = 0.523$ m

Putting in these values gives

$$\sin \theta_1 = \frac{(1)(0.523 \text{ m})}{0.75 \text{ m}} = 0.697$$

$$\theta_1 = 44.2°$$

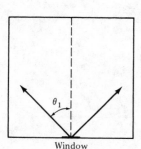

Window

FIG. 21.2

The angular position of the first minimum is shown superimposed on the recreation area, Fig. 21.2, drawn approximately to scale. It is possible that the prisoner did not hear the whistle if he was standing along the line of the first minimum of the diffraction pattern.

Avoiding Pitfalls

1.   The integer n is used to number the bright fringes in a double-slit or grating interference pattern. (It is *not* an index of refraction.) The symbol $n = 0$ refers to the central maximum, because $\theta = 0$ is obtained from $\sin \theta = n\lambda/d$ when n is zero, regardless of wavelength or slit separation. Other integral values of n (called *orders*) number the bright fringes out from the center on either side; their positions depend on wavelength and slit separation as well as on n.

2.   The equations $\sin \theta = n\lambda/d$, giving the angular position of interference maxima, and $\tan \theta = y_n/D$, relating the linear displacement from the central maximum to the screen distance, can be combined into $n\lambda/d \simeq y_n/D$ only when θ is small enough that $\sin \theta \simeq \tan \theta$. Beyond $\theta = 12°$, $\sin \theta$ and $\tan \theta$ do not agree to two significant figures. Since angles greater than 12° are often encountered when gratings are used, the two equations should be used individually for gratings.

3.   Gratings are usually characterized by their number of lines per centimeter. The *reciprocal* of this value is centimeters per line, the quantity d in the grating equation. (And, if you are working a problem in *mks* units, don't forget to change d from centimeters to meters.)

4.   The index of refraction appearing in the thin-film interference equations is *always* the index of the thin film, not the surrounding medium(s). The index of refraction of the surrounding medium is needed to determine whether or not a phase change occurs on reflection, but it does not appear explicitly in the equations.

5.   If a condition for *constructive* interference is determined for a thin film situation, the condition for *destructive* interference differs from it by half a wavelength.

6.   In thin-film interference, a large number of film thicknesses will satisfy the condition for constructive (or destructive) interference of a given wavelength; likewise a large number of wavelengths (not all visible, of course) will undergo constructive (or destructive) interference for a given film thickness.

7.   The condition for interference maxima in a double-slit or grating experiment, $\sin \theta = n\lambda/d$, is very similar in appearance to the condition for minima in a single-slit diffraction pattern, $\sin \theta = n\lambda/w$. Please recall that the first equation locates *bright* fringes, whereas the second equation locates *dark* areas.

1. In a Young's double-slit experiment, the distance between the slits is half a millimeter and the slits are 2 m from the screen. Two interference patterns can be seen on the screen, one due to green light of wavelength 500 nm and the other due to red light of wavelength 650 nm. What is the separation on the screen between the third order interference fringes of the two different patterns?

2. For what visible wavelength(s) will a soap bubble ($n = 1.35$) 350 nm thick produce maximum *destructive* interference? (*Hint:* Since the condition for maximum *constructive* interference is $2t = (m + \frac{1}{2})\lambda/n$, the condition for destructive interference differs by half a wavelength: $2t = m\lambda/n$.)

3. A grating with 3000 lines/cm is illuminated with monochromatic light. The first order maximum occurs at 11.0°. Determine

    (a) the wavelength used,
    (b) the angular position of the second order maximum, and
    (c) the number of orders observed.

22

Electric Force

Terms

Define or describe briefly what is meant by the following terms. If you have difficulty, refer to the textbook section given in parentheses.

electric charge (22.1)

atom (22.1)

electron (22.1)

proton (22.1)

neutron (22.1)

nucleus (22.1)

positive ion (22.1)

negative ion (22.1)

coulomb (22.1)

Coulomb's law (22.2)

dielectric constant (22.4)

permittivity (22.4)

polar molecule (22.5)

electric dipole moment (22.5)

electric field (22.6)

lines of force (22.7)

electromagnetic wave (22.8)

Equation Review

For each equation, be able to state the situation to which it applies, what quantity each symbol represents, and the units for measuring each quantity in some consistent set of units.

$$F = \frac{kq_1q_2}{r^2} \qquad \text{(22.1)}$$

$$K = \frac{F_{vacuum}}{F_{medium}}$$

$$F = \frac{kq_1q_2}{Kr^2} \qquad \text{(22.2)}$$

$$p = qa \qquad \text{(22.3)}$$

$$\vec{E} = \frac{\vec{F}}{q'} \qquad \text{(22.4)}$$

Problems with Solutions and Discussion

PROBLEM: A birthday party balloon of radius 18 cm contains 1 mole of helium gas. How many electrons does the gas contain and what is the total amount of their negative charge? (There are two electrons per helium atom.)

Solution: One mole of helium contains Avogadro's number of helium atoms, or $N_A = 6.023 \times 10^{23}$ atoms. At two electrons per atom, the number of electrons is

$$N_e = (6.023 \times 10^{23} \text{ atoms})(2 \text{ electrons/atom}) = \underline{1.205 \times 10^{24} \text{ electrons}}$$

The electron charge is -1.6×10^{-19} C, so the total charge Q of the electrons is

$$Q_e = (1.205 \times 10^{24} \text{ electrons})(-1.6 \times 10^{-19} \text{ C/electron}) = \underline{-1.93 \times 10^5 \text{ C}}$$

Assuming that the helium gas is electrically neutral, it must also contain 1.205×10^{24} protons, comprising 1.93×10^5 C of positive charge.

PROBLEM: Suppose all the electrons could be removed from the helium in the party balloon and placed on the moon. What would be the force of attraction between them and the positive helium ions remaining on earth?

Solution: From the inside front cover of the textbook, we find the average earth-moon distance is 3.84×10^8 m. Using Coulomb's law to determine the force,

$$F = \frac{(9 \times 10^9 \text{ N} \cdot \text{m}^2/\text{C}^2)(1.93 \times 10^5 \text{ C})^2}{(3.84 \times 10^8 \text{ m})^2} = \underline{2270 \text{ N}}$$

The force of attraction, even at that very great distance of separation, is still 2270 N. Even though the charge on the electron is tiny, macroscopic objects contain inconceivably large numbers of electrons and protons. If they could be separated, the forces between them would be enormous, even at astronomical distances. The electrical force holds electrons and nuclei together as atoms, the basic building blocks of all matter in our universe. Large forces are necessary to remove any but the most loosely bound electrons from atoms, which accounts for the stability of matter and the fact that ordinary matter is electrically neutral.

PROBLEM: In each of the situations in Fig. 22.1, show by vectors the force on the charge at position A due to each of the charges at the other labeled positions. Add the vectors

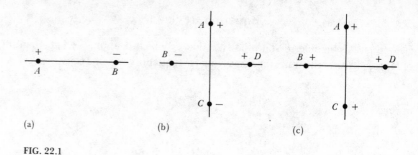

(a) (b) (c)

FIG. 22.1

graphically by a sketch, approximately to scale, and indicate the resultant force. (Assume all charges are equal in magnitude.)

Solution: (a) There is one force; it attracts the charge at A to the charge at B. If released, the charge at A will move toward B (Fig. 22.2a).

(b) The charge at B exerts an attractive force $\vec{F}_B$ (Fig. 22.2b). The charge at D exerts a repulsive force $\vec{F}_D$, equal in magnitude to $\vec{F}_B$ because B and D are equidistant from A. The charge at C exerts an attractive force $\vec{F}_C$, smaller than $\vec{F}_B$ and $\vec{F}_D$ because C is further away. The forces are added as vectors (c) to obtain the resultant force $\vec{R}$. If released, the charge at A would move in the general direction of the charge at B.

(c) The charges at B and D exert repulsive forces $\vec{F}_B$ and $\vec{F}_D$, equal in magnitude and directed as shown in (d). $\vec{F}_C$ is also repulsive but smaller than $\vec{F}_B$ and $\vec{F}_D$ because the charge at C is further away. The resultant force R is large and points in the direction of $\vec{F}_C$ in (e). If released, the charge at A will move away from the other charges along the line joining C and A.

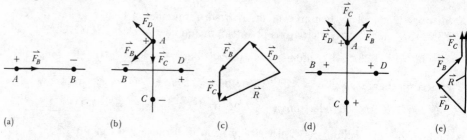

(a) (b) (c) (d) (e)

FIG. 22.2

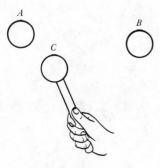

FIG. 22.3

PROBLEM: Two identical metal spheres that have equal charges repel each other with a force F. Suppose a third identical sphere (initially uncharged and with an insulating handle) touches sphere A and then sphere B and is then removed (Fig. 22.3). By what factor does the force between A and B change?

Solution: Metals are good conductors. When two identical spheres touch, the excess charge will be shared equally between them, as each charge tries to move away from all the others. Let q be the initial charge on sphere A and sphere B. The force between them is given by Coulomb's law:

$$F = \frac{kq^2}{r^2}$$

where r is the separation of their centers. When C touches A, the charge q is shared equally. A and C each now have a charge of $\frac{1}{2}q$. When C, with its charge of $\frac{1}{2}q$ touches B with its charge q, again the total excess charge $(q + \frac{1}{2}q = \frac{3}{2}q)$ is shared equally between B and C; therefore each ends up with $\frac{1}{2}(\frac{3}{2}q) = \frac{3}{4}q$. Now sphere C is removed. Sphere A, with charge $\frac{1}{2}q$, remains at the same distance from sphere B with charge $\frac{3}{4}q$. They now repel with a force

$$F' = \frac{k(\frac{1}{2}q)(\frac{3}{4}q)}{r^2} = \frac{3}{8}\frac{kq^2}{r^2} = \frac{3}{8}F$$

The repulsive force is $\frac{3}{8}$ as great as it was initially.

Let us use the idea of conservation of charge to check our results. Initially the total charge of the system was q (on A) + q (on B) = $2q$. At the end of the procedure, we had $\frac{1}{2}q$ (on A) + $\frac{3}{4}q$ (on B) + $\frac{3}{4}q$ (on C) = $2q$. The total amount of charge has not changed.

PROBLEM: An object with a charge of $-2\ \mu C$ and mass 20 g is held at a point 2 m from a fixed charge of 200 μC (Fig. 22.4). (a) What is the initial acceleration of the object when released from rest? (Assume the electrical force is the only force acting.) (b) If the same conditions are repeated in water, what will the initial acceleration be?

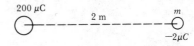

FIG. 22.4

Solution: (a) The electrical force on the object is given by Coulomb's law:

$$F = \frac{kq_1q_2}{r^2} = \frac{(9 \times 10^9\ \text{N}\cdot\text{m}^2/\text{C}^2)(2 \times 10^{-6}\ \text{C})(200 \times 10^{-6}\ \text{C})}{(2\ \text{m})^2} = \underline{0.9\ \text{N}}$$

The initial acceleration, from Newton's second law, is

$$a = \frac{F}{m} = \frac{0.9\ \text{N}}{2 \times 10^{-2}\ \text{kg}} = \underline{45\ \text{m/s}^2}$$

Note that all quantities were expressed in *mks* units in the above equations. The charges are opposite in sign, so the object will accelerate toward the positive charge. Since force is a function of distance, as the object gets closer, the force increases, causing the acceleration to increase. The object's acceleration continues to increase until it reaches the fixed charge.

(b) In water, the force is smaller than in air because the polar water molecules near each charged object align themselves in such a way as to effectively cancel part of its charge. The dielectric constant of water, $K = 80$, gives the reduction of the force in water compared to the force in air (vacuum).

$$K_{\text{water}} = \frac{F_{\text{vacuum}}}{F_{\text{water}}}$$

Therefore

$$F_{\text{water}} = \frac{F_{\text{vacuum}}}{K} = \frac{F_{\text{vacuum}}}{80}$$

If the mass of the object is unchanged but the force on it is reduced by a factor of 80, the acceleration will likewise be reduced by a factor of 80. In water the initial acceleration will be

$$a_{\text{water}} = \frac{a_{\text{air}}}{80} = \frac{45\ \text{m/s}^2}{80} = \underline{0.56\ \text{m/s}^2\ \text{toward the positive charge}}$$

PROBLEM: An ammonia molecule has a dipole moment of 5×10^{-30} C·m. (a) What charge separation between e and $-e$ would result in the same dipole moment? (b) What force is felt by the charges in this equivalent dipole?

Solution: (a) The electric dipole moment p is defined by Eq. (22.3), $p = qa$, where q is the magnitude of each of two opposite charges and a is their separation. In this case,

$$p = 5 \times 10^{-30}\ \text{C}\cdot\text{m, and}$$
$$q = e = 1.6 \times 10^{-19}\ \text{C}$$

Solving Eq. (22.3) for a gives

$$a = \frac{p}{q} = \frac{5 \times 10^{-30}\ \text{C}\cdot\text{m}}{1.6 \times 10^{-19}\ \text{C}} = \underline{3.1 \times 10^{-11}\ \text{m}}$$

The equivalent charge separation in an ammonia molecule is 3.1×10^{-11} m, less than the radius of a hydrogen atom, which is 5.3×10^{-11} m.

(b) The charges are attracted to each other with a force given by Coulomb's law,

$$F = \frac{kq_1q_2}{r^2} = \frac{ke^2}{r^2} = \frac{(9 \times 10^9\ \text{N}\cdot\text{m}^2/\text{C}^2)(1.6 \times 10^{-19}\ \text{C})^2}{(3.1 \times 10^{-11}\ \text{m})^2} = \underline{2.4 \times 10^{-7}\ \text{N}}$$

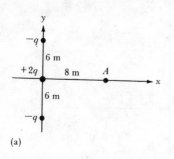

(a)

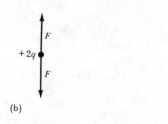

(b)

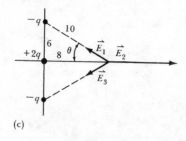

(c)

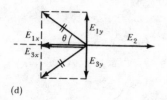

(d)

FIG. 22.5

PROBLEM: Consider the charge distribution shown on the y axis, consisting of *three fixed charges* (Fig. 22.5a): $+2q$ at the origin, $-q$ located 6 m above, and $-q$ located 6 m below the origin. Let $q = 1 \times 10^{-6}$ C. Point A is located on the positive x axis 8 m from the origin. (a) What is the electrostatic *force* on the charge $+2q$ at the origin? (b) Determine the electric *field* at the point A.

Solution: (a) The charge $+2q$ is attracted equally to the charge $-q$ above and the charge $-q$ below (Fig. 22.5b). The forces are opposite in direction; therefore the net force on the charge $+2q$ is zero.

(b) The electric field E is defined by Eq. (22.4) as

$$\vec{E} = \frac{\vec{F}}{q'}$$

We imagine a small *positive* test charge q' at point A and determine the forces acting on it. We may either determine the net resultant force $\vec{F}$ and divide by q' to get the resultant $\vec{E}$; or we may divide each individual force $\vec{F}$ by q' to get the field $\vec{E}$ due to one charge, then add all the $\vec{E}$'s as vectors to get the resultant $\vec{E}$. We will use the latter method. Let $\vec{F}_1$ be the force due to the charge above the origin. This charge will attract the test charge q'. Since $\vec{F}$ and $\vec{E}$ have the same direction [the vector equation (Eq. 22.4) equates magnitudes *and* directions], $\vec{E}_1$ lies as shown in (c). Its magnitude is

$$E_1 = \frac{F_1}{q'} = \frac{kqq'}{r^2 q'} = \frac{kq}{r^2}$$

The distance r is the hypotenuse of a right triangle in which

$$r = \sqrt{(6 \text{ m})^2 + (8 \text{ m})^2} = 10 \text{ m}$$

Substituting these values gives

$$E_1 = \frac{(9 \times 10^9 \text{ N} \cdot \text{m}^2/\text{C}^2)(1 \times 10^{-6} \text{ C})}{(10 \text{ m})^2} = 90 \text{ N/C}$$

By symmetry, $\vec{E}_3$ is also 90 N/C; it points toward the charge below the origin as shown in (c). $\vec{E}_2$, the field due to the charge $+2q$ at the origin is given by

$$E_2 = \frac{F_2}{q'} = \frac{k(2q)(q')}{r^2 q'} = \frac{k(2q)}{r^2} = \frac{(9 \times 10^9 \text{ N} \cdot \text{m}^2/\text{C}^2)(2)(1 \times 10^{-6} \text{ C})}{(8 \text{ m})^2} = 280 \text{ N/C}$$

Since q' and $2q$ are both positive, the force is repulsive. $\vec{F}_2$ and $\vec{E}_2$ both point *away from* $+2q$ as shown.

We must now add $\vec{E}_1$, $\vec{E}_2$, and $\vec{E}_3$ as vectors. Using the method of components, the vectors are shown resolved in (d). By symmetry ($E_1 = E_3$ and the angles are also equal), the y components are equal and opposite, cancelling each other: $\vec{E}_y = 0$. On the x axis we have

$$E = E_{1x} + E_{3x} + E_{2x} = -E_1 \cos\theta - E_3 \cos\theta + E_2$$

where we have taken fields to the left as negative and fields to the right as positive. From the right triangle in the previous diagram,

$$\cos\theta = \frac{8 \text{ m}}{10 \text{ m}} = 0.8$$

$$E = -(90 \text{ N/C})(0.8) - (90 \text{ N/C})(0.8) + 280 \text{ N/C}$$

$$= -72 - 72 + 280 = \underline{136 \text{ N/C to the right}}$$

The resultant electric field at point A is 136 N/C; since we got a positive number, it is directed to the right.

PROBLEM: A charge of $+q$ is located 1 m to the left of point P (Fig. 22.6a). Where should a second charge of $-2q$ be placed to make the electric field zero at point P?

Solution: The electric field at P due to the charge $+q$ is determined by imagining a small positive charge q' at P. Then

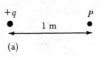

(a)

$$E_1 = \frac{F_1}{q'} = \frac{kqq'}{r_1^2 q'} = \frac{kq}{r_1^2}$$

In this case $r_1 = 1$ m, so

(b)

$$E_1 = \frac{kq}{(1 \text{ m})^2} \qquad (1)$$

The direction of $\vec{E}_1$ is to the right, since the two positive charges repel each other. To make $\vec{E}$ zero at point P, a field $\vec{E}_2$ is required that is equal in magnitude but opposite in direction to $\vec{E}_1$ (Fig. 22.6b). This means the charge $-2q$ must lie to the left of P, to attract the positive q' to the left. To determine its exact distance from P, we use the condition that at point P, $E_2 = E_1$ for the fields to give a resultant $\vec{E} = 0$ there. The field E_2 is given by

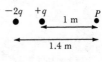

(c)

FIG. 22.6

$$E_2 = \frac{F_2}{q'} = \frac{k(2q)(q')}{r_2^2 q'} = \frac{k(2q)}{r_2^2} \qquad (2)$$

where r_2 is the distance we wish to determine. Equating the magnitudes of E_1 and E_2 from (1) and (2) above gives

$$\frac{kq}{(1 \text{ m})^2} = \frac{k(2q)}{r_2^2}$$

$$r_2^2 = 2(1 \text{ m})^2$$

$$r_2 = \sqrt{2}\,(1 \text{ m}) = \underline{1.4 \text{ m}}$$

If the charge $-2q$ lies 1.4 m to the left of P, the electric field at P will be zero (Fig. 22.6c).

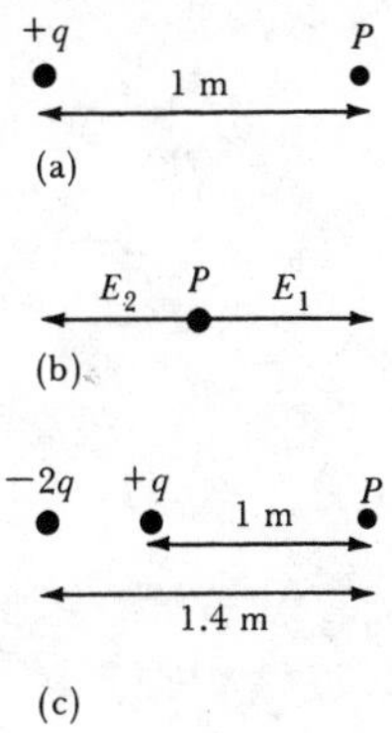

Avoiding Pitfalls

1. The fundamental indivisible unit of electric charge, designated by e, is that carried by the electron $(-e)$ and the proton $(+e)$, where $e = 1.6 \times 10^{-19}$ C. All other electric charges are integral multiples of e.

2. For solids an electrical charge is due to an excess or deficiency of electrons. In liquids and gases, both positive and negative charges (ions) can move.

3. Coulomb's law is used to calculate the *magnitude* of the force between two charged particles. The *direction* of the force is always along the line joining the charges and *toward* each other for unlike charges, *away from* each other for like charges.

4. The force determined by Coulomb's law is treated like any other force vector. It is added with other forces in the equilibrium equations for a stationary charged particle, or in Newton's second law to find the acceleration of a charged particle.

5. To determine the force on a charged particle due to a number of other charged particles, use Coulomb's law to find $\vec{F}_1$ due to q_1, repeat to find $\vec{F}_2$ due to q_2, etc., then add all forces *as vectors*.

6. To avoid erroneously inverting the ratio of forces which defines K ($K = F_{\text{vacuum}}/F_{\text{medium}}$), recall that the presence of a dielectric *always reduces* the forces between charged particles.

7. Do not confuse k, the Coulomb's law constant of proportionality, with K, the dielectric constant of a medium.

8. An electric dipole consists of two equal and opposite charges separated by a certain fixed distance. If the charges are not equal in magnitude or if they have the same sign, the term *dipole* is not appropriate.

9. The electric field $\vec{E}$ is a vector quantity with both magnitude and direction. Its direction is defined as the direction of the force on a *positive charge*. Thus if the charged particle is negative, the force on it is *opposite* to the direction of $\vec{E}$.

10. To calculate the electric field at a point due to several charges, calculate $\vec{E}_1$ due to q_1, $\vec{E}_2$ due to q_2, etc., and then add all the $\vec{E}$'s *as vectors*.

Drill Problems
Answers in Appendix

1. Two charges experience an attractive force of 4 N. If both charges are made three times as large and their separation is doubled, what will the force between them be?

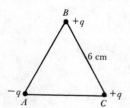

B +q

6 cm

−q   +q
A   C

FIG. 22.7

2. Points A, B, and C are at the vertices of an equilateral triangle of side 6 cm (Fig. 22.7). At A is a charge of $-q$, and at B and C are charges of $+q$. Let $q = 2 \times 10^{-6}$ C.

(a) What force (magnitude and direction) does the charge at C experience due to the charges at A and B?
(b) If this configuration existed in water, what force would be felt by the charge at C?

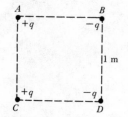

A   B
+q   −q

         1 m

+q   −q
C   D

FIG. 22.8

3. Determine the electric field at the center of the square array of point charges shown in Fig. 22.8. Let $q = 2 \times 10^{-9}$ C. The side of the square has a length of 1 m.

162

Electrical Energy

23

Terms

Define or describe briefly what is meant by the following terms. If you have difficulty, refer to the textbook section given in parentheses.

electrical potential energy (23.1)

electrical potential difference (23.4)

volt (23.4)

ground point (23.4)

absolute potential (23.4)

equipotential surface (23.4)

Equation Review

For each equation, be able to state the situation to which it applies, what quantity each symbol represents, and the units for measuring each quantity in some consistent set of units.

$$\Delta PE_q = kq_1q_2\left(\frac{1}{r} - \frac{1}{r_0}\right) \tag{23.1}$$

$$\Delta PE_q = \sum_i \frac{kq_iq'}{K}\left(\frac{1}{r_i} - \frac{1}{r_{i_0}}\right) \tag{23.3}$$

$$W = \Delta KE + \Delta PE_g + \Delta PE_s + \Delta PE_q + \Delta E_{th} + \cdots \tag{8.10}$$

$$V_{AB} = \frac{PE_{qB} - PE_{qA}}{q'} = \frac{\Delta PE_{qAB}}{q'} \tag{23.4}$$

$$\Delta PE_q = qV \tag{23.5}$$

$$V = \frac{kq}{r} \tag{23.6}$$

$$E_{av} = -\frac{V}{d} \tag{23.8}$$

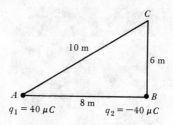

10 m

6 m

A 8 m B

$q_1 = 40\ \mu C$ $q_2 = -40\ \mu C$

FIG. 23.1

Problems with Solutions and Discussion

PROBLEM: Charges q_1 and q_2 of 40 μC and -40 μC respectively are situated as shown in Fig. 23.1. Determine the change in potential energy as q_2 moves from B to C.

Solution: The change in electrical potential energy is given by Eq. (23.1)

$$\Delta PE_q = kq_1q_2\left(\frac{1}{r} - \frac{1}{r_0}\right)$$

where r is the final separation, 10 m in this case, and r_0 is the initial separation of 8 m. Substituting numerical values gives

$$\Delta PE_q = (9 \times 10^9 \text{ N}\cdot\text{m}^2/\text{C}^2)(40 \times 10^{-6} \text{ C})(-40 \times 10^{-6} \text{ C})\left(\frac{1}{10 \text{ m}} - \frac{1}{8 \text{ m}}\right)$$

$$\Delta PE_q = (-14.4)(0.1 - 0.125) \text{ J} = \underline{0.36 \text{ J}}$$

The change in potential energy is 0.36 J. The positive sign tells us that the potential energy of the configuration has increased. To achieve this, external work was done in moving q_2 from B to C. Since q_2 and q_1 are unlike charges, they attract each other; q_2 will *not* move further away from q_1 of its own accord. External work is required amounting to $W = \Delta PE_q$.

PROBLEM: Consider again the original configuration in the previous problem. Determine the change in potential energy as q_1 moves from A to C.

Solution: Again,

$$\Delta PE_q = kq_1q_2\left(\frac{1}{r} - \frac{1}{r_0}\right)$$

where $r_0 = 8$ m and $r = 6$ m. Substituting numerical values gives

$$\Delta PE_q = (9 \times 10^9 \text{ N}\cdot\text{m}^2/\text{C}^2)(40 \times 10^{-6} \text{ C})(-40 \times 10^{-6} \text{ C})\left(\frac{1}{6 \text{ m}} - \frac{1}{8 \text{ m}}\right)$$

$$\Delta PE_q = (-14.4)(0.167 - 0.125) = \underline{-0.60 \text{ J}}$$

The change in potential energy is negative, -0.60 J; the potential energy has decreased. In moving from A to C, charge q_1 is "rolling downhill," being attracted to q_2. No *external* work is required for this move; the system loses electrical potential energy which will be converted to some other form of energy, such as kinetic energy or perhaps thermal energy.

PROBLEM: A charge q_1 of 5 μC is located at one corner of an equilateral triangle of side 4 m (Fig. 23.2a). Determine the change in potential energy when a charge q_2 of 1 μC moves from point B to point A.

Solution: According to Eq. (23.1)

$$\Delta PE_q = kq_1q_2\left(\frac{1}{r} - \frac{1}{r_0}\right)$$

where r is the final separation between q_1 and q_2, and r_0 is their initial separation. In this case $r = 4$ m and $r_0 = 4$ m. Thus,

$$\Delta PE_q = (9 \times 10^9 \text{ N}\cdot\text{m}^2/\text{C}^2)(5 \times 10^{-6} \text{ C})(1 \times 10^{-6} \text{ C})\left(\frac{1}{4 \text{ m}} - \frac{1}{4 \text{ m}}\right) = 0$$

Because the separation of the two charges does not change, the potential energy also <u>does not change</u>. Note that q_2 may move to any point on the circumference of a circle of radius 4 m centered on q_1 with no change in electrical potential energy (Fig. 23.2b).

An alternate way of solving the same problem is to ask how much work is done in moving q_2 from A to B. Since the work done is equal to the change in electrical potential energy (assuming all other forms of energy remain constant), the answer is again <u>zero</u>.

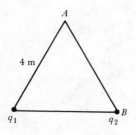

A

4 m

q_1 B

q_2

(a)

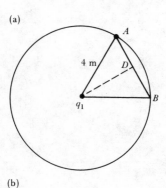

A

4 m D

q_1 B

(b)

FIG. 23.2

164

Work is put into the system (positive work) to move q_2 from B to D, since the positive charges are getting closer to each other, but the system does work (negative work) in moving q_2 from D to A, since the positive charges repel each other. These two works cancel, leaving a total work of zero. Similarly, it requires no work to move charge q_2 anywhere on the circle shown in the last diagram. This can be deduced by the fact that $W = \Delta PE_q$ and $\Delta PE_q = 0$; or from the definition of work, $W = F\,\Delta r \cos\theta$. In moving on the circle $\Delta \vec{r}$ and $\vec{F}$ are always perpendicular and $\cos 90° = 0$.

PROBLEM: Two 1-g particles, each with a charge of 30 μC are shot at each other from a very large distance, each with a speed of 500 m/s. How close are they when they (momentarily) come to rest? Ignore any frictional effects.

Solution: As the particles approach each other they repel. Their kinetic energy is converted into electrical potential energy of repulsion.

$$\text{Loss of } KE = \text{Gain of } PE_q$$

$$\cancel{2}\left(\frac{1}{\cancel{2}}mv^2\right) = kq_1q_2\left(\frac{1}{r} - \frac{1}{r_0}\right)$$

where each has an initial kinetic energy of $\frac{1}{2}mv^2$, their initial separation is $r_0 = \infty$, and their final separation at the distance of closest approach is r, the unknown we wish to determine. Solving for r and putting in numbers gives

$$r = \frac{kq_1q_2}{mv^2} = \frac{(9 \times 10^9 \text{ N}\cdot\text{m}^2/\text{C}^2)(30 \times 10^{-6} \text{ C})^2}{(1 \times 10^{-3} \text{ kg})(500 \text{ m/s})^2} = \underline{3.24 \times 10^{-2} \text{ m}}$$

The particles are brought to rest by their mutual repulsion at a separation of 3.2 cm, all kinetic energy having been transformed into electrical potential energy. They have rolled as far as possible up an "electrical hill." The electrical force of repulsion will now cause them to separate again, and if there are no other forces acting, they will regain at a large distance all the kinetic energy they started with.

PROBLEM: Determine the electric field and the absolute electric potential at a point halfway between two equal like charges (Fig. 23.3a).

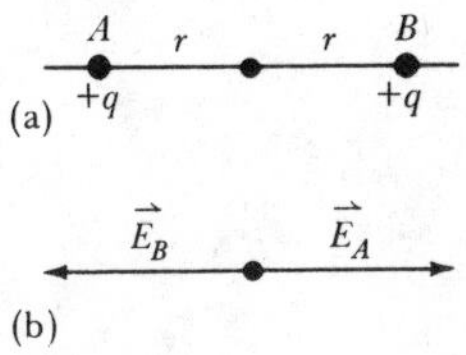

(a)

(b)

FIG. 23.3

Solution: The electric *field* $\vec{E}$ at a distance r from a charge q is given by

$$\vec{E} = \frac{\vec{F}}{q'} = \frac{kqq'}{r^2q'} = \frac{kq}{r^2}$$

where we imagine a small positive test charge q' to be located at the point of interest. Since the charges at A and B are equal and equidistant from the center point, $\vec{E}_A$ and $\vec{E}_B$ are equal and opposite (Fig. 23.3b); the resultant $\vec{E}$ is zero.

The absolute electric *potential* at a distance r from a charge q is given by Eq. (23.6), $V = kq/r$. With two charges in the vicinity,

$$V = \frac{kq}{r} + \frac{kq}{r} = \underline{\frac{2kq}{r}}$$

The voltage is *not* zero. So there can be points in space where $\vec{E} = 0$ but V is not zero. This is consistent with the idea that the $V = 0$ reference level, here taken at $r = \infty$, can in fact be taken at any convenient location.

PROBLEM: Determine the electric field and the absolute electric potential at a point halfway between two equal unlike charges (Fig. 23.4).

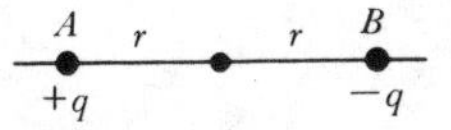

FIG. 23.4

Solution: Here it is the electric *potential* (voltage) that is zero:

$$V = \frac{kq}{r} + \frac{k(-q)}{r} = \underline{0}$$

165

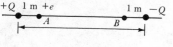

FIG. 23.5

whereas the electric *field* is non-zero. A small positive test charge q' placed midway between A and B would be attracted to B and repelled from A. The net electric field is thus $\underline{2kq/r^2 \text{ to the right.}}$

PROBLEM: Two charges of 5×10^{-9} C are located 6 m from each other as shown in Fig. 23.5. If a positron (a particle with the mass of an electron and charge $+e$) is released from point A, one meter to the right of the positive charge, what will its speed be when it reaches point B, one meter to the left of the negative charge?

Solution: Our first thought might be to determine the force on the positron at A, and from it an acceleration, and then use a constant acceleration equation to get the speed at B. However, the acceleration of the positron between A and B is not constant, because the force changes, so this approach would give an incorrect result. When nonconstant forces act, the energy approach will almost always be a simpler way to solve the problem.

The positron loses electrical potential energy and gains kinetic energy; if no other energy changes occur, conservation of energy says the loss and gain must be equal in magnitude: $\Delta PE_q = \frac{1}{2}mv^2$. ΔPE_q may be calculated by the multiple-charge Eq. (23.3), but let us use instead a slightly different approach embodied in Eq. (23.5):

$$\Delta PE_q = qV_{AB}$$

where V_{AB} is the change in voltage from A to B. The voltage (absolute potential) at A is the algebraic sum of the potentials due to the two charges at their respective distances.

$$V_A = \frac{k(+Q)}{1 \text{ m}} + \frac{k(-Q)}{5 \text{ m}} = kQ\left(\frac{1}{1 \text{ m}} - \frac{1}{5 \text{ m}}\right)$$
$$= (9 \times 10^9 \text{ N}\cdot\text{m}^2/\text{C}^2)(5 \times 10^{-9} \text{ C})(1 - 0.2) \text{ m}^{-1}$$
$$V_A = +36 \text{ V}$$

At B,

$$V_B = \frac{k(+Q)}{5 \text{ m}} + \frac{k(-Q)}{1 \text{ m}} = kQ\left(\frac{1}{5} - \frac{1}{1}\right)$$
$$= (9 \times 10^9 \text{ N}\cdot\text{m}^2/\text{C}^2)(5 \times 10^{-9} \text{ C})(-0.8 \text{ m}^{-1})$$
$$V_B = -36 \text{ V}$$

The voltage has *decreased* from $+36$ V to -36 V, a change of -72 V. Putting this into Eq. (23.5) gives

$$\Delta PE_q = qV_{AB} = (+1.6 \times 10^{-19} \text{ C})(-72 \text{ V}) = -1.15 \times 10^{-17} \text{ J}$$

We may now solve the energy conservation equation, $|\Delta PE_q| = \frac{1}{2}mv^2$, for v:

$$v = \sqrt{\frac{2|\Delta PE_q|}{m}} = \sqrt{\frac{2(1.15 \times 10^{-17} \text{ J})}{9.1 \times 10^{-31} \text{ kg}}}$$
$$= \sqrt{2.53 \times 10^{13}} \text{ m/s} = 5 \times 10^6 \text{ m/s}$$

A positron released from point A would reach point B with a speed of 5×10^6 m/s.

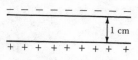

FIG. 23.6

PROBLEM: A voltage difference of 200 V is applied between two parallel plates separated by 1 cm of air (Fig. 23.6). (a) What is the electric field between the plates? (b) What is the electric field between the plates if the space there is filled with water?

Solution: (a) The electric field is given by Eq. (23.8):

$$|E| = \frac{V}{d} = \frac{200 \text{ V}}{0.01 \text{ m}} = \underline{20,000 \text{ V/m}}$$

(We have used E rather than E_{av} because between parallel plates the field is essentially constant.) The direction of E is the direction of the force on a positive charge, or $\underline{\text{upward}}$.

(b) The presence of water reduces the force between charged particles by a factor equal to the dielectric constant. For water, $K = 80$, from Table 22.1 in the textbook. The electric field is defined as $\vec{E} = \vec{F}/q'$. If the force on a charged particle q' is reduced, the electric field is reduced by the same factor. Therefore the electric field with the water present is

$$E_{water} = \frac{E_{air}}{K} = \frac{20{,}000 \text{ V/m}}{80} = \underline{250 \text{ V/m}}$$

The direction of the electric field of 250 V/m is again <u>upward</u>.

PROBLEM: Consider a lightning stroke which lasts 0.2 s and transfers a total of 30 C of charge across a potential difference of 10^8 V between cloud and ground. (a) What is the average power developed during this stroke? (b) If all the electrical energy released could be stored and utilized, how long would it operate a 100-W light bulb?

Solution: (a) Power is work or energy per unit time. The energy release of the stroke can be obtained from Eq. (23.5), $\Delta PE_q = q\,\Delta V$, where $q = 30$ C and $\Delta V = 10^8$ V.

$$\Delta PE_q = (30 \text{ C})(10^8 \text{ V}) = 3 \times 10^9 \text{ J}$$

The electrical energy released as the charge moves through a potential difference of 10^8 volts is 3×10^9 J. This energy is released during a time period $t = 0.2$ s, so the power developed is

$$P = \frac{\Delta PE_q}{t} = \frac{3 \times 10^9 \text{ J}}{0.2 \text{ s}} = \underline{1.5 \times 10^{10} \text{ W}}$$

This power is equivalent to that of 150 million 100-watt light bulbs. Such a tremendous power output lasts for only a fraction of a second, however.

(b) The total energy released is 3×10^9 J, from part (a). A 100-watt light bulb utilizes energy at the rate of $P = E/t$. Solving for t gives

$$t = \frac{E}{P} = \frac{3 \times 10^9 \text{ J}}{100 \text{ W}} = \underline{3 \times 10^7 \text{ s}} = 3 \times 10^7 \text{ s} \left(\frac{1 \text{ day}}{86400 \text{ s}}\right) = 347 \text{ days}$$

The electrical energy from a 0.2-s flash of lightning is enough to operate a 100-watt light bulb day and night for almost a year.

Avoiding Pitfalls

1. Electrical potential energy and electrical potential (voltage) are both *scalar* quantities which have magnitude but no direction, unlike the electric field which is a vector quantity with both magnitude and direction.

2. In *every equation* in the Equation Review which involves charge q, the signs of the various charges must be included in the calculations.

3. Changes in potential energy and electric potential may be negative because the sign of the charges involved may be positive or negative. A negative sign on a change in electrical potential energy means it has decreased. A negative sign on a voltage means it is less than the voltage of the reference level arbitrarily taken as $V = 0$.

4. The electric forces on charged particles are always such as to make them move in the direction of decreasing potential energy. An external force is required to make them move in the direction of increasing potential energy.

5. Positive and negative work have the same meaning in electrical contexts that they have always had. Positive work is done by an external force, increasing the energy of the system. Negative work is work done by the system, thereby decreasing its energy.

6. As with gravitational potential energy, the reference point where electrical potential (voltage) is considered zero is chosen at a point convenient to the particular problem. This is because it is *changes* in potential that are physically meaningful. The same is true of electrical potential energy.

7. Electric fields may be measured in units of newtons per coulomb or volts per meter, since 1 N/C = 1 V/m. The choice is determined by which unit is most similar to the units of other quantities in the problem.

8. The equation $E_{av} = -V/d$ is not a vector equation because V is not a vector quantity. The meaning of the negative sign is that the direction in which V is increasing is opposite to the direction of the electric field.

Drill Problems
Answers in Appendix

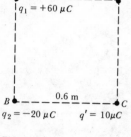

FIG. 23.7

1. Electrical charges are placed at two corners, A and B, of a square whose side is 0.6 m long as shown in Fig. 23.7. How much work is required to move a $+10$-μC charge between C and D?

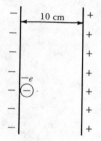

FIG. 23.8

2. An electron, released from a negatively charged plate, is attracted to a plate with an equal positive charge 10 cm away (Fig. 23.8). The difference in potential between the plates is 12 V.

(a) In what direction is the electric field between the plates?
(b) What is the electron's speed when it has traveled 6 cm in the direction of the positive plate?

3. An electric field exists in the earth's atmosphere. At the surface of the earth, its average value is 120 V/m and it is directed vertically downward.

(a) What difference in potential exists between the head and feet of a person 2 m tall, due to the earth's field?
(b) What potential energy (gravitational plus electrical) would a proton lose in falling through this distance in a vacuum? The proton's mass is 1.67×10^{-27} kg.

Electric Current and Ohm's Law

24

Terms

Define or describe briefly what is meant by the following terms. If you have difficulty, refer to the textbook section given in parentheses.

battery (24.1)

electrodes (24.1)

electrolyte (24.1)

terminals (24.1)

source of emf (24.1)

electric current (24.2)

ion current (24.2)

ampere (24.2)

direct current (24.2)

alternating current (24.2)

drift velocity (24.2)

free electron density (24.2)

Ohm's law (24.3)

resistivity (24.3)

conductivity (24.3)

electrical resistance (24.3)

ohm (24.3)

bilateral impedance rheography (24.4)

temperature coefficient of resistance (24.5)

thermistor (24.5)

kilowatt-hour (24.6)

Equation Review

For each equation, be able to state the situation to which it applies, what quantity each symbol represents, and the units for measuring each quantity in some consistent set of units.

$$I = \frac{\Delta q}{\Delta t} \tag{24.1}$$

$$I = -(n_e e A)v_d \tag{24.3}$$

$$V = IR \tag{24.4}$$

$$R = \rho \frac{L}{A} \tag{24.6}$$

$$R_T = R_0(1 + \alpha \, \Delta T) \tag{24.9}$$

$$P = \frac{\Delta PE_q}{\Delta t} \tag{24.10}$$

$$P = IV = I^2 R = \frac{V^2}{R} \tag{24.11}$$

Problems with Solutions and Discussion

PROBLEM: A certain 100-watt light bulb draws a current of 0.8 ampere. How long does it take for 48 coulombs of charge to pass through the light bulb? How many electrons have passed through?

Solution: Equation (24.1) relates charge, current, and time. Solving for Δt gives

$$\Delta t = \frac{\Delta q}{I} = \frac{48 \text{ C}}{0.8 \text{ A}} = \underline{60 \text{ s}}$$

48 coulombs of charge flow through the bulb in one minute of operation. Since one electron has a (negative) charge of 1.6×10^{-19} C, the number of electrons in 48 C is

$$\frac{48 \text{ C}}{1.6 \times 10^{-19} \text{ C/electron}} = \underline{3 \times 10^{20} \text{ electrons}}$$

PROBLEM: Preceding a flash of lightning, the bottom of a thundercloud is negative with respect to the earth below. Although the entire flash may last several tenths of a second, the *main stroke* lasts only about 100 μs and carries a current of the order of 10,000 A. How many coulombs of charge are transported during the main stroke?

Solution: Current is defined as the charge flowing past a certain point per unit time,

$$I = \frac{\Delta q}{\Delta t}$$

Solving for Δq gives

$$\Delta q = I \, \Delta t = (10,000 \text{ A})(100 \times 10^{-6} \text{ s}) = 1 \text{ C}$$

One coulomb of charge is transported during the main stroke. A typical lightning flash consists of 3 or 4 strokes and a continuing current, the whole event having a total duration of 0.2 s and transferring a total charge of 25 C.

170

PROBLEM: The drift velocity of electrons in a silver wire 2 mm in diameter is half a millimeter per second. If there are 5.8×10^{28} electrons/m³ which are free to move, what current is the wire carrying?

Solution: Equation (24.3) relates the drift velocity of the electrons v_d to the current I:

$$I = -(n_e e A) v_d$$

where

n_e = free electron density = 5.8×10^{28} electrons/m³

$e = 1.6 \times 10^{-19}$ C

$A = \pi r^2$, where $r = 1 \times 10^{-3}$ m; $A = \pi(1 \times 10^{-3} \text{ m})^2 = 3.1 \times 10^{-6}$ m²

$v_d = 0.5$ mm/s $= 5 \times 10^{-4}$ m/s

Substituting these values gives

$$I = -\left(5.8 \times 10^{28}\, \frac{\text{electrons}}{\text{m}^3}\right)\left(1.6 \times 10^{-19}\, \frac{\text{C}}{\text{electron}}\right)$$
$$\times (3.1 \times 10^{-6} \text{ m}^2)(0.5 \times 10^{-3} \text{ m/s})$$
$$I = \underline{-14\text{A}}$$

The wire is carrying a current of 14 A. The negative sign indicates that the positive current flows in the opposite direction to the electron drift velocity.

PROBLEM: Twenty grams of copper (density 8900 kg/m³) is drawn into a wire 1 mm in diameter. What is the resistance of the wire?

Solution: The resistance of an object as a function of its shape and dimensions is given by Eq. (24.6), $R = \rho L / A$. The cross-sectional area A of a wire 1 mm in diameter is πr^2, where $r = 0.5$ mm $= 5 \times 10^{-4}$ m. Thus,

$$A = \pi(5 \times 10^{-4} \text{ m})^2 = 7.85 \times 10^{-7} \text{ m}^2$$

The length is not known but can be determined because 20 g of copper constitutes a particular volume, given by the density relationship, density $= m/V$.

$$V = \frac{m}{\text{density}} = \frac{0.02 \text{ kg}}{8900 \text{ kg/m}^3} = 2.25 \times 10^{-6} \text{ m}^3$$

Assuming the wire to be cylindrical in shape, $V = AL$, where L is the length. Thus

$$L = \frac{V}{A} = \frac{2.25 \times 10^{-6} \text{ m}^3}{7.85 \times 10^{-7} \text{ m}^2} = 2.87 \text{ m}$$

Putting these values into Eq. (24.6), along with the resistivity of copper, $\rho = 1.7 \times 10^{-8}\ \Omega \cdot$m, obtained from text Table 24.1 gives

$$R = \rho \frac{L}{A} = (1.7 \times 10^{-8}\ \Omega \cdot \text{m}) \frac{2.87 \text{ m}}{7.85 \times 10^{-7} \text{ m}^2} = \underline{6.2 \times 10^{-2}\ \Omega}$$

The resistance of almost 3 m of this copper wire is 0.062 Ω, less than a tenth of an ohm. In most electrical circuit problems, we can consider the resistance of short pieces of connecting wire to be negligible.

PROBLEM: To compare the electrical properties of insulators and conductors, consider an aluminum wire of a certain cross-sectional area. How long would the wire have to be to have the same resistance to the flow of electric current as a one-millimeter long piece of ordinary glass of the same cross-sectional area?

Solution: For the aluminum wire,

$$R_A = \frac{\rho_A L_A}{A}, \text{ where } \rho_A = 2.8 \times 10^{-8}\ \Omega \cdot \text{m, from Table 24.1}$$

For the glass filament, $R_G = \frac{\rho_G L_G}{A}$, where $\rho_G = 9 \times 10^{11} \ \Omega \cdot m$, also from Table 24.1.

If aluminum and glass are to have equal resistance, $R_A = R_G$,

$$\frac{\rho_A L_A}{\cancel{A}} = \frac{\rho_G L_G}{\cancel{A}}$$

$$L_A = \frac{\rho_G L_G}{\rho_A} = \frac{(9 \times 10^{11} \ \Omega \cdot m)(1 \times 10^{-3} \ m)}{2.8 \times 10^{-8} \ \Omega \cdot m} = \underline{3.2 \times 10^{16} \ m}$$

This is truly an astronomical distance. A wire this long could reach to the sun and back 100,000 times! The properties of conductors and insulators are indeed miles apart.

PROBLEM: Ventricular fibrillation (text Fig. 24.7, page 488) is rapid uncoordinated twitching of the heart muscles, stopping its pumping action. What minimum potential difference between a person's hands would cause ventricular fibrillation, if the resistance is 1500 Ω?

Solution: Text Fig. 24.7 shows that the minimum current for the onset of ventricular fibrillation is 100 mA $= 100 \times 10^{-3}$ A. We can determine the voltage that would supply this current through a 1500-Ω resistance using Ohm's law:

$$V = IR = (100 \times 10^{-3} \ A)(1500 \ \Omega) = \underline{150 \ V}$$

The resistance between a person's hands is highly variable; 1500 Ω corresponds to "sweaty palms." In this circumstance 150 V is sufficient for the onset of ventricular fibrillation.

PROBLEM: It is observed that the resistance of a certain barbed wire is twice as great at 240°C as it is at 20°C. What is the temperature coefficient of resistance of this wire and of what material might it be made?

Solution: Since we have two different sets of conditions, one at 20°C and one at 240°C, we may write equations for each and solve simultaneously for the unknown α. Eq. (24.9),

$$R_T = R_0(1 + \alpha \, \Delta T)$$

gives the resistance R_T as a function of temperature change ΔT, where R_0 and α are constants for a particular substance. R_0 is the value of R at 0°C, so ΔT is the temperature difference from 0°C.

$$\text{At } 20°C, \qquad R_T = R_0[1 + \alpha(20)]$$
$$\text{At } 240°C, \text{ the resistance} = 2R_T: \quad 2R_T = R_0[1 + \alpha(240)]$$

Substituting the value for R_T from the first equation into the second equation gives

$$2\cancel{R_0}[1 + \alpha(20)] = \cancel{R_0}[1 + \alpha(240)]$$
$$2 + (40)\alpha = 1 + (240)\alpha$$
$$1 = (200 \ C°)\alpha$$
$$\alpha = \underline{0.005 \ (C°)^{-1}}$$

The temperature coefficient of resistance is 0.005 $(C°)^{-1}$. Referring to Table 24.2, we see that the barbed wire is probably made of <u>iron</u>.

PROBLEM: The tungsten filament of a certain 100-watt light bulb has a temperature of about 2500°C when operating at its steady state. (a) What is its resistance? (b) What is its resistance the moment it is turned on, when the filament temperature is 20°C? (c) What is its (momentary) power output when first turned on?

Solution: (a) Of the various forms of the electrical power equation, Eq. (24.11), we choose

$$P = \frac{V^2}{R}$$

172

because we know that $P = 100$ W and $V = 120$ V (assuming a standard household circuit), and we wish to find R. Solving for R gives

$$R = \frac{V^2}{P} = \frac{(120 \text{ V})^2}{100 \text{ W}} = \underline{144 \ \Omega}$$

At $T = 2500°C$, the filament resistance is 144 Ω.

(b) The resistance of the filament at 20°C can be determined from Eq. (24.9):

$$R_T = R_0(1 + \alpha T)$$

if we know R_0, the resistance at 0°C. This can be obtained by applying Eq. (24.9) at the higher temperature, where $R_T = 144 \ \Omega$; $\Delta T = 2500 - 0 = 2500$ C°; and $\alpha = 0.0045$ (C°)$^{-1}$, from Table 24.2.

$$R_0 = \frac{R_T}{1 + \alpha \ \Delta T} = \frac{144 \ \Omega}{1 + (0.0045 \text{ C°}^{-1})(2500 \text{ C°})} = \frac{144 \ \Omega}{12.25} = 11.8 \ \Omega$$

The resistance at $T = 20°C$ can now be determined:

$$R_T = R_0(1 + \alpha \ \Delta T) = (11.8 \ \Omega)[1 + (0.0045 \text{ C°}^{-1})(20 \text{ C°})]$$
$$= (11.8 \ \Omega)(1.09) = \underline{13 \ \Omega}$$

(c) The momentary power output when the voltage is 120 V and the resistance is only 13 Ω is

$$P = \frac{V^2}{R} = \frac{(120 \text{ V})^2}{13 \ \Omega} = \underline{1100 \text{ W}}$$

This surge of power heats up the filament so that it very quickly becomes luminous. As it heats, the resistance increases, reducing the power until it reaches the steady state power output of 100 watts. The initial power surge also accounts partially for the fact that very frequently a light bulb burns out as it is being turned on.

PROBLEM: An electric heating coil operating from a 120-V line is submerged in a cup of water to heat it up for coffee. It takes 3 minutes to bring the water, of mass 300 g, from 40°C to a boil. Determine the current drawn by the heating coil and its resistance.

Solution: If we know the power supplied by the coil to the water, we may use Eq. (24.11), $P = IV$, to find the current it draws. The power supplied can be determined because we can calculate the heat energy necessary to heat the water to a boil, and we know the time it takes to do this.

$$P = \frac{Q}{t}, \text{ where } Q = mc \ \Delta T$$

and

$$m = 300 \text{ g} = 0.3 \text{ kg};$$
$$\Delta T = 100 - 40 = 60 \text{ C°};$$
$$c = \text{specific heat capacity of water} = 4180 \text{ J/kg} \cdot \text{C°};$$

and

$$t = 3 \text{ min} \left(\frac{60 \text{ s}}{1 \text{ min}} \right) = 180 \text{ s}$$

Then,

$$P = \frac{mc \ \Delta T}{t} = \frac{(0.3 \text{ kg})(4180 \text{ J/kg} \cdot \text{C°})(60 \text{ C°})}{180 \text{ s}} = 420 \frac{\text{J}}{\text{s}} = 420 \text{ W}$$

Solving $P = IV$ for I gives

$$I = \frac{P}{V} = \frac{420 \text{ W}}{120 \text{ V}} = \underline{3.5 \text{ A}}$$

Then from Ohm's law,

$$R = \frac{V}{I} = \frac{120 \text{ V}}{3.5 \text{ A}} = \underline{34 \ \Omega}$$

The heating coil draws a current of 3.5 amperes and has a resistance of 34 ohms.

PROBLEM: Compare the cost of baking a potato in a microwave oven and in a conventional oven. The microwave tube draws 1500 watts and operates continually for 4 minutes. The conventional oven element draws 3000 watts. The oven preheats for 10 minutes and is then on 30 percent of the total cooking time of 1 hour. Electricity costs 7¢ per kilowatt-hour.

Solution: We will determine the energy in each case, in kilowatt-hours. The equation defining power as energy per unit time, $P = E/t$, can be solved for energy: $E = Pt$. To get the energy in kilowatt-hours, we may calculate energy in joules and use the conversion relation, $3.6 \times 10^6 \text{ J} = 1 \text{ kW} \cdot \text{h}$. Or (and we shall use this method), we can immediately express power in kilowatts and time in hours; then E will automatically have units of kilowatt-hours.
Microwave oven:

$$P = 1500 \text{ W} = 1.5 \text{ kW}$$

$$t = 4 \text{ min} \left(\frac{1 \text{ h}}{60 \text{ min}} \right) = 0.067 \text{ h}$$

$$E = Pt = (1.5 \text{ kW})(0.067 \text{ h}) = 0.1 \text{ kW} \cdot \text{h}$$

At 7¢ per kW·h, $(0.1 \text{ kW} \cdot \text{h})(7\text{¢}/\text{kW} \cdot \text{h}) = \underline{0.7\text{¢}}$. It will cost 0.7¢ to bake the potato in the microwave oven.
Conventional oven:

$$P = 3000 \text{ W} = 3 \text{ kW}$$

$$t = 10 \text{ min (preheat)} + 60 \text{ min}(0.3) \text{ (cooking)} = 10 + 18 = 28 \text{ min}$$

$$t = 28 \text{ min} \left(\frac{1 \text{ h}}{60 \text{ min}} \right) = 0.467 \text{ h}$$

$$E = Pt = (3 \text{ kW})(0.467 \text{ h}) = \underline{1.4 \text{ kW} \cdot \text{h}}$$

The cost will thus be $(1.4 \text{ kW} \cdot \text{h})(7\text{¢}/\text{kW} \cdot \text{h}) = \underline{9.8\text{¢}}$. It costs 9.8¢ to bake the potato in a conventional oven.

Comparing the two costs, $9.8/0.7 = 14$, we see that it costs <u>14 times as much</u> to bake the potato in a conventional oven as in a microwave oven. (This does not take into consideration the initial cost of buying the microwave oven!)

Avoiding Pitfalls

1. The direction of a current is defined as the direction positive charge carriers would flow, opposite in direction to the electron flow.

2. A free electron in a conductor has an enormous *average speed* ($\sim 10^6$ m/s) but its *average velocity* is zero (because of numerous random collisions, there is no preferred direction of motion) until an electric field is applied. In the presence of an electric field, the electrons have an average *drift velocity* ($\sim$millimeters per second) through the conductor. It is this drift velocity that constitutes the electric current.

3. A source of emf is generally considered to supply a *constant* potential difference to a particular electrical circuit.

4. The *resistivity* ρ of a substance depends on its intrinsic properties, not on its geometry; the *resistance* R depends on both the intrinsic properties (ρ) and the length and cross-sectional area of the sample.

5. In the equation $R_T = R_0(1 + \alpha \, \Delta T)$, R_0 is the resistance at some reference temperature T_0 (here taken as 0°C). The temperature interval ΔT is then $(T - T_0)$. If the

resistance is known at any particular temperature, this pair of values may be used as R_0 and T_0, the reference situation from which the resistance R_T at other temperatures may be determined.

6. The equation for electrical power, $P = IV$, is used to calculate the power used up or dissipated by a circuit element, such as a household appliance; it may also be used to calculate the power *supplied* to an electrical circuit by a source of emf.

7. A quantity measured in kilowatt-hours is neither a power nor a time, but an *energy* ($E = Pt$) for which the proper *mks* unit is the *joule*.

Drill Problems
Answers in Appendix

1. A 1.55-V battery operates a certain electronic watch continuously for a year.

    (a) What average current does the watch draw if the battery is capable of supplying a total charge of 0.36 C during the year?

    (b) What is the resistance of the watch, assuming it to follow Ohm's law?

2. A copper wire 6 m long with a cross-sectional area of 2 mm^2 is connected to a 6-V battery. Determine

    (a) the wire's resistance,

    (b) the current flowing,

    (c) the electric field in the wire, and

    (d) the power dissipated by the wire.

3. An electric heater connected to a 120-V potential difference operates 16 hours a day during winter. Its resistance is 10 ohms when in operation.

    (a) What current does it draw?

    (b) What is its power output?

    (c) At 7¢ per kilowatt-hour, how much does it cost to operate the heater for the month of February (28 days)?

25 Direct-Current Electric Circuits

Terms

Define or describe briefly what is meant by the following terms. If you have difficulty, refer to the textbook section given in parentheses.

Kirchhoff's loop rule (25.1)

Kirchhoff's junction rule (25.2)

equivalent resistance (25.4)

series resistors (25.4)

parallel resistors (25.4)

internal resistance (25.5)

ammeter (25.5)

voltmeter (25.5)

Equation Review

For each equation, be able to state the situation to which it applies, what quantity each symbol represents, and the units for measuring each quantity in some consistent set of units.

$$\Sigma I = I_1 + I_2 + \ldots = 0 \tag{25.1}$$

$$R_{eq} = R_1 + R_2 + R_3 + \ldots \tag{25.4}$$

$$\frac{1}{R_{eq}} = \frac{1}{R_1} + \frac{1}{R_2} + \frac{1}{R_3} + \ldots \tag{25.5}$$

Problems with Solutions and Discussion

PROBLEM: Apply Kirchhoff's loop rule to determine the current flowing in the circuit shown in Fig. 25.1a.

Solution: We first assign a direction and symbol to the current in the circuit: I, going in the clockwise direction, since the polarity of the 30-V source of emf will determine the current direction. Kirchhoff's loop rule states that the sum of the voltage changes around any closed loop in an electric circuit is zero. We assume the resistance of the connecting wires is negligible so that the voltage changes occur across the sources of emf and the resis-

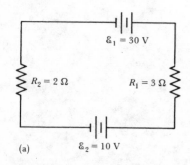

(a)

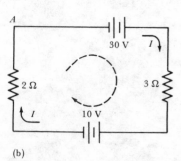

(b)

FIG. 25.1

176

tors. Starting at point A and progressing clockwise around the loop [see dotted line in (b)], the voltage changes are as follows:

$$\mathscr{E}_1 - IR_1 - \mathscr{E}_2 - IR_2 = 0$$

Note that, as in the case of $\mathscr{E}_2$, whenever we go through a source of emf "backward" (from the high voltage side to the low voltage side), the voltage drops; thus $\mathscr{E}_2$ is negative in the loop equation. Putting in numbers and solving for I, we get

$$30 - I(3) - 10 - I(2) = 0$$
$$-5I = -20$$
$$I = \underline{4\ A}$$

The current in the loop is 4 amperes. What would happen if we ignored the relative sizes of the sources of emf and assumed I to go in the counterclockwise direction? Kirchhoff's loop rule, this time starting at A and going *counterclockwise* along with the assumed current, would give the equation

$$-2I + 10 - 3I - 30 = 0$$

In this loop, it is $\mathscr{E}_1$ that we went through "backward" and which therefore has a negative sign indicating a voltage drop. Solving for I gives

$$-5I = 20$$
$$I = \underline{-4\ A}$$

The same current value results; the negative sign indicates we chose the wrong direction for the current. The real current is 4 A in the *clockwise* direction. So one should not spend too much time trying to figure out which way the actual current flows in a given branch. If the incorrect direction is assumed, the current will have a minus sign, but the numerical value will be correct.

PROBLEM: Consider the junction shown in Fig. 25.2a, which is part of a larger electrical circuit. Use Kirchhoff's junction rule to determine the current I in the branch to the right.

Solution: The junction rule states that the algebraic sum of the currents flowing into and out of any branch point is zero, $\Sigma I = 0$, where currents entering the junction are positive and currents leaving the junction are negative. For this junction,

$$2\ A - 5\ A + 6\ A + I = 0$$
$$3\ A + I = 0$$
$$I = \underline{-3\ A}$$

The current is 3 A; since it has a negative sign, it is *leaving* the junction (Fig. 25.2b).

PROBLEM: For the circuit shown in Fig. 25.3a, use Kirchhoff's rules to write two loop equations and one junction equation. Solve for the current in each branch.

Solution: Using the procedure outlined in the text, we first assign a direction and symbol to the current in each branch, as shown in Fig. 25.3b. Next we apply the junction rule, $\Sigma I = 0$, to the junction marked A.

$$I_1 + I_2 - I_3 = 0 \qquad\qquad (1)$$

where currents entering the junction are positive and currents leaving it are negative. (Since all currents appear in this equation, a second point equation would give no new information.)

  Next we apply the loop rule twice to get two more equations. The loops chosen are indicated by the dotted lines on the diagram.

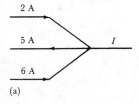

(a)

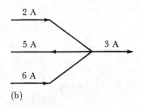

(b)

FIG. 25.2

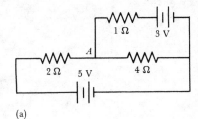

(a)

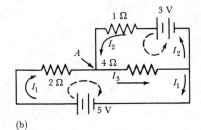

(b)

FIG. 25.3

177

Upper loop, circulating counterclockwise:

$$3\text{ V} - I_2(1\ \Omega) - I_3(4\ \Omega) = 0 \qquad (2)$$

Lower loop, circulating clockwise:

$$5\text{ V} - I_1(2\ \Omega) - I_3(4\ \Omega) = 0 \qquad (3)$$

The three equations (1), (2) and (3), are sufficient to determine the three unknown currents. Solving (1) for I_1 gives $I_1 = I_3 - I_2$, and substituting it into (3) gives

$$5 - 2(I_3 - I_2) - 4I_3 = 0 \qquad \text{(where we have dropped the units)}$$

$$5 - 6I_3 + 2I_2 = 0 \qquad (3a)$$

Equations (2) and (3a) can be solved simultaneously by multiplying Eq. (2) by 2 and adding it to Eq. (3a) in order to eliminate I_2.

$$(2) \times 2: \quad 6 - 2I_2 - 8I_3 = 0$$
$$(3a): \quad \underline{5 + 2I_2 - 6I_3 = 0}$$
$$11 \qquad\quad - 14I_3 = 0$$
$$I_3 = \underline{{}^{11}\!/_{14}\text{ A}}$$

Substituting this value of I_3 into Eq. (2) gives

$$3 - I_2 - 4({}^{11}\!/_{14}) = 0$$

$$I_2 = 3 - \frac{44}{14} = \frac{42 - 44}{14} = \underline{-\frac{2}{14}\text{ A}}$$

I_2 has a negative sign; this means we chose the wrong direction for I_2. It actually goes clockwise in the upper branch—the 5-V source of emf is large enough to make current flow "backward" through the 3-V source of emf. Since the three equations were written before we knew I_2 had the incorrect direction, we must continue I_2 through the solution as a negative number. We do not "change horses midstream"; that is, alter assumed directions in the middle of a solution. We may obtain I_1 from Eq. (1):

$$I_1 = I_3 - I_2 = {}^{11}\!/_{14} - (-{}^{2}\!/_{14}) = \underline{{}^{13}\!/_{14}\text{ A}}$$

These then are the three currents in the circuit. I_1 and I_3 go in the direction originally assumed; I_2 goes opposite in direction to that shown in Fig. 25.3b.

As a check on our results, let us write a third loop equation, this time around the perimeter of the circuit (Fig. 25.4), starting at the lower right corner and progressing counterclockwise:

$$5 - 2I_1 + 1I_2 - 3 = 0$$

The signs of the emfs show we are going through the 5-V battery in the "forward" direction and through the 3-V battery "backward." Notice also that our direction around the loop is *opposite* the direction of the assumed current going through the 1-Ω resistor, so this voltage change is an increase, $+IR$, not the usual (negative) voltage drop that occurs when passing through a resistor in the same direction as the assumed current. Substituting the values of $I_1 = {}^{13}\!/_{14}$ A and $I_2 = -{}^{2}\!/_{14}$ A, we get

$$5 - 2({}^{13}\!/_{14}) + (1)(-{}^{2}\!/_{14}) - 3 = 5 - {}^{26}\!/_{14} - {}^{2}\!/_{14} - 3$$
$$= 5 - {}^{28}\!/_{14} - 3 = 5 - 2 - 3 = 0$$

This is a true statement, confirming that our original solution is correct.

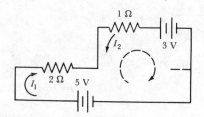

FIG. 25.4

PROBLEM: Consider a circuit consisting of an 18-V source of emf and two resistors of 3 Ω and 6 Ω (Fig. 25.5a). (a) What current flows through each resistor when connected in parallel across the source of emf? (b) What voltage drop occurs over each resistor when connected in series across the source of emf?

178

Solution: (a) Both resistors have the same voltage drop, 18 V. Applying Ohm's law to each, we get

$$3\ \Omega: \quad I_1 = \frac{V}{R_1} = \frac{18\ \text{V}}{3\ \Omega} = \underline{6\ \text{A}}$$

$$6\ \Omega: \quad I_2 = \frac{V}{R_2} = \frac{18\ \text{V}}{6\ \Omega} = \underline{3\ \text{A}}$$

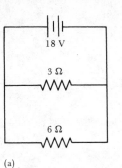

(a)

By Kirchhoff's junction rule, a total of 9 A must be supplied to the circuit. Notice that the resistances are in the ratio of 1 : 2, the currents through them in the ratio of 2 : 1. This is always true for a parallel arrangement: the current flowing in is divided among the parallel branches in the inverse ratio of the resistance of the branches, the highest current going through the branch of lowest resistance, etc. This idea holds for more than two branches. If the resistances of four parallel branches are in the ratio of 1 : 1 : 2 : 4, the currents through the branches will be in the ratio of 4 : 4 : 2 : 1, respectively.

(b) The same current flows through both resistors—this is the nature of a series connection (Fig. 25.5b). It is determined by the equivalent resistance of the combination. By Eq. (25.4)

$$R_{eq} = R_1 + R_2 = 3\ \Omega + 6\ \Omega = 9\ \Omega$$

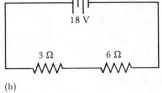

(b)

FIG. 25.5

Ohm's law then gives the current:

$$I = \frac{V}{R_{eq}} = \frac{18\ V}{9\ \Omega} = 2\ \text{A}$$

Now we can determine the voltage drop over each resistance from Ohm's law.

$$3\ \Omega: \quad V_1 = IR_1 = (2\ \text{A})(3\ \Omega) = \underline{6\ \text{V}}$$

$$6\ \Omega: \quad V_2 = IR_2 = (2\ \text{A})(6\ \Omega) = \underline{12\ \text{V}}$$

The sum of the voltages across the series resistors equals the voltage of the source of emf, 18 V. The voltage is divided in direct proportion to the resistances, the larger resistance having the larger voltage drop. The resistances are in the ratio of 1 : 2; so are the voltage drops. This idea holds for all direct current series circuits.

PROBLEM: In the circuit shown in Fig. 25.6a, (a) what is the equivalent resistance between points A and B? (b) Find the current through R_1. (c) Find the voltage drop across R_4.

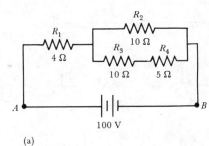

(a)

Solution: (a) Starting in the parallel loop and working outward, we get

Series (Fig. 25.6b): $\quad R_{3,4} = 10\ \Omega + 5\ \Omega = 15\ \Omega$

Parallel (Fig. 25.6c): $\quad \dfrac{1}{R_{2,3,4}} = \dfrac{1}{10\ \Omega} + \dfrac{1}{15\ \Omega} = \dfrac{3+2}{30} = \dfrac{5}{30} = \dfrac{1}{6}$

$$R_{2,3,4} = 6\ \Omega$$

Series (Fig. 25.6d): $\quad R_{eq} = 4\ \Omega + 6\ \Omega = \underline{10\ \Omega}$

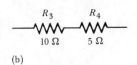

(b)

The equivalent resistance between A and B is 10 ohms.

(b) The total current in the circuit is determined by Ohm's law:

$$I_t = \frac{V}{R_{eq}} = \frac{100\ V}{10\ \Omega} = 10\ \text{A}$$

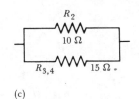

(c)

Since this current does not branch before passing through R_1, the current through R_1 equals the total current of $\underline{10\ \text{A}}$.

(c) To find the voltage drop over a resistor "buried" in a series-parallel network, we work our way in to it using Ohm's law for each circuit element or group of elements to determine the current or voltage, whichever is needed.

Across R_1,

$$V_1 = IR_1 = (10\ \text{A})(4\ \Omega) = 40\ \text{V}$$

(d)

FIG. 25.6

179

If 40 V of the total 100-V drop occurs over R_1, the remainder occurs over $R_{2,3,4}$, the parallel resistors:

$$V_{2,3,4} = 100 - 40 = 60 \text{ V}$$

Thus, $V_2 = V_{3,4} = 60$ V, since the identifying characteristic of a parallel arrangement is that each branch has the same voltage drop.

Across $R_{3,4}$,

$$I_{3,4} = \frac{V_{3,4}}{R_{3,4}} = \frac{60 \text{ V}}{15 \ \Omega} = 4 \text{ A}$$

Thus, 4 A flows through the bottom branch (leaving $10 - 4 = 6$ A to flow through the top branch). This 4-A current flows through both R_3 and R_4, since they are connected in series.

We now have enough information to get V_4 by Ohm's law:

$$V_4 = I_4 R_4 = (4 \text{ A})(5 \ \Omega) = \underline{20 \text{ V}}$$

The voltage drop over the 5-Ω resistor is 20 V.

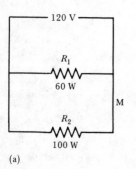

(a)

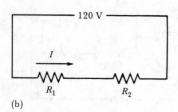

(b)

FIG. 25.7

PROBLEM: Two light bulbs, labeled 60 W and 100 W respectively, are designed to operate in a 120-V household circuit. (a) How much current does each draw (at steady state) when connected in parallel? (b) How much current does each draw when connected in series to the 120-V line, and what is the power output of each? (Assume the steady-state resistance of the bulbs is constant.)

Solution: (a) When connected in parallel, the voltage across each bulb is 120 V (Fig. 25.7a). According to $P = VI$, the bulb with the larger power output will draw the larger current.

$$\text{60-W bulb:} \quad I_1 = \frac{P_1}{V} = \frac{60 \text{ W}}{120 \text{ V}} = \underline{0.5 \text{ A}}$$

$$\text{100-W bulb:} \quad I_2 = \frac{P_2}{V} = \frac{100 \text{ W}}{120 \text{ V}} = \underline{0.83 \text{ A}}$$

(b) When connected in series (Fig. 25.7b), the same current flows through each bulb and is determined by the equivalent resistance of the combination. To get R_{eq}, we need the individual resistances of the bulbs. These can be obtained from Ohm's law and from part (a).

$$\text{60-W bulb:} \quad R_1 = \frac{V}{I_1} = \frac{120 \text{ V}}{0.5 \text{ A}} = 240 \ \Omega$$

$$\text{100-W bulb:} \quad R_2 = \frac{V}{I_2} = \frac{120 \text{ V}}{0.83 \text{ A}} = 144 \ \Omega$$

In series, $R_{eq} = 240 \ \Omega + 144 \ \Omega = 384 \ \Omega$. So the current through the combination will be

$$I = \frac{V}{R_{eq}} = \frac{120 \text{ V}}{384 \ \Omega} = \underline{0.31 \text{ A}}$$

The power of each bulb can be obtained from $P = I^2R$. (We do not use $P = V^2/R$ or $P = VI$, because the 120 V is now applied over the combination, not over each one individually.)

$$\text{60-W bulb:} \quad P_1 = I^2R_1 = (0.31 \text{ A})^2(240 \ \Omega) = 23 \text{ W}$$

$$\text{100-W bulb:} \quad P_2 = I^2R_2 = (0.31 \text{ A})^2(144 \ \Omega) = 14 \text{ W}$$

A very strange thing has occurred: the 100-W bulb shines less brightly than the 60-W bulb in the series arrangement. The designations of power output or consumption on light bulbs and other household appliances *always* assume they will be connected *in parallel,* directly across the stated voltage.

180

PROBLEM: Figure 25.8a shows a galvanometer which has a resistance of 20 Ω and requires a current of 1 mA for full-scale deflection. (a) What is the potential difference across the galvanometer at full-scale deflection? (b) How would you connect a resistor to this galvanometer to convert it to an ammeter for measuring currents up to 500 mA? (c) How much current passes through this resistor? (d) What should the resistance of this resistor be?

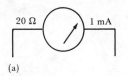

(a)

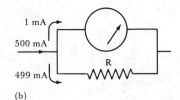

(b)

FIG. 25.8

Solution: (a) Since full-scale deflection is caused by a current of 1 mA = 1×10^{-3} A, the potential difference, by Ohm's law, is

$$V = IR = (1 \times 10^{-3} \text{ A})(20 \text{ Ω}) = \underline{0.02 \text{ V}}$$

(b) The resistor should be connected <u>in parallel</u>, so that no more than 1 mA goes through the galvanometer and the rest can go through the resistor (Fig. 25.8b).

(c) For a maximum current of 500 mA, since 1 mA goes through the galvanometer, 499 mA must pass through the resistor. This is an example of Kirchhoff's junction rule, $\Sigma I = 0$:

$$500 \text{ mA} - 1 \text{ mA} - 499 \text{ mA} = 0$$

(d) Since the meter and resistor are in parallel, they have the same voltage drop, 0.02 V at full scale. The resistance R, by Ohm's law, must therefore be:

$$R = \frac{V}{I} = \frac{0.02 \text{ V}}{0.499 \text{ A}} = \underline{0.04 \text{ Ω}}$$

We have designed an ammeter for measuring currents up to 500 mA by determining the value of the very low resistance (called the *shunt resistance*) that should be connected in parallel with the galvanometer. The ammeter is connected in series in a circuit. When 500 mA flow, all but 1 mA goes through the shunt, the 1 mA through the meter causing it to deflect full scale, indicating 500 mA. When 200 mA flow, this current divides at the branch point in the same ratio as before, so that the meter deflects ⅖ of full scale, the rest of the current passing through the shunt.

Avoiding Pitfalls

1. In using Kirchhoff's rules, strict attention to algebraic signs is necessary. In the loop rule, voltage increases are positive and voltage drops are negative. In the junction rule, currents entering the junction are positive and currents leaving the junction are negative.

2. When a negative current results from a solution using Kirchhoff's rules, it means the true direction is opposite to the assumed direction; the magnitude of the current is unaffected by the negative sign.

3. Current is unchanged in a particular branch of a circuit. Its value does not change when going through a resistor. The only way a current can increase or decrease is at a branch point when currents from other branches combine with it.

4. If a circuit contains combinations of resistors in series and parallel, the equivalent resistance technique will usually be most efficient in analyzing the circuit. If the resistors are not in series or parallel networks (for example, if they are separated by sources of emf), Kirchhoff's rules will be the easiest way to analyze the circuit.

5. For series-parallel networks of resistors, the total current flowing is determined by the equivalent resistance of the circuit, using Ohm's law: $I = V/R_{eq}$. When this current branches, current through individual circuit elements also follows Ohm's law, $I = V/R$, where I is the current in the branch, and V and R are the potential difference and resistance for the particular element of interest.

6. For resistors in series, the same current flows through all and the voltage drop over each is directly proportional to its resistance (larger resistors have larger voltage drops). For resistors in parallel, the same potential difference exists across all and the current in each is inversely proportional to its resistance (less current flows through larger resistances).

7. When determining the equivalent resistance of resistors in parallel, do not forget the final step of inverting $\frac{1}{R_{eq}}$ to get R_{eq}.

181

8. Resistors should be connected in series to maximize the equivalent resistance, which will be larger than the largest of the resistances. Resistors should be connected in parallel to minimize the equivalent resistance, which will be less than the smallest of the resistance in parallel.

9. Voltmeters are connected in parallel with the element whose voltage is being measured. Ammeters are connected in series in the circuit branch whose current is being measured.

Drill Problems
Answers in Appendix

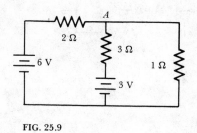

FIG. 25.9

1. For the circuit in Fig. 25.9, assign currents and apply Kirchhoff's junction rule to junction A and Kirchhoff's loop rule for the right loop and the outside perimeter. Solve your equations to determine the current in each branch of the circuit. (*Note:* Other equations are possible and will yield correct results. Using those suggested will allow direct comparison with the solution in the Appendix.)

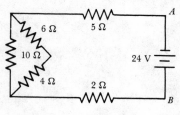

FIG. 25.10

2. In the circuit in Fig. 25.10,

   (a) what is the equivalent resistance between A and B?
   (b) Find the current through the 5-Ω resistor.
   (c) Find the voltage drop across the 4-Ω resistor.

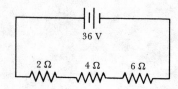

FIG. 25.11

3. For the circuit in Fig. 25.11, determine

   (a) the power supplied by the source of emf, and
   (b) the power dissipated by each resistor.
   (c) Show that the power supplied is equal to the power dissipated.

182

Capacitors

Terms

Define or describe briefly what is meant by the following terms. If you have difficulty, refer to the textbook section given in parentheses.

capacitor (26.1)

capacitance (26.2)

farad (26.2)

dielectric breakdown (26.4)

dielectric strength (26.4)

RC series circuit (26.6)

time constant (26.6)

Equation Review

For each equation, be able to state the situation to which it applies, what quantity each symbol represents, and the units for measuring each quantity in some consistent set of units.

$$C = \frac{K}{4\pi k}\frac{A}{d} \tag{26.1}$$

$$q = CV \tag{26.2}$$

$$\Delta PE_q = \frac{1}{2}qV = \frac{1}{2}CV^2 = \frac{1}{2}\frac{q^2}{C} \tag{26.3}$$

$$i = i_0 e^{-t/RC} \tag{26.4}$$

$$q = q_f(1 - e^{-t/RC}) \tag{26.5}$$

$$\tau = RC \tag{26.6}$$

$$q = q_0 e^{-t/RC} \tag{26.7}$$

$$C_{eq} = C_1 + C_2 + C_3 + \ldots \tag{26.8}$$

$$\frac{1}{C_{eq}} = \frac{1}{C_1} + \frac{1}{C_2} + \frac{1}{C_3} + \ldots \tag{26.9}$$

Problems with Solutions and Discussion

PROBLEM: A capacitor is made of two pieces of tinfoil, each 80 cm × 40 cm, with a piece of wax paper 0.004-cm thick in between. The dielectric constant of the wax paper is 2.5. (a) What is the capacitance of this capacitor? (b) Determine the charge on it when connected to a 12-V battery.

Solution: (a) The capacitance depends on the geometry of the capacitor and the intrinsic properties of the dielectric. It is given by Eq. (26.1)

$$C = \frac{K}{4\pi k}\frac{A}{d}$$

where in this case

$K = 2.5$,

A = area of each "plate" = 80 cm × 40 cm = 0.8 m × 0.4 m = 0.32 m²

d = separation between plates = 0.004 cm = 4×10^{-3} cm = 4×10^{-5} m

and k = Coulomb's law force constant = 9×10^9 N·m²/C²

Substituting these values gives

$$C = \frac{(2.5)(0.32\ \text{m}^2)}{4\pi(9 \times 10^9\ \text{N·m}^2/\text{C}^2)(4 \times 10^{-5}\ \text{m})} = 1.8 \times 10^{-7}\ \text{F} = \underline{0.18\ \mu\text{F}}$$

(b) According to Eq. (26.2), $q = CV$. If $V = 12$ V, then

$$q = CV = (1.8 \times 10^{-7}\ \text{F})(12\ \text{V}) = 2.2 \times 10^{-6}\ \text{C} = \underline{2.2\ \mu\text{C}}$$

The tinfoil and wax paper capacitor has a capacitance of 0.18 μF and holds a charge of 2.2 μC when the charging voltage is 12 V.

PROBLEM: What is the maximum voltage that could be applied to the tinfoil capacitor in the previous problem, if the dielectric strength of the wax paper is 2×10^7 V/m?

Solution: The dielectric strength is defined as the value of the *electric field* in the dielectric at which breakdown occurs (the dielectric molecules ionize, and it becomes a conductor). By Eq. (23.8),

$$E = \frac{V}{d}$$

where d is the separation across which the potential difference V exists. Therefore,

$$V = Ed = (2 \times 10^7\ \text{V/m})(4 \times 10^{-5}\ \text{m}) = \underline{800\ \text{V}}$$

This capacitor will break down if a voltage of 800 V or greater is applied to it.

PROBLEM: Two parallel plates separated by half a millimeter of air are charged to a potential difference of 450 V, then disconnected from the source. A sheet of glass half a millimeter thick is then slid between the plates. If the potential difference is now 90 V, what is the dielectric constant of the glass?

Solution: Since the capacitance of a capacitor depends on the dielectric constant of the material between the plates, the capacitance will change when the glass is slid in. The voltage also changes. The initial charge on the capacitor, from Eq. (26.2), is $q_1 = C_1 V_1$, and the final charge is $q_2 = C_2 V_2$, where the subscripts $_1$ and $_2$ represent the initial and final conditions, respectively. But the *charge on the plates remains constant.* (Where can it go? The source of emf has been removed and there is no longer a circuit whereby charge from one plate can get to the other.) If $q_1 = q_2$, then $C_1 V_1 = C_2 V_2$.

We now substitute the expression for capacitance from Eq. (26.1),

$$\left(\frac{K_1}{4\pi k}\frac{A}{d}\right)V_1 = \left(\frac{K_2}{4\pi k}\frac{A}{d}\right)V_2$$

The only change in the capacitor is its dielectric material. Thus, $K_1 V_1 = K_2 V_2$. Solving for K_2, the unknown dielectric constant of the glass, we get

$$K_2 = \frac{K_1 V_1}{V_2} = \frac{(1)(450\ \text{V})}{90\ \text{V}} = \underline{5}$$

The dielectric constant of the glass is 5. The voltage drops by the same ratio that the dielectric constant increases. The electric field in the space between the plates polarizes the molecules of the dielectric. These dipoles align themselves antiparallel to the electric field, effectively reducing it and thereby reducing the potential difference between the plates.

PROBLEM: A 0.6 μF capacitor consists of two metal plates separated by 1 mm of Plexiglas. It is found that breakdown and sparking will occur when the voltage between the plates exceeds 10^5 V. (a) What is the dielectric strength of Plexiglas? (b) What is the maximum amount of charge that can be stored by this capacitor?

Solution: (a) The dielectric strength is the value of the electric field at which breakdown occurs. Electric field is related to voltage by Eq. (23.8), $E = V/d$. Since breakdown occurs above 10^5 V, and $d = 1\ \text{mm} = 10^{-3}\ \text{m}$,

$$E_{\text{max}} = \frac{10^5\ \text{V}}{10^{-3}\ \text{m}} = \underline{10^8\ \text{V/m}}$$

The dielectric strength of Plexiglas is about 10^8 V/m.

(b) Charge stored at the maximum voltage of 100,000 V is

$$q = CV = (0.6 \times 10^{-6}\ \text{F})(10^5\ \text{V}) = 6 \times 10^{-2}\ \text{C} = \underline{0.06\ \text{C}}$$

PROBLEM: A parallel plate capacitor is charged to a certain voltage. While still connected to the battery, the plates are pulled apart, increasing their separation distance. State *qualitatively* what happens to the (a) voltage, (b) capacitance, (c) electric field, and (d) charge on the capacitor plates. Justify each response briefly.

Solution: (a) The voltage remains constant because the plates are connected throughout to a constant source of emf.

(b) The capacitance decreases, because it is inversely proportional to the plate separation; the larger the separation, the smaller the capacitance.

(c) The electric field is given by $E = V/d$. If V remains constant and d increases, the electric field decreases.

(d) The charge is related to the voltage and capacitance by $q = CV$. If V remains constant and C decreases, the charge on each plate decreases. Current flows through the source of emf during the separation process as some electrons flow from the negative plate (reducing its charge) over to the positive plate (neutralizing an equal amount of its excess charge). Another way of thinking about the negative plate is that when the capacitor is fully charged, the Coulomb force of attraction to the positive plate is just balanced by the Coulomb repulsion of like charges on the negative plate. When the plates are pulled apart, the force of attraction decreases (as the distance squared) and the negative plate then repels electrons through the external circuit until the attraction and repulsion forces are again balanced.

PROBLEM: A capacitor has plates 50 cm on a side, separated by 3 cm of air. It is initially charged to 15,000 V, then disconnected from the voltage source. How much work must be done to separate them to 5 cm, if no charge is allowed to escape?

Solution: According to the work-energy equation, $W = \Delta PE_q$, the work done will be equal to the change in electrical potential energy, provided no other energy changes are occurring. The electrical potential energy stored in a capacitor with charge q is

$$PE_q = \frac{1}{2}\frac{q^2}{C}$$

185

We choose this form rather than $\frac{1}{2}CV^2$ or $\frac{1}{2}qV$ because when the plates are separated, they are no longer attached to a voltage source; there is no assurance that V will be unchanged. We do know, however, that q *will not change* because no charges can escape.

To get the potential energy initially stored, we need the capacitance. Let C_1 be the initial capacitance, given by Eq. (26.1)

$$C_1 = \frac{K}{4\pi k} \frac{A}{d_1}$$

where $K = 1$; $A = 50$ cm $\times$ 50 cm $= 0.5$ m $\times$ 0.5 m $= 0.25$ m²; and $d_1 =$ initial plate separation $= 3$ cm $= 0.03$ m. Then:

$$C_1 = \frac{(1)(0.25 \text{ m}^2)}{4\pi(9 \times 10^9 \text{ N} \cdot \text{m}^2/\text{C}^2)(0.03 \text{ m})} = 7.4 \times 10^{-11} \text{ F} = 74 \text{ pF}$$

The final capacitance, when the plate separation is 5 cm, $= 0.05$ m, is

$$C_2 = \frac{(1)(0.25 \text{ m}^2)}{4\pi(9 \times 10^9 \text{ N} \cdot \text{m}^2/\text{C}^2)(0.05 \text{ m})} = 4.4 \times 10^{-11} \text{ F} = 44 \text{ pF}$$

The charge is determined from Eq. (26.2), $q = CV$. Since we know the initial voltage is 15,000 V, then:

$$q = C_1 V_1 = (74 \times 10^{-12} \text{ F})(15,000 \text{ V}) = 1.1 \times 10^{-6} \text{ C}$$

This charge remains unchanged during the separation process. From the above relationship, we see that as the capacitance decreases, the voltage between the plates *must increase* to maintain a constant charge q.

Returning to the work-energy equation, we find that

$$W = \Delta PE_q = (PE_q)_2 - (PE_q)_1$$

$$W = \frac{1}{2} \frac{q^2}{C_2} - \frac{1}{2} \frac{q^2}{C_1}$$

$$W = \frac{(1.1 \times 10^{-6} \text{ C})^2}{2(44 \times 10^{-12} \text{ F})} - \frac{(1.1 \times 10^{-6} \text{ C})^2}{2(74 \times 10^{-12} \text{ F})} = 1.38 \times 10^{-2} \text{ J} - 0.82 \times 10^{-2} \text{ J}$$

$$W = 0.56 \times 10^{-2} \text{ J} = \underline{0.0056 \text{ J}}$$

5.6 millijoules of work is required to separate the plates from 3 cm to 5 cm. The plates, being oppositely charged, attract each other and work is required to pull them further apart.

PROBLEM: A series circuit consists of a 4-μF capacitor, a 2-Megohm resistor, a 60-V battery and a switch (Fig. 26.1). At $t = 0$ the switch is closed. (a) What is the initial charge on the capacitor? (b) What is the initial current? (c) What is the time constant of the circuit?

Solution: (a) Initially the capacitor is uncharged, $\underline{q = 0}$; it is part of an open circuit. This can also be seen from Eq. (26.5):

$$q = q_f(1 - e^{-t/RC})$$

When $t = 0$, $e^{-0} = \frac{1}{e^0} = \frac{1}{1}$, since any quantity raised to the zero power is equal to 1.

$$q = q_f(1 - 1) = 0$$

(b) The current at any time t is given by Eq. (26.4):

$$i = i_0 e^{-t/RC}$$

where

$$i_0 = \frac{V}{R} = \frac{60 \text{ V}}{2 \times 10^6 \text{ }\Omega} = 30 \times 10^{-6} \text{ A} = \underline{30 \text{ }\mu\text{A}}$$

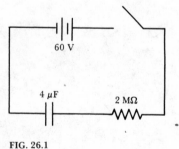

60 V

4 μF   2 MΩ

FIG. 26.1

186

At $t = 0$,

$$i = (30 \ \mu A)e^{-0} = (30 \ \mu A)(1) = 30 \ \mu A$$

(c) The time constant of the circuit is the quantity RC.

$$RC = (2 \times 10^6 \ \Omega)(4 \times 10^{-6} \ F) = \underline{8 \ s}$$

The circuit has a time constant of 8 seconds. That RC has units of seconds is proved in the text. It can also be seen by observing that the exponent of e in Eqs. (26.4) and (26.5) is $^{-t/RC}$. Since an exponent must be unitless and the numerator t has units of seconds, the denominator RC must also have units of seconds.

PROBLEM: For the previous circuit, evaluate the following quantities at $t = 4$ s: (a) the charge on the capacitor, (b) the current in the circuit, and (c) the voltage drop across the resistor. (d) Show that Kirchhoff's loop rule is satisfied at $t = 4$ s.

Solution: (a) The charge at any time t is given by

$$q = q_f(1 - e^{-t/RC})$$

where

$$q_f = CV = (4 \times 10^{-6} \ F)(60 \ V) = 240 \times 10^{-6} \ C = 240 \ \mu C$$

At $t = 4$ s

$$q = (240 \ \mu C)(1 - e^{-4/8})$$

where $RC = 8$ s from the previous problem. Now $e^{-4/8} = e^{-1/2} = \dfrac{1}{e^{1/2}} = \dfrac{1}{\sqrt{e}} =$

$\dfrac{1}{\sqrt{2.72}} = \dfrac{1}{1.649} = 0.606$. Therefore,

$$q = (240 \ \mu C)(1 - 0.606) = (240 \ \mu C)(0.394) = \underline{94.6 \ \mu C}$$

(b) The current at any time t is

$$i = i_0 e^{-t/RC}$$

where $i_0 = 30 \ \mu A$, from the previous problem. At $t = 4$ s,

$$i = (30 \ \mu A)(e^{-4/8}) = (30 \ \mu A)(0.606) = \underline{18.2 \ \mu A}$$

(c) The voltage drop across the resistor is given by Ohm's law, as usual, $V_R = iR$. We need to recall, however, that since the current i is a function of time, so is V_R. At $t = 4$ s,

$$V_R = i_{4s}R = (18.2 \times 10^{-6} \ A)(2 \times 10^6 \ \Omega) = 36.4 \ V$$

Four seconds after the switch is closed, the charge on the capacitor is 94.6 μC, the current in the circuit is 18.2 μA, and the voltage drop across the resistor is 36.4 V.

(d) Kirchhoff's loop rule, starting at the top left corner and circulating clockwise, gives

$$\mathcal{E} - V_R - V_C = 0$$

where V_C, the voltage drop across the capacitor, is given by q/C. At $t = 4$ s, $q = 94.6 \ \mu C$. Putting in numerical values,

$$60 \ V - 36.4 \ V - \frac{94.6 \times 10^{-6} \ C}{4 \times 10^{-6} \ F} = 0$$

$$60 \ V - 36.4 \ V - 23.6 \ V = 0$$

This is a true statement; Kirchhoff's loop rule is satisfied. As t gets larger, V_R decreases (as i decreases) and V_C increases (as q increases). After a very long time, there is no voltage drop over the resistor (because $i = 0$) and the entire voltage of the battery appears across the capacitor plates, $\mathcal{E} = V_C$. The capacitor is now fully charged.

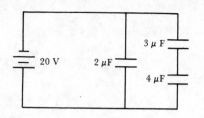

FIG. 26.2

PROBLEM: How much energy is stored in the capacitor network shown in Fig. 26.2?

Solution: We may calculate the energy stored by each capacitor and add them up, but an easier method is to determine the equivalent capacitance of the network and use it to obtain the stored energy.

The capacitors in series are added according to Eq. (26.9):

$$\frac{1}{C_{eq}} = \frac{1}{C_1} + \frac{1}{C_2} = \frac{1}{3} + \frac{1}{4} = \frac{4+3}{12} = \frac{7}{12}$$

$$C_{eq} = \text{\textonehalf} \text{\%} = 1\text{\%} \ \mu F$$

The parallel branches are added according to Eq. (26.8):

$$C_{eq} = C_1 + C_2 = 2 \ \mu F + 1\text{\%} \ \mu F = 3\text{\%} \ \mu F = 3.7 \ \mu F$$

The equivalent capacitance of the above network is 3.7 μF. The potential energy stored by the network will be equal to that stored by C_{eq}; that is the meaning of an *equivalent* capacitance. The voltage across C_{eq} is 20 V, so

$$PE_q = \text{\textonehalf} CV^2 = \text{\textonehalf}(3.7 \times 10^{-6} \ F)(20 \ V)^2 = \underline{7.4 \times 10^{-4} \ J}$$

The network of capacitors stores 0.74 millijoules of energy.

Avoiding Pitfalls

1. The capacitance of a given capacitor depends on its geometrical and material properties, not on the charge it is storing nor on the voltage between its plates.

2. The equivalent capacitance for capacitors in series and in parallel is determined by reversing the methods used for determining equivalent resistances of resistors in series and parallel.

3. The equations $V = iR$ and $q = CV$ apply both in steady-state condition and also during the charging and discharging of a capacitor through a resistor, when i and q are functions of time.

4. For an RC series circuit with a given resistance and capacitance, the time constant for charging the capacitor is the same as the time constant for discharging the capacitor.

Drill Problems
Answers in Appendix

1. Suppose a capacitor is made of two parallel metal plates, each 60 m wide and 100 m long (bigger than a football field). The plates are 3 cm apart, with air between them.

   (a) What is the capacitance of this capacitor?
   (b) State three different ways you could increase its capacitance.
   (c) If a potential difference of 1000 V is applied between the plates, how much energy is stored in the capacitor?

2. A parallel plate capacitor is charged to a certain voltage, then disconnected from the charging source and insulated so that no charge can escape. The plates are then pulled apart, increasing their separation distance. State *qualitatively* what happens to

  (a) the charge on the capacitor plates,
  (b) the capacitance,
  (c) the voltage between the plates,
  (d) the electric field, and
  (e) the energy stored in the capacitor as the plates are separated.
       Justify each response briefly.

3. A series circuit consists of a $6\text{-}\mu\text{F}$ capacitor, initially charged to a potential diffference of $V_0 = 100$ V, a 1-megohm resistor, and a switch (Fig. 26.3). At time $t = 0$, the switch is closed.

  (a) What is the initial charge on the capacitor?
  (b) What is the initial current in the circuit?
  (c) What is the time constant of the circuit? After one time constant has passed, determine
  (d) the current in the circuit,
  (e) the voltage drop over the resistor, and
  (f) the charge on the capacitor.

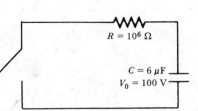

FIG. 26.3

189

27

Magnetic Forces and Fields

Terms

Define or describe briefly what is meant by the following terms. If you have difficulty, refer to the textbook section given in parentheses.

magnetic field (27.1)

tesla (27.1)

weber/m² (27.1)

gauss (27.1)

right-hand rule (27.2)

magnetohydrodynamics (27.2)

magnetic mirror (27.3)

plasma (27.3)

magnetic bottle (27.3)

mass spectrometer (27.3)

electric motor (27.5)

armature (27.5)

commutator (27.5)

Biot-Savart law (27.7)

permeability (27.7)

magnetic dipole moment (27.8)

paramagnetic atoms (27.8)

ferromagnetic materials (27.8)

domains (27.8)

Equation Review

For each equation, be able to state the situation to which it applies, what quantity each symbol represents, and the units for measuring each quantity in some consistent set of units.

$$F_m = qvB \sin \theta \tag{27.1}$$

$$qvB = \frac{mv^2}{r} \tag{27.5}$$

$$\tau = IAB \cos \theta \tag{27.7}$$

$$B = \frac{\mu}{4\pi} \frac{qv \sin \theta}{r^2} \tag{27.8}$$

$$B = \frac{\mu}{2\pi} \frac{I}{a} \tag{27.9}$$

$$B = \frac{N\mu I}{2a} \tag{27.11}$$

$$B = \mu \frac{N}{L} I \tag{27.12}$$

Problems with Solutions and Discussion

PROBLEM: A particle with a charge of 60 μC and traveling at 500 m/s enters a magnetic field of 2 T at an angle of 30° as shown in Fig. 27.1a. What magnetic force acts on the charged particle?

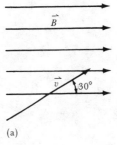

(a)

Solution: The magnetic force on a charged particle has a magnitude given by Eq. (27.1):

$$F_m = qvB \sin \theta = (60 \times 10^{-6} \text{ C})(500 \text{ m/s})(2 \text{ T}) \sin 30 = (0.06 \text{ N})(0.5) = \underline{0.03 \text{ N}}$$

The direction of the force is given by the right-hand rule: When the fingers (of the right hand) point in the direction of the magnetic field and the thumb in the direction of motion of the charge, the palm pushes in the direction of the magnetic force on a positive charge (b). Here the force is <u>into the page</u>, perpendicular to the plane determined by $\vec{B}$ and $\vec{v}$. The magnetic force on the charged particle is thus 0.03 N, into the page.

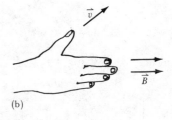

(b)

FIG. 27.1

PROBLEM: Consider a segment of current-carrying wire of length $L = 10$ cm, lying at a 30°-angle in the magnetic field of the previous problem, $B = 2$ T (Fig. 27.2). If this segment of wire feels the same force as the charged particle of the previous problem, what current flows through it?

Solution: Since currents are in fact charges in motion, we predict that a current-carrying wire will experience a force in a magnetic field given by

$$F_m = qvB \sin \theta$$

The current in the wire is related to the charge flowing through the length L of wire by $I = q/t$, where t is the time taken for q to travel the distance L. Thus, $q = It$. Also the speed v of the moving charges is just L/t. Substituting these into the magnetic force equation gives

$$F_m = (It)\left(\frac{L}{t}\right) B \sin \theta = ILB \sin \theta$$

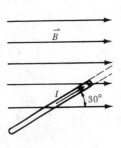

FIG. 27.2

191

This equation is equivalent to Eq. (27.1), except that the moving charges are expressed in terms of current through a length of conductor.

If the 10-cm wire feels the same force as the moving charge in the previous problem, $F_m = 0.03$ N, the current must be

$$I = \frac{F_m}{BL \sin \theta} = \frac{0.03 \text{ N}}{(2 \text{ T})(0.1 \text{ m})(0.5)} = \underline{0.3 \text{ A}}$$

The wire carries a current of 0.3 A. As a check, let us verify that the motion of the charged particle is equivalent to a 0.3-A current. We had 60 μC of charge moving at 500 m/s; at this speed, how long would it take to travel 10 cm?

$$t = \frac{L}{v} = \frac{0.1 \text{ m}}{500 \text{ m/s}} = 2 \times 10^{-4} \text{ s}$$

Therefore the equivalent current is

$$I = \frac{q}{t} = \frac{60 \times 10^{-6} \text{ C}}{2 \times 10^{-4} \text{ s}} = \underline{0.3 \text{ A}}$$

Since the current and the moving charge are equivalent, the magnetic field exerts the same force on both.

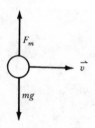

FIG. 27.3

PROBLEM: A small sphere of mass 10 g and charge -500 μC is projected horizontally into a magnetic field with a speed of 300 m/s. Determine the magnitude and direction of the magnetic field such that the sphere does not fall under the influence of gravity but continues moving horizontally. Assume that the direction of motion is perpendicular to the magnetic field.

Solution: The magnetic *force* must be equal and opposite to the weight of the sphere for it to continue horizontally without falling (Fig. 27.3):

$$F_m - mg = 0, \quad \text{or} \quad F_m = mg$$

Using Eq. (27.1) for F_m,

$$qvB = mg \; (\sin \theta = 1)$$

and solving for B gives

$$B = \frac{mg}{qv}$$

Putting in numerical values we get

$$B = \frac{(0.01 \text{ kg})(9.8 \text{ m/s}^2)}{(500 \times 10^{-6} \text{ C})(300 \text{ m/s})} = \underline{0.65 \text{ T}}$$

We know the direction of the magnetic force must be upward; we have yet to determine the direction of the *magnetic field* that will result in this upward force. Orienting the right hand so that the palm pushes up and the thumb points horizontally, the fingers, representing B, point into the page. However, the right-hand rule is formulated for a *positive* charge; since this sphere carries a *negative* charge, the required direction of $\bar{B}$ is out of the page. The magnetic field that will just support the sphere's weight is 0.65 teslas out of the page.

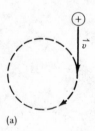

(a)

(b)

(c)

FIG. 27.4

PROBLEM: (a) In what direction is the magnetic field that causes a proton to move in the circular path shown in Fig. 27.4a? (b) In what direction would you apply an *electric field*, in the same region as the magnetic field, so that the proton would move through the region in a straight line?

Solution: (a) Since the proton moved in a circle, the direction of the force on it must be toward the center of the circle. Looking at the point farthest right, the centripetal force points left. $\vec{F}$ and $\vec{v}$ determine a plane (in this case the plane of the page) and $\vec{B}$ must be perpendicular to this plane, either out of the page or into the page (Fig. 27.4b). We may use the right-hand rule to determine which is correct. Orienting the palm of the right hand so that it pushes to the left while the thumb points downward in the direction of the proton's velocity at that point, results in the fingers pointing <u>out of the page</u>. This must be the direction of the magnetic field $\vec{B}$.

(b) Since the magnetic force $\vec{F}_m$ is to the left as the proton enters the region, the electric force $\vec{F}_q$ must be to the right, equal and opposite, cancelling $\vec{F}_m$ so that the proton continues in a straight line (Fig. 27.4c). The direction of the electric field is defined as the direction of the force on a positive charge, so $\vec{E}$ also points <u>to the right</u>.

PROBLEM: An alpha particle (mass 6.64×10^{-27} kg and charge $+2e$) enters a magnetic field B of 0.5 T, directed out of the page as shown in Fig. 27.5a, with a speed of 2×10^6 m/s. How far from its entrance point does it leave the field and what is its velocity as it leaves?

Solution: As it enters the field, the alpha particle experiences a *downward* force, perpendicular to both $\vec{B}$ and $\vec{v}$. (Confirm this by applying the right-hand rule.) A force perpendicular to the velocity vector causes circular motion; the alpha particle takes the path indicated (b). Since the magnetic force, $F_m = qvB$, supplies the centripetal force which causes this circular motion, we may write Eq. (27.5):

$$qvB = \frac{mv^2}{r}$$

Solving for r, the radius of the circle, gives

$$r = \frac{mv^2}{qvB} = \frac{mv}{qB}$$

The distance d is just the diameter of the circular path, or

$$d = 2r = \frac{2mv}{qB}$$

where

$$q = 2e = 2(1.6 \times 10^{-19}\ \text{C}) = 3.2 \times 10^{-19}\ \text{C}$$

Substituting numerical values gives

$$d = \frac{2(6.64 \times 10^{-27}\ \text{kg})(2 \times 10^6\ \text{m/s})}{(3.2 \times 10^{-19}\ \text{C})(0.5\ \text{T})} = 0.166\ \text{m} = \underline{16.6\ \text{cm}}$$

The alpha particle leaves the region of the magnetic field 16.6 cm from where it entered. A centripetal force, because it is always perpendicular to the velocity vector, cannot speed up or slow down an object, it can only change its direction. Since the alpha particle entered the field with a velocity of 2×10^6 m/s to the right, it exits with a velocity of $\underline{2 \times 10^6}$ $\underline{\text{m/s to the left}}$.

PROBLEM: How much energy does a particle with charge q gain (or lose) when moving through a magnetic field B at speed v?

Solution: The energy gained will be equal to the work done, by the work-energy equation $W = \Delta E$. The basic definition of work is $Fx \cos \theta$. But since the magnetic force is always perpendicular to the motion, $\cos \theta$ is always 0 and <u>no work is done</u> on the particle. Consequently the <u>energy of the charged particle remains constant</u>. A particle cannot gain or lose energy in a magnetic field the way it can in an electric field, provided the magnetic field is not changing. (In a particle accelerator like the cyclotron, the magnetic field serves to confine the charged particles by keeping them moving in circular paths; the increase in speed is obtained by intermittent application of an electric field to the charged particles.)

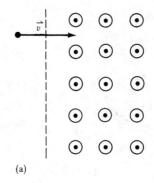

(a)

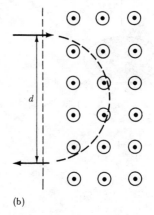

(b)

FIG. 27.5

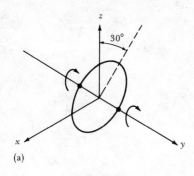

(a)

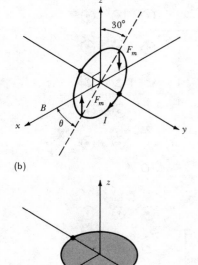

(b)

(c)

FIG. 27.6

PROBLEM: A circular loop of wire with a radius of 8 cm can rotate around the *y* axis. It is held such that its plane makes an angle of 30° with the vertical (*z* axis) as shown in Fig. 27.6a. A magnetic field *B* of 0.5 T is everywhere parallel to the *x* axis. (a) What is the torque on the loop if a current of 10 A flows in it? (b) What would be the torque if the loop were replaced by a coil made up of 20 circular loops each of radius 8 cm, all other conditions remaining the same? (c) In what position would the loop experience the maximum torque?

Solution: (a) The torque on a current-carrying loop in a magnetic field is given by Eq. (27.7): $\tau = IAB \cos \theta$, where *A* is the cross-sectional area of the loop and θ is the angle between the *plane* of the loop and the magnetic field direction. In this case, $\theta = 60°$. The area of the loop is $A = \pi r^2 = \pi(0.08 \text{ m})^2 = 0.02 \text{ m}^2$. The torque on the loop is therefore

$$\tau = (10 \text{ A})(0.02 \text{ m}^2)(0.5 \text{ T})(0.5) = \underline{0.05 \text{ N} \cdot \text{m}}$$

The direction of the torque is indicated by showing the magnetic force vectors on the top and bottom of the loop when the current is in the direction shown (Fig. 27.6b).

(b) If instead of 1 loop, there were 20 circular loops in the coil, each loop would feel a torque of 0.05 N·m, so the total torque on the coil would be

$$\tau = 20(0.05 \text{ N} \cdot \text{m}) = \underline{1.0 \text{ N} \cdot \text{m}}$$

Thus, for *N* loops, the torque is *N* times as great as on a single loop.

(c) The maximum torque will be experienced when $\cos \theta = 1$. This occurs when $\theta = 0$; that is, when the plane of the loop is parallel to the direction of *B* as shown (Fig. 27.6c). This torque decreases as θ increases until the loop plane is perpendicular to the direction of *B*. Here $\cos \theta = 0$, so $\tau = 0$, and the loop, if released, will not rotate.

PROBLEM: Show that the units on the right side of Eq. (27.7), $\tau = IAB \cos \theta$, reduce to the proper *mks* units for torque.

Solution: Torque is defined as a force (N) times a perpendicular moment arm (m), so the proper *mks* unit for torque is N·m. The units of the right-hand side of Eq. (27.7) are

$$(\text{A})(\text{m}^2)(\text{T})$$

where an ampere (A) is a coulomb per second (C/s). Substituting this in gives

$$\left(\frac{\text{C}}{\text{s}} \right) (\text{m}^2)(\text{T})$$

The unit for magnetic field, the tesla, can be expressed in terms of other units by recalling the defining equation for magnetic force, $F_m = qvB \sin \theta$; solving this for *B* gives

$$B = \frac{F}{qv \sin \theta} \quad \text{so} \quad \text{T} = \frac{\text{N}}{(\text{C})(\text{m/s})} = \frac{\text{N} \cdot \text{s}}{\text{C} \cdot \text{m}}$$

Substituting this gives

$$\left(\frac{\cancel{\text{C}}}{\cancel{\text{s}}} \right) (\text{m}^2) \left(\frac{\text{N} \cdot \cancel{\text{s}}}{\cancel{\text{C}} \cdot \text{m}} \right) = \text{N} \cdot \text{m}$$

Thus the quantity $IAB \cos \theta$ has units of N·m, the proper *mks* unit for torque.

PROBLEM: A solenoid is made by winding a 110-m long wire around a cardboard tube that is 40 cm long and 5 cm in diameter. If a magnetic field of 0.02 T is produced inside the solenoid, what current is flowing through the wire?

Solution: Equation (27.12) relates the properties of a solenoid to the magnetic field inside it:

$$B = \mu_0 \frac{N}{L} I$$

where μ_0 (the permeability of air) $= 4\pi \times 10^{-7}$ T·m/A, *N* is the number of turns of

194

wire, and L is the length of the solenoid. To determine N for this solenoid, we note that the length of wire needed for each turn is equal to πD, the circumference of the tube. Therefore,

$$N = \text{number of turns} = \frac{110 \text{ m}}{\pi D} = \frac{110 \text{ m}}{\pi(0.05 \text{ m})} = 700 \text{ turns}$$

Solving Eq. (27.12) for I and putting in numerical values gives

$$I = \frac{BL}{\mu_0 N} = \frac{(0.02 \text{ T})(0.4 \text{ m})}{(4\pi \times 10^{-7} \text{ T·m/A})(700)} = \underline{9.1 \text{ A}}$$

A current of 9.1 amperes flowing through this solenoid produces a magnetic field of 0.02 teslas.

PROBLEM: A proton is moving with a speed of 4×10^7 m/s straight toward a long wire carrying a current of 50 A upward (Fig. 27.7a). What force does the proton experience when it is 5 cm from the wire?

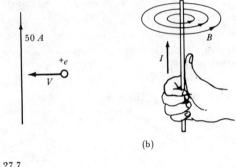

(a)   (b)

FIG. 27.7

Solution: The proton feels a magnetic force given by Eq. (27.1): $F_m = qvB \sin\theta$, where B is the magnetic field of the wire. Here the magnitude of $\vec{B}$ can be determined by Eq. (27.9):

$$B_{\text{wire}} = \frac{\mu_0 I}{2\pi a}$$

where the distance from the wire, $a = 5 \text{ cm} = 0.05 \text{ m}$.

$$B_{\text{wire}} = \frac{(4\pi \times 10^{-7} \text{ T·m/A})(50 \text{ A})}{2\pi(0.05 \text{ m})} = 2 \times 10^{-4} \text{ T}$$

The direction of $\vec{B}$ is given by the *right-hand rule:* the wire is grasped with the right hand so that the thumb points in the direction of I; the fingers then curl around the wire in the direction of the concentric lines of the field $\vec{B}$ (Fig. 27.7b). They go *into the page* to the right of the wire and come *out of the page* to the left of the wire. If $\vec{B}$ goes into the page at the point of interest, then the angle θ between $\vec{B}$ and $\vec{v}$ is 90°.

$$F_m = qvB \sin\theta = (1.6 \times 10^{-19} \text{ C})(4 \times 10^7 \text{ m/s})(2 \times 10^{-4} \text{ T})(1)$$

$$F_m = \underline{1.28 \times 10^{-15} \text{ N}}$$

You may wish to confirm, by the right-hand rule for the magnetic force, that the direction of $\vec{F}$ is <u>downward</u>, perpendicular to both $\vec{B}$ and $\vec{v}$.

PROBLEM: A very long wire carrying a current of 40 A is straight except where it has a circular kink as shown in Fig. 27.8. The circular loop lies in the plane of the page and has a 10-cm radius. Determine the magnetic field at C, the center of the loop.

Solution: The magnetic field at the center of the loop can be thought of as consisting of two components, the field due to the straight wire, $\vec{B}_{\text{wire}}$, and the field due to the circular loop, $\vec{B}_{\text{loop}}$:

$$\vec{B} = \vec{B}_{\text{wire}} + \vec{B}_{\text{loop}}$$

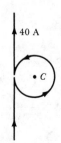

FIG. 27.8

195

The point C is 10 cm = 0.01 m from the long wire. Eq. (27.9) gives the magnetic field at a distance a from a long straight wire carrying a current I:

$$B_{wire} = \frac{\mu_0}{2\pi}\frac{I}{a} = \frac{(4\pi \times 10^{-7} \text{ T}\cdot\text{m/A})(40 \text{ A})}{2\pi(0.1 \text{ m})}$$

$$= 8 \times 10^{-5} \text{ T} \otimes \text{ (symbol for into the page)}$$

As in the previous problem, the direction of B_{wire} is *into the page* at point C, by the right-hand rule.

The magnetic field due to a loop of current is given by Eq. (27.10):

$$B_{loop} = \frac{\mu_0 I}{2a}$$

where a is the radius of the loop, here 0.1 m, and I is the current it carries, 40 A.

$$B_{loop} = \frac{(4\pi \times 10^{-7} \text{ T}\cdot\text{m/A})(40 \text{ A})}{2(0.1 \text{ m})} = 2.5 \times 10^{-4} \text{ T} = 0.25 \text{ mT}$$

The direction of $\vec{B}_{loop}$ at the center can be determined by the right-hand rule of the previous problem, or by examination of Fig. 27.23(b) in the text. Here we note that a current traveling counterclockwise when viewed from the top of the loop, produces a magnetic field at the loop's center which is directed perpendicularly upward from the plane of the loop. Thus $\vec{B}_{loop}$ in our problem is *out of the page* (symbol $\odot$).

We now add $\vec{B}_{wire}$ and $\vec{B}_{loop}$ as vectors. They are in the same plane but opposite in direction. Taking *out of the page* as positive we get

$$B = B_{loop} \odot - B_{wire} \otimes = 25 \times 10^{-5} \text{ T} - 8 \times 10^{-5} \text{ T}$$

$$\underline{B = 17 \times 10^{-5} \text{ teslas out of the page, } \odot}$$

Avoiding Pitfalls

1.   Magnetic forces act on charges in motion. There is no magnetic force on a stationary charge.

2.   Only charged particles with a component of velocity perpendicular to the magnetic field direction experience a magnetic force. A particle whose velocity is parallel (or antiparallel) to the magnetic field experiences no magnetic force.

3.   The *direction* of the magnetic force is always perpendicular to both the magnetic field $\vec{B}$ and the velocity of the charge, $\vec{v}$. The right-hand rule for determining the direction of the perpendicular force is for *positively* charged particles. Negatively charged particles experience a magnetic force in the opposite direction.

4.   If a charged particle's velocity is perpendicular to a uniform magnetic field, circular motion results. The circular path lies in the plane determined by the magnetic *force* and the velocity vectors.

5.   There is no net force on a flat current loop in a uniform magnetic field; however there is in general a torque, given by $\tau = IAB \cos\theta$, where θ is the angle between the *plane* of the loop and the direction of the magnetic field.

6.   Magnetic fields are produced by moving electric charges and currents. A current does not feel a force due to its own magnetic field.

7.   Since the magnetic field is a vector quantity, vector addition techniques must be used to obtain the resultant of more than one magnetic field.

Drill Problems
Answers in Appendix

1.   (a) Imagine that you are sitting in a room with your back to one wall and that a proton beam is traveling horizontally from the wall behind you toward the opposite wall in front of you. If the beam is deflected to your left, what is the *direction* of the magnetic field that exists in the room? (b) Figure 27.9a shows part of a conductor carrying a current I in a

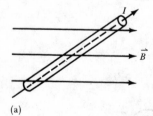

(a)

(b)

FIG. 27.9

magnetic field *B*. What is the *direction* of the force on this conductor? (c) Figure 27.9b shows the cross section of a wire carrying current *out of the page*. Sketch in a few lines representing the magnetic field near this wire.

2. As shown in Fig. 27.10, a particle with a charge of $+3e$ undergoes circular motion of radius 30 cm when it enters a region where a magnetic field of 0.1 T exists.

   (a) Determine the direction of the magnetic field in this region.
   (b) What is the momentum of the particle?
   (c) If the particle is a lithium nucleus of mass 10^{-27} kg, how long does it take to complete one revolution?

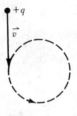

FIG. 27.10

3. Two current-carrying wires form two concentric circular loops as shown in Fig. 27.11. The larger loop has a radius of 30 cm and carries a current of 50 A in the clockwise direction. The inner loop has a radius of 20 cm and carries 20 A in the counterclockwise direction. Determine the magnetic field at the center of the loops. (Ignore the effect of the lead wires through which the currents enter the loops.)

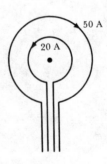

FIG. 27.11

28

Induced Voltage

Terms

Define or describe briefly what is meant by the following terms. If you have difficulty, refer to the textbook section given in parentheses.

electromagnetic induction (28.1)

magnetic flux (28.2)

Faraday's law (28.2)

Lenz's law (28.3)

electric generator (28.4)

peak voltage (28.4)

mutual inductance (28.5)

henry (28.5)

self-induced emf (28.6)

self inductance (28.6)

back emf (28.6)

transformer (28.7)

primary coil (28.7)

secondary coil (28.7)

Equation Review

For each equation, be able to state the situation to which it applies, what quantity each symbol represents, and the units for measuring each quantity in some consistent set of units.

$$\Phi = AB \cos \phi \tag{28.1}$$

$$\mathscr{E} = -N\frac{\Delta\Phi}{\Delta t} = -N\frac{\Phi - \Phi_0}{t - t_0} \tag{28.2}$$

$$\mathcal{E} = \mathcal{E}_0 \sin 2\pi ft = (NBA2\pi f) \sin 2\pi ft \qquad (28.5)$$

$$M = \frac{N_2 \Phi_{21}}{i_1} \qquad (28.6)$$

$$\mathcal{E}_2 = -M \frac{\Delta i_1}{\Delta t} \qquad (28.7)$$

$$L = \frac{N\Phi}{i} \qquad (28.9)$$

$$\mathcal{E} = -L \frac{\Delta i}{\Delta t} \qquad (28.10)$$

$$\frac{\mathcal{E}_1}{\mathcal{E}_2} = \frac{N_1}{N_2} \qquad (28.11)$$

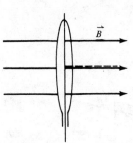

(a)

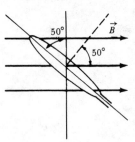

(b)

Problems with Solutions and Discussion

PROBLEM: A circular loop of 40-cm radius is held with its plane perpendicular to a magnetic field of 0.2 T as shown in Fig. 28.1a. (a) What is the magnetic flux passing through the loop? (b) What flux passes through the loop if it is tilted 50° from the vertical? (c) What is the flux through the loop if it is tilted 90° from the vertical?

Solution: The flux through the loop is given by Eq. (28.1):

$$\Phi = AB \cos \phi$$

where A is the cross-sectional area of the loop and is the same for all cases: $A = \pi r^2 = \pi(0.4 \text{ m})^2 = 0.5 \text{ m}^2$. B is the magnetic field and also does not change: $B = 0.2$ T. The only difference in the three cases is the *orientation* of the loop in the field, specified by the angle ϕ between $\vec{B}$ and the line *perpendicular to the cross-sectional area* of the loop. It is shown by a dashed line in each diagram.

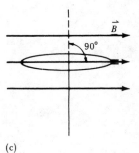

(c)

FIG. 28.1

(a) Here the perpendicular line and the field $\vec{B}$ are parallel ($\phi = 0°$):

$$\Phi = AB \cos 0° = (0.5 \text{ m}^2)(0.2 \text{ T})(1) = \underline{0.1 \text{ T} \cdot \text{m}^2}$$

(b) Here the perpendicular line makes a 50° angle with the field $\vec{B}$:

$$\Phi = AB \cos 50° = (0.5 \text{ m}^2)(0.2 \text{ T})(0.643) = \underline{0.064 \text{ T} \cdot \text{m}^2}$$

Because the loop is tilted, fewer magnetic field lines can pass through, so the flux is less than in case (a).

(c) Here the perpendicular line and the field $\vec{B}$ are perpendicular ($\phi = 90°$):

$$\Phi = AB \cos 90° = (0.5 \text{ m}^2)(0.2 \text{ T})(0) = 0$$

No flux can pass through the loop in this orientation.

PROBLEM: A Boy Scout pitches a tent with its length in the direction toward the magnetic north pole. At this location the field has a magnitude of 5.5×10^{-5} T and a downward inclination of 65° as shown in Fig. 28.2. The front of the tent makes an equilateral triangle of side 1.5 m. Determine the magnetic flux through the front of the tent.

Solution: The magnetic flux is determined by Eq. (28.1): $\Phi = AB \cos \phi$. A is the area of interest, in this case the triangular tent front. A triangle's area is given by

$$\tfrac{1}{2}(\text{base})(\text{height}) = \tfrac{1}{2}(1.5 \text{ m})(1.5 \text{ m} \sin 60°) = 0.97 \text{ m}^2$$

Now ϕ is the angle between the *perpendicular to the area* and the magnetic field. This perpendicular is horizontal, in the direction of the horizontal component of the earth's field.

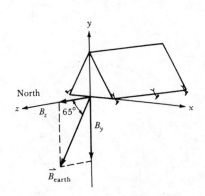

FIG. 28.2

199

The angle ϕ is thus 65°. This can also be seen by noting that the door is in the x-y plane, so its perpendicular lies on the z axis, and $\vec{B}$ makes an angle of 65° with the z axis. The flux through the tent door is therefore

$$\Phi = AB \cos \phi = (0.97 \text{ m}^2)(5.5 \times 10^{-5} \text{ T})(\cos 65°) = \underline{2.3 \times 10^{-5} \text{ T} \cdot \text{m}^2}$$

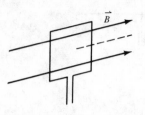

FIG. 28.3

PROBLEM: A coil of wire having 150 turns and a cross-sectional area of 0.18 m² lies with its plane perpendicular to a uniform magnetic field (Fig. 28.3). If the magnetic field is turned on in 0.05 s, an emf of 12 V is measured in the coil. What is the magnitude of the magnetic field?

Solution: The emf induced in the coil as the magnetic flux through the coil changes is given by Eq. (28.2): $\mathscr{E} = -N \Delta\Phi/\Delta t$. Φ, the magnetic flux, is given by Eq. (28.1): $\Phi = AB \cos \phi$.

The flux can change as a result of a changing magnetic field $\vec{B}$, a changing cross-sectional area A, or a changing orientation of the loop in the field, specified by ϕ, the angle between the normal to the area and the magnetic field. In this case, $\phi = 0$, so $\cos \phi = 1$, and $A = 0.18$ m² and neither of these quantities changes. The changing flux is due to the change in B:

$$\Delta\Phi = \Phi - \Phi_0 = BA \cos \phi - B_0 A \cos \phi = (B - B_0)A \cos \phi$$

$$= (B - 0)(0.18 \text{ m}^2)(1) = (0.18 \text{ m}^2)B$$

Substituting this into Eq. (28.2), along with $N = 150$ turns, $\Delta t = 0.05$ s, and $\mathscr{E} = 12$ V gives

$$12 \text{ V} = -\frac{(150)(0.18 \text{ m}^2)B}{0.05 \text{ s}}$$

Solving for B we find that

$$B = \frac{(12 \text{ V})(0.05 \text{ s})}{(150)(0.18 \text{ m}^2)} = \underline{0.022 \text{ T}}$$

Here we have omitted the negative sign, whose sole function is to remind us of the method for determining the polarity of the induced voltage.

PROBLEM: A circular loop of wire is positioned half in and half out of the region of uniform magnetic field shown in Fig. 28.4a. Determine the direction (clockwise or counterclockwise) of the induced current in the loop in each of the following situations: (a) the loop moves in the $+x$ direction; (b) the loop moves in the $-x$ direction; (c) the loop moves in the $+y$ direction; (d) the magnetic field increases; (e) the magnetic field decreases.

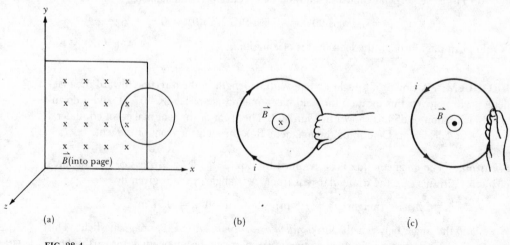

(a) (b) (c)

FIG. 28.4

Solution: Lenz's law states that the induced current will be in such a direction as to *oppose the change* which caused it.

(a) The loop, in moving to the right, experiences a decreasing field *B*, pointing into the page. To oppose this, its current must *strengthen B into the page.* A <u>clockwise current</u> will produce a magnetic field into the page, strengthening the decreasing external magnetic flux through the loop (Fig. 28.4b).

(b) When the loop moves to the left, the magnetic flux through it increases. To oppose this increase, the loop's current must produce a magnetic field opposite in direction or *out of the page.* A <u>counterclockwise current</u> will achieve this and thus oppose the change in flux (Fig. 28.4c).

(c) When the loop moves up, the amount of magnetic flux it encloses does not change; therefore there is <u>no induced current</u>.

(d) If the magnetic field into the page increases, the loop's induced current must oppose this increase by producing a magnetic field *out of the page;* therefore a <u>counterclockwise current</u> will flow.

(e) If the magnetic field into the page decreases, the loop's current must oppose this *decrease;* that is, it must *strengthen* the field into the page. To produce a field into the page, the induced current will flow <u>clockwise</u>.

PROBLEM: A square loop of side 0.5 m lies with its plane perpendicular to a uniform magnetic field of 0.08 T as shown in Fig. 28.5a. What is the induced emf when the loop is quickly (0.2 s) pulled out into a circular shape? In what direction does the induced current flow?

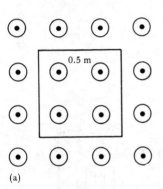

(a)

Solution: An emf is induced if the magnetic flux through the loop changes, according to Eq. (28.2): $\mathcal{E} = -N\,\Delta\Phi/\Delta t$. Since magnetic flux is defined as $\Phi = AB\cos\phi$, Φ will change if there is a change in the loop area *A*, or the magnetic field $\vec{B}$, or the orientation of the loop to the field, indicated by ϕ. In this case it is the loop *area* that changes, changing Φ accordingly.

$$\Delta\Phi = \Phi - \Phi_0 = B(A - A_0) \quad (\phi = 0, \text{ so } \cos\phi = 1)$$

The initial area is the area of the square, $A_0 = 0.5 \text{ m} \times 0.5 \text{ m} = 0.25 \text{ m}^2$. The final area is the area of the circle which has the same perimeter as the square had. The square's perimeter is 4(0.5 m) = 2 m, so the circle's circumference is 2 m.

$$2\pi r = 2 \text{ m}$$

Solving for *r*,

$$r = \frac{2 \text{ m}}{2\pi} = 0.318 \text{ m}$$

(b)

FIG. 28.5

The circle's area is thus

$$A = \pi r^2 = \pi(0.318 \text{ m})^2 = 0.32 \text{ m}^2$$

Substituting these values into Eq. (28.2), along with *N* = 1, gives

$$\mathcal{E} = -N\frac{B(A - A_0)}{\Delta t} = -(1)\frac{(0.08 \text{ T})(0.32 \text{ m}^2 - 0.25 \text{ m}^2)}{0.2 \text{ s}} = \underline{-0.028 \text{ V}}$$

An emf of 28 mV is induced in the loop as its area changes. The negative sign reminds us that the polarity of the induced emf is such as to cause a current which will induce a magnetic field that opposes the changing flux. In this case the change was an *increase* in flux out of the page (due to an increase in loop area). To oppose this increasing flux, the loop must induce a current which produces a magnetic field *into the page.* This will result from a <u>clockwise current</u>.

PROBLEM: An electric generator rotates at a rate of 20 cycles per second, producing a peak emf of 30 V. The rectangular coil is 30 cm × 50 cm and has 100 turns. (a) What is the magnitude of the magnetic field in which the coil turns? (b) If this same coil is rotated in a magnetic field of 0.08 T, what peak emf will it generate?

Solution: (a) The induced emf produced by a generator is given by Eq. (28.5):

$$\mathcal{E} = (NBA2\pi f) \sin 2\pi ft$$

where the quantity $(NBA2\pi f) = \mathcal{E}_0$, the amplitude or peak voltage of the sinusoidally varying emf. For this problem

$$\mathcal{E}_0 = 30 \text{ V}$$
$$N = 100 \text{ turns}$$
$$A = 30 \text{ cm} \times 50 \text{ cm} = 0.3 \text{ m} \times 0.5 \text{ m} = 0.15 \text{ m}^2$$
$$f = 20 \text{ s}^{-1}$$

Solving for B and substituting numerical values gives

$$B = \frac{\mathcal{E}_0}{NA2\pi f} = \frac{30 \text{ V}}{(100)(0.15 \text{ m}^2)(2\pi)(20 \text{ s}^{-1})} = \underline{0.016 \text{ T}}$$

The magnetic field in which this coil rotates is 0.016 teslas.

(b) If the coil is rotated in a 0.08-T magnetic field, the peak voltage will be

$$\mathcal{E}_0 = NBA2\pi f = (100)(0.08 \text{ T})(0.15 \text{ m}^2)(2\pi)(20 \text{ s}^{-1}) = \underline{150 \text{ V}}$$

This result can also be obtained by noting that this magnetic field is five times larger than the original one. Since $\mathcal{E}_0$ is directly proportional to B, it will generate five times the voltage, or $5 \times 30 \text{ V} = \underline{150 \text{ V}}$.

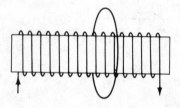

FIG. 28.6

PROBLEM: A solenoid of 10-cm^2 cross-sectional area and 30-cm length has 90 turns. It is encircled by a large coaxial loop of 20 cm^2 cross-sectional area with 2 turns (Fig. 28.6). What is the mutual inductance of this system?

Solution: Mutual inductance is given by Eq. (28.6):

$$M = \frac{N_2 \Phi_{21}}{i_1}$$

In this case we will choose the subscript $_2$ to represent the large loop and $_1$ the solenoid. Thus, Φ_{21} is the flux through the loop produced by the solenoid. (Choosing Φ_{21} to be the flux through the solenoid produced by the loop would not make sense, since the loop carries no primary current which could produce a magnetic flux through the solenoid. Even if it did carry a current, we would still not make this choice, because its magnetic field would not produce a *constant* flux through all the solenoid loops. Even to get some sort of average flux through the solenoid from the loop's field would be an extremely difficult problem.) So Φ_{21} is the flux through the large loop produced by the current in the solenoid. From Eq. (28.1) we know that

$$\Phi = AB \cos \phi$$

where B = field inside solenoid = $\mu_0 \dfrac{N}{L} i$, and A = cross-sectional area of the solenoid, not of the loop. A solenoid has the property of producing a magnetic field that is constant and coaxial inside, and essentially zero elsewhere. Therefore the total flux through the loop is the flux inside the solenoid. If we multiplied B by the loop's area, we would calculate more flux than actually passes through the loop. The angle between the perpendicular to the loop's plane and the magnetic field B of the solenoid is 0°; both are coaxial. Therefore, $\cos \phi = 1$, and

$$\Phi_{21} = \left(\mu_0 \frac{N_1}{L} i_1 \right) (A_1)(1)$$

Substituting this into Eq. (28.6) above, we find that

$$M = \frac{N_2 \Phi_{21}}{i_1} = \frac{N_2}{\dot{x}_1} \left(\mu_0 \frac{N_1}{L} \dot{x}_1 \right) A_1 = \frac{\mu_0 N_2 N_1 A_1}{L}$$

where $\mu_0 = 4\pi \times 10^{-7}$ T·m/A; $N_2 = 2$ turns; $N_1 = 90$ turns; $A_1 = 10$ cm^2 = 0.001 m^2; and $L = 30$ cm = 0.3 m. Substituting these values gives

$$M = \frac{(4\pi \times 10^{-7}\ \text{T·m/A})(2)(90)(0.001\ \text{m}^2)}{0.3\ \text{m}} = \underline{7.54 \times 10^{-7}\ \text{H}}$$

Note that M is a function only of the geometrical and material properties of the system. Equation (28.6) seems to imply that M depends on the current i, but this is not true; Φ depends on i, so the seeming dependence on current cancels out of the expression.

PROBLEM: A transformer for charging a pocket calculator from a 120-V line supplies 5.7 V to the calculator. (a) What is the ratio N_1/N_2 of turns in the primary coil to turns in the secondary? (b) If the same transformer were used as a step-up transformer operating from a 120-V line, what voltage would the secondary supply?

Solution: (a) The emfs of the primary and secondary coils of a transformer are directly proportional to the number of turns in each coil, provided the same magnetic flux passes through both coils. (This is often achieved by winding the primary and secondary coils on the same iron core.) Eq. (28.11) shows this proportionality:

$$\frac{\mathcal{E}_1}{\mathcal{E}_2} = \frac{N_1}{N_2}$$

In this case,

$$\frac{N_1}{N_2} = \frac{\mathcal{E}_1}{\mathcal{E}_2} = \frac{120\ \text{V}}{5.7\ \text{V}} = \underline{21}$$

In this step-down transformer there are 21 turns on the primary for each turn on the secondary, thereby stepping down the voltage from 120 V to $^{120}/_{21} = 5.7$ V.

(b) The transformer can be used to step up the input voltage by reversing the roles of the primary and the secondary. If the coil with the smaller number of turns is connected to the 120-V line, it becomes the primary. In this case,

$$\frac{N_p}{N_s} = \frac{1}{21} = \frac{\mathcal{E}_p}{\mathcal{E}_s}$$

If $\mathcal{E}_p = 120$ V, then $\mathcal{E}_s = \mathcal{E}_p(21) = 120\ \text{V}(21) = \underline{2500\ \text{V}}$. This transformer can step up 120 V to 2500 V if the coil with fewer turns is used as the primary. Many transformers dissipate considerable thermal energy (they get quite warm when operating). If a transformer is reversed from the way it was designed to operate, care must be taken to determine whether the windings are capable of dissipating the thermal energy which is generated in the reverse configuration.

Avoiding Pitfalls

1. In the equation for calculating magnetic flux, $\Phi = AB \cos \phi$, ϕ is the angle between the magnetic field direction and the direction of the line drawn *perpendicular to A*, the cross-sectional area of interest. [Do not confuse Φ (flux) with ϕ (angle specifying orientation of the area to $\vec{B}$).]

2. Faraday's law, $\mathcal{E} = -N \Delta\Phi/\Delta t$, is used to determine the *magnitude* of the induced emf in a conductor. The only function of the negative sign is as a reminder that the polarity of the induced emf (and the direction of the resulting current in a closed circuit) is such as to *oppose the change* in flux which caused it.

3. Always use Lenz's law to determine the direction of the induced current in a circuit within which a magnetic flux is changing. The negative sign in Faraday's law does not indicate an explicit direction because no positive direction has been agreed on.

4. The induced emf is large, not when the magnetic flux is large, but when there is a *large change* in the magnetic flux.

5. Steady state conditions do not produce induced voltages and currents. Thus electromagnetic induction is not important in dc circuits except at the instant the current starts or stops flowing. Induction is very important in ac circuits where currents and their associated magnetic fields are continually changing in both magnitude and direction.

6. A coil rotating at a given frequency in a uniform magnetic field produces an induced emf that is a sinusoidal function of time.

7. Self-inductance L and mutual inductance M are independent of current and depend only on the geometry of the coil(s) and the properties of any nearby or enclosed magnetic material.

Drill Problems
Answers in Appendix

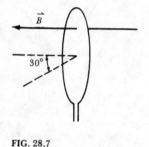

FIG. 28.7

1. A circular wire loop of 25-cm radius is initially oriented so that its plane is perpendicular to a 0.2-T magnetic field as shown in Fig. 28.7. In a time period of 0.15 s the loop is tilted 30° and the magnetic field doubles in magnitude. What emf is induced in the loop?

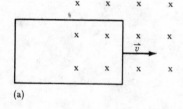

(a)

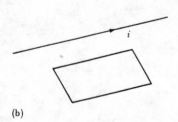

(b)

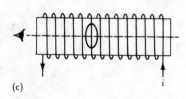

(c)

FIG. 28.8

2. (a) What is the direction of the induced current in the square loop as it enters a region of uniform magnetic field B directed into the page as shown in Fig. 28.8a? (b) The loop and the long straight wire lie side by side in the same plane (Fig. 28.8b). What is the direction of the induced current in the square loop if the current i in the wire suddenly drops to zero. (c) The solenoid and the small loop inside it are coaxial (Fig. 28.8c). What is the direction of the induced current in the small loop, as seen by an observer to the left, when the solenoid current i is increased?

3. Twenty turns of wire are wrapped around a core to make a coil with a self-inductance of 50 mH.

    (a) If the current through this coil changes from $+2$ A to -2 A in 0.02 s, how large an emf is induced in the coil?
    (b) Suppose a secondary coil of 60 turns is wound onto the same core. How large an emf is induced in it during the current change in (a)?
    (c) What is the mutual inductance of the two-coil system?

29

Alternating-Current Circuits

Terms

Define or describe briefly what is meant by the following terms. If you have difficulty, refer to the textbook section given in parentheses.

ac generator (29.1)

cycle (29.1)

peak voltage (29.1)

peak current (29.2)

root-mean-square current (29.3)

impedance (29.4, 29.7)*

inductive reactance (29.4)

capacitive reactance (29.5)

phase angle (29.6)

RLC series circuit (29.7)

power factor (29.8)

resonant frequency (29.9)

Equation Review

For each equation, be able to state the situation to which it applies, what quantity each symbol represents, and the units for measuring each quantity in some consistent set of units.

$$v = v_0 \sin (2\pi ft) \tag{29.1}$$

$$i = i_0 \sin (2\pi ft) \tag{29.2}$$

$$p = [i_0^2 \sin^2(2\pi ft)]R = i^2 R \tag{29.3}$$

$$I_{rms} = 0.707 i_0 \tag{29.4a}$$

$$V_{rms} = 0.707 v_0 \tag{29.4b}$$

$$X_L = 2\pi f L \tag{29.6}$$

$$I = \frac{V}{X_L} \tag{29.7}$$

$$X_C = \frac{1}{2\pi f C} \tag{29.8}$$

$$I = \frac{V}{X_C} \tag{29.9}$$

$$V = \sqrt{V_R^2 + (V_L - V_C)^2} \tag{29.10}$$

$$\tan \phi = \frac{V_L - V_C}{V_R} = \frac{X_L - X_C}{R} \tag{29.11}$$

$$Z = \sqrt{R^2 + (X_L - X_C)^2} \tag{29.12}$$

$$I = \frac{V}{Z} \tag{29.13}$$

$$P = VI \cos \phi \tag{29.14}$$

$$f_0 = \frac{1}{2\pi \sqrt{LC}} \tag{29.16}$$

Problems with Solutions and Discussion

PROBLEM: An ac generator produces a voltage given by $v = 300 \sin 377t$, where v is in volts and t in seconds. (a) What is the peak voltage produced by this generator? (b) What is its frequency? (c) If the generator delivers a peak current of 200 mA to a load resistor, what is its resistance?

Solution: (a) The general expression for the voltage as a function of time for an ac generator can be written as $v = v_0 \sin (2\pi f t)$. Comparing this with the above expression, we can identify 300 V as the peak voltage, or the amplitude of the voltage function.

(b) Further comparison of the two expressions above reveals that the argument of the sine is

$$2\pi f t = 377t$$
$$f = \frac{377}{2\pi} = 60 \text{ Hz}$$

(c) The peak current and peak voltage are related by Ohm's law, $i_0 = v_0/R$, where $i_0 = 200 \text{ mA} = 0.2 \text{ A}$. Therefore,

$$R = \frac{v_0}{i_0} = \frac{300 \text{ V}}{0.2 \text{ A}} = 1500 \text{ }\Omega$$

The load resistor has a resistance of 1500 ohms.

PROBLEM: For the ac generator above, determine the voltage and current in the resistor at $t = 0.02$ s.

Solution: Voltage as a function of time is given by $v = 300 \sin 377t$. Technically this is the generator voltage, but if the resistor is the only other circuit element (and we neglect the resistance of connecting wires as usual), it is also the voltage drop over the load resistor.

207

At $t = 0.02$ s,

$$v = 300 \sin [377(0.02)] = 300 \sin 7.54$$

Note that since f has units of s^{-1} and t has units of s, the quantity $2\pi ft$ is *unitless*, which indicates it is an angle *measured in radians*. The sine of 7.54 *radians* is 0.951. Thus,

$$v = 300(0.951) = \underline{285 \text{ V}}$$

The current through the resistor may be obtained from Ohm's law,

$$i = \frac{v}{R} = \frac{285 \text{ V}}{1500 \text{ } \Omega} = \underline{0.19 \text{ A}}$$

Alternatively, we may note that if voltage varies sinusoidally, the current through a resistance does also, with the same frequency and in phase:

$$i = i_0 \sin 2\pi ft$$

where $i_0 = 0.2$ A from the previous problem. So at $t = 0.02$ s,

$$i = (0.2 \text{ A})(\sin 7.54) = (0.2 \text{ A})(0.951) = \underline{0.19 \text{ A}}$$

PROBLEM: An ac source supplies a peak voltage of 200 V to a 100-Ω resistor. What average power is dissipated by the resistor as thermal energy?

Solution: The power dissipated in a resistor is given by $P = I^2R = V^2/R = VI$, provided we understood that for ac circuits, I and V are the *rms values* of these quantities. The rms voltage of the source is given by

$$V_{rms} = V = 0.707 v_0 = 0.707(200 \text{ V}) = 141 \text{ V}$$

Thus the power dissipated is

$$P = \frac{V^2}{R} = \frac{(141 \text{ V})^2}{100} = \underline{199 \text{ W}}$$

PROBLEM: A travel iron is rated at 600 watts when used with a 120-V ac line. (a) What peak current flows through it? (b) What peak current would flow through it if connected to a 240-V line? (c) In which case is more power utilized? (Assume that the resistance of the iron is constant.)

Solution: (a) Current, voltage, and power are related by $P = VI$ for a resistor, so

$$I = \frac{P}{V} = \frac{600 \text{ W}}{120 \text{ V}} = 5 \text{ A}$$

Recall, however, that for ac circuits, these are rms quantities, so 5 A is the rms current. It is related to i_0, the peak current, by Eq. (29.4):

$$I = 0.707 i_0$$

Solving for i_0,

$$i_0 = \frac{I}{0.707} = \frac{5 \text{ A}}{0.707} = \underline{7.1 \text{ A}}$$

(b) If connected to a 240-V line, the power drawn is no longer 600 watts; that designation assumed a 120-V line. We shall need the resistance of the iron to determine the new current. Ohm's law for the 120-V case gives

$$R = \frac{V}{I} = \frac{120 \text{ V}}{5 \text{ A}} = 24 \text{ } \Omega$$

Therefore, when connected to the 240-V line,

$$I = \frac{V}{R} = \frac{240 \text{ V}}{24 \text{ } \Omega} = 10 \text{ A}$$

A 10-A current is drawn; again this is the rms current. The peak current is

$$i_0 = \frac{I}{0.707} = \frac{10 \text{ A}}{0.707} = \underline{14 \text{ A}}$$

(c) In the first case the power utilized is given as 600 watts. In the second case the power is

$$P = VI = (240 \text{ V})(10 \text{ A}) = 2400 \text{ watts}$$

The power drawn is 4 times as great when connected to the 240-V line as it is for the 120-V line connection. This can also be seen from the relationship $P = I^2R$. If R remains constant, when V doubles, so does I; and when I doubles, the power dissipated increases as I^2, so it quadruples.

PROBLEM: At what frequency will the capacitive reactance of a 30-μF capacitor be the same as the inductive reactance of a 30-mH coil?

Solution: We want the frequency at which $X_C = X_L$. These reactances depend on frequency as given by Eqs. (29.8) and (29.6) respectively. Using these expressions for the reactances gives

$$\frac{1}{2\pi fC} = 2\pi fL$$

$$(2\pi fC)(2\pi fL) = 1 \quad \text{so} \quad f^2 = \frac{1}{4\pi^2 LC}$$

$$f = \frac{1}{2\pi \sqrt{LC}} = \frac{1}{2\pi \sqrt{(30 \times 10^{-3} \text{ H})(30 \times 10^{-6} \text{ F})}} = \underline{168 \text{ Hz}}$$

At a frequency of 168 Hz, the resistance to current flow of each of these circuit elements is the same. (You may have noted that the equation we obtained for f is in fact the same as for the *resonant frequency* of an RLC circuit. At the resonant frequency, the current flowing in a series circuit is higher than at any other frequency. This occurs when inductor and capacitor offer identical resistances to the flow of current—but at differing times during the current cycle.)

PROBLEM: Consider a 450-Ω resistor, a coil with inductance of 2 H, and a 6-μF capacitor. They are connected in series across a 60-Hz, 120-V supply. Determine (a) the impedance of the circuit, (b) the current in the circuit, (c) the phase angle between the voltage and current, and (d) the rms voltage across each circuit element.

Solution: (a) The impedance Z is given by Eq. (29.12):

$$Z = \sqrt{R^2 + (X_L - X_C)^2}$$

X_L, the inductive reactance, is calculated from Eq. (29.6):

$$X_L = 2\pi fL = 2\pi(60 \text{ s}^{-1})(2 \text{ H}) = 754 \text{ }\Omega$$

X_C, the capacitive reactance, is calculated from Eq. (29.8):

$$X_C = \frac{1}{2\pi fC} = \frac{1}{2\pi(60 \text{ s}^{-1})(6 \times 10^{-6} \text{ F})} = 442 \text{ }\Omega$$

Therefore:

$$Z = \sqrt{(450)^2 + (754 - 442)^2} = \sqrt{(450)^2 + (312)^2} = \underline{548 \text{ }\Omega}$$

The series circuit has an impedance of 548 Ω.

(b) The current is obtained from Eq. (29.13), which is essentially Ohm's law with resistance replaced by impedance:

$$I = \frac{V}{Z} = \frac{120 \text{ V}}{548 \text{ }\Omega} = \underline{0.22 \text{ A}}$$

(c) The phase angle ϕ is related to the resistive properties of the circuit by Eq. (29.11)

$$\tan \phi = \frac{X_L - X_C}{R}$$

Before doing the calculation, we can note that the inductive reactance (754 Ω) is greater than the capacitive reactance (442 Ω), so the voltage across the coil (which leads the current by 90°) dominates the voltage across the capacitor (which lags the current by 90°). Therefore we predict a positive phase angle indicating that the voltage over the entire series combination leads the current. Substituting numerical values we get

$$\tan \phi = \frac{754 \ \Omega - 442 \ \Omega}{450 \ \Omega} = \frac{312}{450} = 0.693$$

$$\phi = \underline{34.7°}$$

As predicted, the voltage <u>leads</u> the current by about 35°.

(d) We use an Ohm's law type of equation to get the rms voltage over the various circuit elements.

Resistor: $V_R = IR = (0.22 \text{ A})(450 \ \Omega) = \underline{99 \text{ V}}$

Capacitor: $V_C = IX_C = (0.22 \text{ A})(442 \ \Omega) = \underline{97 \text{ V}}$

Inductor: $V_L = IX_L = (0.22 \text{ A})(754 \ \Omega) = \underline{166 \text{ V}}$

PROBLEM: A solenoid has a resistance of 10 Ω and an inductance of 160 mH. It is connected across an 80-V, 60-Hz power supply. (a) What current flows in the solenoid? (b) What is the voltage drop across it? (c) Determine the power it absorbs.

Solution: Even though the resistance and inductance reside in the same circuit element, the solenoid is treated as a resistance and inductance in series, as shown in Fig. 29.1a. The circuit is an RLC series circuit in which there is no capacitive reactance, or an RL circuit.

(a) To determine the current, we must first find the impedance from Eq. (29.12):

$$Z = \sqrt{R^2 + (X_L - X_C)^2}$$

where $X_L = 2\pi fL = 2\pi(60 \text{ s}^{-1})(0.16 \text{ H}) = 60 \ \Omega$, and $X_C = 0$. Thus,

$$Z = \sqrt{(10)^2 + (60)^2} = \sqrt{3700} = 61 \ \Omega$$

The current, by the Ohm's law analog for ac circuits, is

$$I = \frac{V}{Z} = \frac{80 \text{ V}}{61 \ \Omega} = \underline{1.31 \text{ A}}$$

(b) The voltage drop across the coil will be the voltage drop across R and L together, since it is impossible to separate them physically. The voltage across R is

$$V_R = IR = (1.31 \text{ A})(10 \ \Omega) = 13 \text{ V}$$

The voltage across L, from Eq. (29.7), is

$$V_L = IX_L = (1.31 \text{ A})(60 \ \Omega) = 79 \text{ V}$$

We cannot simply add the two voltages together because they are out of phase by 90° (and if we did, we would get $13 + 79 = 92$ V, an impossible result since the source supplies only 80 V). We use the vector method of addition (Fig. 29.1b), which results in Eq. (29.10):

$$V_{RL} = \sqrt{V_R^2 + V_L^2} = \sqrt{(13)^2 + (79)^2} = \underline{80 \text{ V}}$$

Of course, looking at a circuit diagram (Fig. 29.1c) in which the voltmeter is shown measuring the desired voltage, we see immediately that V_{RL} must be equal to the source voltage, 80 V. The above calculation was unnecessary. However, had there been other circuit elements besides the solenoid, such as a capacitor and another resistor, the above technique would be necessary to determine the voltage across the solenoid (including both its resistance and inductance) and that voltage would *not* be equal to the entire source voltage.

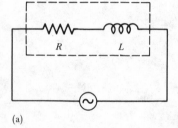

(a)

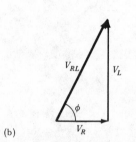

(b)

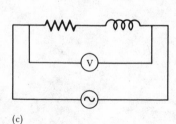

(c)

FIG. 29.1

210

(c) The power dissipated by the solenoid is given by Eq. (29.14):

$$P = VI \cos \phi$$

where ϕ is the phase angle between the voltage and the current. Since the current is in phase with V_R, we may determine ϕ from the vector diagram in part (b).

$$\cos \phi = \frac{V_R}{V_{RL}} = \frac{13 \text{ V}}{80 \text{ V}} = 0.1625$$

Thus,

$$P = (80 \text{ V})(1.3 \text{ A})(0.1625) = \underline{17 \text{ W}}$$

This result can also be obtained by considering only the resistance, since it alone dissipates power:

$$P = I^2R = (1.3 \text{ A})^2(10 \text{ }\Omega) = 17 \text{ W}$$

The solenoid dissipates 17 watts of power. As for the inductance, it absorbs power as the magnetic field within the solenoid builds up and returns the power to the circuit when the field collapses; there is no net power loss associated with this process.

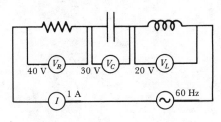

PROBLEM: Suppose you use an rms voltmeter to measure the voltages in the series RLC circuit shown in Fig. 29.2: $V_R = 40$ V; $V_C = 30$ V; $V_L = 20$ V. The current measured by the rms ammeter is 1 A. (a) Determine the source voltage and the phase angle between the voltage and the current. (b) Determine the values of R, C, and L.

FIG. 29.2

Solution: (a) The voltages in an ac circuit simply cannot be added together as scalars, because they are not in phase with one another. They may, however, be added using the vector method outlined in the text. We shall plot V_R on the $+x$ axis, V_L on the $+y$ axis since it leads V_R by 90°, and V_C on the $-y$ axis since it lags V_R by 90°. These vectors are shown in Fig. 29.3, followed by the vector addition of the three.

We find that

$$V = \sqrt{V_R^2 + V_{LC}^2} = \sqrt{(40)^2 + (-10)^2} = \underline{41 \text{ V}}$$

The voltage across all three elements is 41 V (not 90 V, which would result by simple addition). This 41 V is also the voltage of the source, since the circuit has no other elements. Referring to the last diagram, we can find the phase angle ϕ from:

$$\tan \phi = \frac{-10 \text{ V}}{40 \text{ V}} = -0.25$$

$$\phi = \underline{-14°}$$

The source voltage *lags* the current by 14°.

(b) We now obtain the values of the individual circuit elements.

Resistance: By Ohm's law, $I = \dfrac{V}{R}$, so

$$R = \frac{V_R}{I} = \frac{40 \text{ V}}{1 \text{ A}} = 40 \text{ }\Omega$$

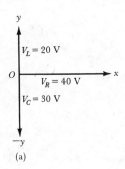

(a)

Capacitance: By Eq. (29.9), $I = \dfrac{V_C}{X_C} = \dfrac{V_C}{1/(2\pi fC)} = V_C(2\pi fC)$

Solve for C: $C = \dfrac{I}{2\pi fV_C} = \dfrac{1 \text{ A}}{2\pi(60)(30)} = 88 \times 10^{-6} \text{ F} = 88 \text{ }\mu\text{F}$

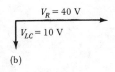

(b)

Inductance: By Eq. (29.7), $I = \dfrac{V_L}{X_L} = \dfrac{V_L}{2\pi fL}$

Solve for L: $L = \dfrac{V_L}{2\pi fI} = \dfrac{20}{2\pi(60)(1)} = 0.053 \text{ H} = 53 \text{ mH}$

As a check on our results, let us calculate the impedance and see if it produces the observed current when the source voltage is 41 V. The impedance is $Z = \sqrt{R^2 + (X_L - X_C)^2}$ where $X_L = 2\pi fL = 20 \text{ }\Omega$, and $X_C = 1/(2\pi fC) = 30 \text{ }\Omega$.

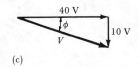

(c)

FIG. 29.3

211

$$Z = \sqrt{(40)^2 + (20 - 30)^2} = \sqrt{(40)^2 + (-10)^2} = 41 \ \Omega$$

Therefore

$$I = \frac{V}{Z} = \frac{41 \ V}{41 \ \Omega} = 1 \ A$$

Our results for the circuit elements and the source voltage are consistent with the measured current.

PROBLEM: In a certain RLC series circuit, the source emf of 100 V lags the current of 0.3 A by a phase angle of 40°. (a) What average power is delivered by the source? (b) What changes could be made to increase the power factor of this circuit?

Solution: (a) The average power delivered to an ac circuit is given by $VI \cos \phi$, where ϕ is the angle between the current and the source voltage, and $\cos \phi$ is called the power factor.

$$P = (100 \ V)(0.3 \ A) \cos (-40°) = (30)(0.766) = \underline{23 \ W}$$

By way of comparison, a 100-V dc source of emf delivers power to a circuit carrying 0.3 A of current at a constant rate of

$$P = VI = (100 \ V)(0.3 \ A) = 30 \ W$$

The power factor is the factor by which the power supplied to an ac circuit differs from that to the corresponding dc circuit. In this case the power factor is 0.766. Less power is delivered because the voltage and current are *not* in phase (they do not reach their maximum values at the same time).

(b) The power factor, $\cos \phi$, can be increased by reducing the phase angle ϕ. If $\phi = 0$, then $\cos \phi = 1$, the maximum possible power factor. How can $|\phi|$ be made smaller? Recall that

$$\tan \phi = \frac{X_L - X_C}{R}$$

and $\tan \phi$ increases as ϕ increases. Recall also that in this circuit, $\phi = -40°$ (voltage *lags* the current), so X_C is larger than X_L. To bring ϕ to zero, we wish to have $X_L = X_C$. This can be achieved either by increasing X_L or decreasing X_C.

Since $X_L = 2\pi fL$, X_L can be made larger by increasing L or increasing f. Since $X_C = 1/(2\pi fC)$, X_C can be made smaller by increasing C or increasing f. Therefore, the power factor of this circuit can be increased in any of three ways; by increasing C, L, or f until $X_L = X_C$. (Further increases would unbalance the circuit again, making $X_L > X_C$, and once again introducing a power factor less than 1).

PROBLEM: Suppose you wish to make a resonating circuit. You have at your disposal a coil of 5-mH inductance and a 20-μF capacitor. What resonant frequency can be obtained?

Solution: The resonant frequency of an RLC circuit is given by Eq. (29.16):

$$f_0 = \frac{1}{2\pi \sqrt{LC}}$$

For $L = 5 \ mH = 5 \times 10^{-3} \ H$, and $C = 20 \ \mu F = 20 \times 10^{-6} \ F$, then

$$f_0 = \frac{1}{2\pi \sqrt{(5 \times 10^{-3} \ H)(20 \times 10^{-6} \ F)}} = \frac{1}{2\pi \sqrt{10^{-7}}} = \underline{500 \ Hz}$$

The resonant frequency of this circuit is 500 Hz.

Avoiding Pitfalls

1. The current in an ac series circuit varies sinusoidally with the same frequency as the source of emf, but it is in general *not* in phase with the varying voltage of the source of emf.

2. It is common to use rms values in discussing ac circuits, since it is the *average* power utilization that is of interest in most situations. The rms value A of any quantity is related to its peak value a_0 by $A = 0.707a_0$.

3. Ohm's law and equations derived using Ohm's law are applicable to ac circuit elements other than resistors, provided resistance is replaced by reactance and the quantities in each equation are consistent (all rms values, or all peak values, etc.).

4. Resistance, capacitive reactance, inductive reactance, and impedance all have units of *ohms*.

5. Whereas the resistance of circuit elements is independent of the frequency of the ac current, the capacitive and inductive reactances of circuit elements depend very strongly on frequency and no one value can be assigned to them. They must be individually calculated for a given frequency.

6. Capacitors offer *low* resistance to the flow of *high*-frequency current in an ac circuit; inductors offer *high* resistance to *high*-frequency currents. At *low* frequencies, capacitors offer *high* resistance to current flow; inductors offer *low* resistance.

7. The rms voltages across the individual elements in an ac series circuit *do not* in general add up to the rms voltage of the source, because the individual voltages are not in phase with each other. It is correct (but very cumbersome!) to add instantaneous voltages algebraically. A method using the techniques of *vector addition* (outlined in the text) is a convenient way to correctly combine the rms voltages across the various circuit elements.

8. For a resistor, voltage and current are in phase. They are not in phase for capacitors and inductors. For an inductor, the voltage leads the current by 90°; for a capacitor, the voltage lags the current by 90°.

9. The power used by an ac circuit is *not* given by the dc formula $P = VI$, because the voltage and current are *not* necessarily in phase. If the phase angle between them is ϕ, the power is given by $P = VI \cos \phi$. If they are in phase, this expression reduces to the dc expression for power.

10. Since the voltage across an inductor is 180° out of phase with the voltage across a capacitor, the resistive effects of these two elements tend to cancel one another. When this cancellation is complete, the circuit is said to be *in resonance*.

11. The reactance (resistance to the flow of alternating current) of inductors and capacitors is a function of frequency, so resonance for a given circuit occurs at a particular *resonant frequency*.

12. The resonant *frequency* of an RLC series circuit depends only on L and C, not on the resistance R in the circuit. The *current* flowing at resonance, however, depends on the resistance and is independent of L and C.

Drill Problems
Answers in Appendix

1. A series RLC circuit contains a 140-V, 60-Hz source of emf, a 1000-Ω resistor, a coil of 6-H inductance, and a capacitance that can be varied.

   (a) What value should C have in order for the source voltage to lead the current by 30°?
   (b) At this value of C, find the current in the circuit.
   (c) What is the voltage across the coil?

2. A 240-V, 60-Hz source of emf delivers 130 watts of power to a circuit of 300-Ω impedance.

(a) What is the power factor of this circuit?
(b) What is the peak voltage of the source of emf?

3. A series RLC circuit is connected to a 40-V, 200-Hz source of emf. R is 150 Ω and C is 20 μF.

(a) What must L be for the circuit to resonate at this frequency?
(b) If R, L, and C are all doubled, what will be the resonant frequency of the circuit?
(c) What is the current in the circuit in part (a) at resonance?

214

Special Relativity

30

Terms

Define or describe briefly what is meant by the following terms. If you have difficulty, refer to the textbook section given in parentheses.

inertial reference frame (30.1)

postulates of special relativity (30.2)

ether (30.2)

time dilation (30.3)

proper reference frame (30.3)

proper time (30.3)

length contraction (30.4)

proper length (30.4)

relativistic mass (30.5)

rest mass (30.5)

rest-mass energy (30.6)

relativistic kinetic energy (30.6)

positron (30.7)

pair production (30.7)

Equation Review

For each equation, be able to state the situation to which it applies, what quantity each symbol represents, and the units for measuring each quantity in some consistent set of units.

$$\Delta t = \frac{\Delta t_0}{\left(1 - \dfrac{v^2}{c^2}\right)^{1/2}} \tag{30.1}$$

$$L = L_0 \left(1 - \frac{v^2}{c^2} \right)^{1/2} \qquad\qquad (30.3)$$

$$m = \frac{m_0}{\left(1 - \frac{v^2}{c^2} \right)^{1/2}} \qquad\qquad (30.4)$$

$$E = mc^2 \qquad\qquad (30.5)$$

$$E_0 = m_0 c^2 \qquad\qquad (30.6)$$

$$KE = E - E_0 = mc^2 - m_0 c^2 \qquad\qquad (30.7)$$

Problems with Solutions and Discussion

PROBLEM: A light on your spaceship flashes. Then, 24 hours later (according to a clock on the spaceship) it flashes again. How fast is the ship moving relative to the earth if an earth observer measures 96 hours between flashes?

Solution: The time between flashes measured by you in your spaceship is the proper time Δt_0. The corresponding time measured by an earth observer who is in motion relative to your spaceship is Δt. The relationship between the time intervals is given by Eq. (30.1):

$$\Delta t = \frac{\Delta t_0}{\left(1 - \frac{v^2}{c^2} \right)^{1/2}}$$

where Δt_0 = proper time you measure on your spaceship = 24 hr = 1 day, and Δt = corresponding "dilated" time on earth = 96 hr = 4 days.

$$4 \text{ days} = \frac{1 \text{ day}}{\left(1 - \frac{v^2}{c^2} \right)^{1/2}}$$

$$\left(1 - \frac{v^2}{c^2} \right)^{1/2} = \frac{1 \text{ day}}{4 \text{ days}}$$

Squaring,

$$1 - \frac{v^2}{c^2} = \frac{1}{16}$$

Solving for v,

$$v^2 = {}^{15}\!/\!_{16} c^2$$

$$v = \frac{\sqrt{15}}{4} c = \underline{0.968c}$$

Your spaceship travels with a speed of $0.968c$ relative to the earth. Or we may say that the earth travels with a speed of $0.968c$ relative to your spaceship.

PROBLEM: Traveling in your spaceship that is 100 m long, you pass your friend in an identical spaceship traveling in the opposite direction at a velocity of $0.8c$ relative to you. (a) How long do you think his spaceship is? (b) How long does he think your spaceship is?

Solution: (a) His spaceship, being identical to yours, has a length of 100 m when measured in its own reference frame. (This would therefore be a *proper length, L_0.*) Its length when in motion with respect to an observer is contracted according to Eq. (30.3):

$$L = L_0 \left(1 - \frac{v^2}{c^2} \right)^{1/2}$$

where v is the relative velocity between object and observer, in this case $0.8c$.

$$L = (100 \text{ m}) \left(1 - \frac{(0.8c)^2}{c^2} \right)^{1/2} = (100 \text{ m})(1 - 0.64)^{1/2}$$

$$= (100 \text{ m})(0.36)^{1/2} = (100 \text{ m})(0.6) = \underline{60 \text{ m}}$$

His spaceship will appear to you to be 60 m long.

(b) The proper length of your spaceship is 100 m; he sees you passing by him at a relative velocity of $0.8c$. He will therefore see a length that is shorter than the length of your spaceship at rest, according to the equation

$$L = L_0 \left(1 - \frac{v^2}{c^2} \right)^{1/2}$$

The situation is identical to that described in (a) except that the direction of the relative velocity is reversed. Here $v = -0.8c$, since we took the first velocity as positive. But since v is squared in the length contraction equation, its negative sign has no effect on the calculation. The result is that he thinks your spaceship is $\underline{60 \text{ m}}$ long.

PROBLEM: Muons, whose average lifetime is 2.2×10^{-6} s, are formed when cosmic rays strike atmospheric nuclei high above the earth's surface, say 10,000 m or more. A typical speed for such a muon is $0.999c$. (a) Without considering relativistic effects, calculate how far down a muon could penetrate, if it decayed after 2.2×10^{-6} s. (b) Using relativistic ideas, determine the lifetime of the above muon as seen by an earth observer, and calculate how far the muon could travel in that time. (c) From the muon's point of view, the length of its path through the atmosphere will be contracted because it is moving at a tremendous speed. How long will it take the muon to traverse the atmosphere?

Solution: (a) At a speed of $0.999c$, the distance traveled by the muon in 2.2×10^{-6} s is

$$x = vt = (0.999)(3 \times 10^8 \text{ m/s})(2.2 \times 10^{-6} \text{ s}) = \underline{660 \text{ m}}$$

Using classical physics, it appears that this muon cannot reach the earth's surface; it decays at a depth of 660 m, still more than 9000 m above the ground.

(b) The muon's lifetime of 2.2×10^{-6} s is a proper time interval Δt_0 because it is measured in the muon's own reference frame. The dilated time as seen by an earth observer moving with a speed of $0.999c$ relative to the muon is given by Eq. (30.1):

$$\Delta t = \frac{\Delta t_0}{\left(1 - \frac{v^2}{c^2} \right)^{1/2}} = \frac{2.2 \times 10^{-6} \text{ s}}{\left(1 - \frac{(0.999c)^2}{c^2} \right)^{1/2}} = \frac{2.2 \times 10^{-6} \text{ s}}{0.045} = \underline{4.9 \times 10^{-5} \text{ s}}$$

During this time the muon can travel a distance of

$$x = vt = (0.999)(3 \times 10^8 \text{ m/s})(4.9 \times 10^{-5} \text{ s}) = \underline{14{,}700 \text{ m}}$$

Since the atmosphere's depth is only about 10,000 m, the muon has no difficulty in reaching the earth's surface before it decays.

(c) In the muon's frame, its lifetime is 2.2×10^{-6} s, but the length of its path through the atmosphere is contracted relative to the proper length of 10,000 m measured by the earth observer. By Eq. (30.3) the contracted length is

$$L = L_0 \left(1 - \frac{v^2}{c^2} \right)^{1/2} = (10{,}000 \text{ m}) \left(1 - \frac{(0.999c)^2}{c^2} \right)^{1/2}$$

$$= (10{,}000 \text{ m})(0.045) = 450 \text{ m}$$

The muon can travel this distance in

$$t = \frac{x}{v} = \frac{450 \text{ m}}{(0.999)(3 \times 10^8 \text{ m/s})} = \underline{1.5 \times 10^{-6} \text{ s}}$$

With a lifetime of 2.2×10^{-6} s, the muon has plenty of time to reach the earth's surface.

Notice that from both the earth observer's point of view (part b) and the muon's point of view (part c), the muon can reach the earth's surface if relativistic effects are taken into account. An abundance of cosmic ray muons do in fact reach the earth's surface, and physicists using newtonian kinematics were unable to explain how this could occur. The success of relativistic kinematics in explaining this cosmic ray phenomenon was one of the earliest experimental confirmations of the theory of relativity.

PROBLEM: (a) If you were riding on a train traveling at 100 mph next to another train traveling at 100 mph in the same direction, what speed would the other train appear to have, relative to you? (b) If you were traveling at the speed of light (assume you are massless) next to a beam of light, what speed would the beam have relative to you?

Solution: (a) Since both trains have the same speed and direction, the train next to you would appear to be at rest with respect to you; the relative velocity is <u>zero</u>.

(b) It is very tempting, particularly with the analogy of the two trains close at hand, to conclude that the beam would appear to be at rest with respect to you, since your speed is also *c*. *However,* this result is in direct contradiction to Einstein's second postulate of relativity. That postulate states that the speed of light (in a vacuum) is always measured to be *c* independent of the state of motion of source or observer. In other words, *all* observers, even ones traveling at speed *c*, will see light moving at a speed of 3×10^8 m/s. The beam will still appear to move away from you at speed *c*.

This result defies our intuition about what should happen in this case. Certainly water waves and other waves which require a material medium do not exhibit this behavior. But in the case of light (which does *not* require a material medium) our intuition fails us. We may feel moved, as did many scientists in the past, to reject the special theory of relativity because the behavior of light is too strange for us to accept. But before doing this, we should at least admit that we have never traveled at relativistic speeds, and we therefore really have no basis to predict what could or could not happen at those speeds.

PROBLEM: Consider an electron traveling at the following speeds: $0.01c$, $0.1c$, $0.5c$, $0.9c$, and $0.99c$. For each speed compare the electron's classical kinetic energy and its relativistic kinetic energy.

Solution: We will work through in detail the case of $v = 0.5c$; the method of calculation is identical for the other cases.

Classically:

$$KE = \tfrac{1}{2}mv^2$$

For $v = 0.5c = 0.5(3 \times 10^8 \text{ m/s}) = 1.5 \times 10^8$ m/s, and for electron mass $m = 9.1 \times 10^{-31}$ kg,

$$KE = \tfrac{1}{2}(9.1 \times 10^{-31} \text{ kg})(1.5 \times 10^8 \text{ m/s})^2 = \underline{1.02 \times 10^{-14} \text{ J}}$$

Relativistically:

$$KE = mc^2 - m_0c^2$$

where

$$m_0 = \text{electron rest mass} = 9.1 \times 10^{-31} \text{ kg};$$

and

$$m = \text{relativistic mass} = \frac{m_0}{\left(1 - \dfrac{v^2}{c^2}\right)^{1/2}}$$

218

$$KE = \frac{m_0 c^2}{\left(1 - \dfrac{v^2}{c^2}\right)^{1/2}} - m_0 c^2 = m_0 c^2 \left[\frac{1}{\left(1 - \dfrac{v^2}{c^2}\right)^{1/2}} - 1\right]$$

$$= (9.1 \times 10^{-31}\ \text{kg})(3 \times 10^8\ \text{m/s})^2 \left[\frac{1}{\left(1 - \dfrac{(0.5c)^2}{c^2}\right)^{1/2}} - 1\right]$$

$$= (8.19 \times 10^{-14}\ \text{J}) \left[\frac{1}{(0.75)^{1/2}} - 1\right]$$

$$= (8.19 \times 10^{-14}\ \text{J})(1.155 - 1) = \underline{1.27 \times 10^{-14}\ \text{J}}$$

Comparing the two results, we get

$$\frac{KE_{\text{rel}}}{KE_{\text{cl}}} = \frac{1.27 \times 10^{-14}\ \text{J}}{1.02 \times 10^{-14}\ \text{J}} = 1.25$$

These results and the results for the other speeds are shown in the table below.

v	KE_{cl}	KE_{rel}	$KE_{\text{rel}}/KE_{\text{cl}}$
$0.01c$	4.10×10^{-18} J	4.10×10^{-18} J	1.00
$0.1c$	4.10×10^{-16} J	4.13×10^{-16} J	1.01
$0.5c$	1.02×10^{-14} J	1.27×10^{-14} J	1.25
$0.9c$	3.32×10^{-14} J	1.06×10^{-13} J	3.19
$0.99c$	4.01×10^{-14} J	4.99×10^{-13} J	12.4
c	4.10×10^{-14} J	∞	∞

As the electron speed increases, the discrepancy between the classical and relativistic kinetic energy becomes greater and greater. At relativistic speeds the kinetic energy is increasing very rapidly. At $0.99c$, it is more than 12 times what classical physics predicts. This means it takes over 12 times as much work as we would have expected to accelerate the electron from rest to that speed. More work is required because the mass is becoming larger and larger. To accelerate the electron to the speed of light requires an infinite amount of work because the mass becomes infinitely large. Thus the speed limit imposed by nature—no material particle can be accelerated to the speed of light.

PROBLEM: At what speed is an object moving if its kinetic energy is twice its rest mass energy?

Solution: We are given that $KE = 2m_0 c^2$, where the rest mass energy is defined by Eq. (30.6), $E_0 = m_0 c^2$. The kinetic energy for relativistic cases is given by Eq. (30.7):

$$KE = mc^2 - m_0 c^2$$

Substituting the KE value gives:

$$mc^2 - m_0 c^2 = 2m_0 c^2$$
$$mc^2 = 3m_0 c^2$$

Putting in the expression for relativistic mass in terms of rest mass from Eq. (30.4) gives

$$\frac{m_0 c^2}{\left(1 - \dfrac{v^2}{c^2}\right)^{1/2}} = 3m_0 c^2 \quad \text{or} \quad \left(1 - \frac{v^2}{c^2}\right)^{1/2} = \frac{1}{3}$$

Squaring we get

$$1 - \frac{v^2}{c^2} = \frac{1}{.9} = 0.111$$

$$\frac{v^2}{c^2} = 1 - 0.111 = 0.899 \quad \text{so} \quad v^2 = 0.899c^2$$

$$v = 0.943c = 0.943(3 \times 10^8 \text{ m/s}) = \underline{2.83 \times 10^8 \text{ m/s}}$$

At a speed of $0.943c$, or 2.8×10^8 m/s, an object's kinetic energy is twice its rest mass energy.

PROBLEM: (a) What is the relativistic mass of a positron accelerated from rest across a potential difference of 80,000 V? (b) What is its final speed?

Solution: (a) Recall that a positron has the same rest mass as an electron and a charge $+e$. Thus, $m_0 = 9.1 \times 10^{-31}$ kg, and $q = 1.6 \times 10^{-19}$ C. Energy conservation predicts that the electrical potential energy loss equals the kinetic energy gain, or

$$KE \text{ gained} = \Delta PE_q \quad \text{where} \quad \Delta PE_q = q \, \Delta V$$

Thus,

$$KE = q \, \Delta V$$

Using the relativistic expression for kinetic energy given by Eq. (30.7) gives

$$mc^2 - m_0 c^2 = q \, \Delta V$$

Solving for m, the relativistic mass, we get

$$m = \frac{q \, \Delta V + m_0 c^2}{c^2} = \frac{q \, \Delta V}{c^2} + m_0$$

Substituting numerical values gives

$$m = \frac{(1.6 \times 10^{-19} \text{ C})(80,000 \text{ V})}{(3 \times 10^8 \text{ m/s})^2} + 9.1 \times 10^{-31} \text{ kg}$$

$$m = 1.42 \times 10^{-31} \text{ kg} + 9.1 \times 10^{-31} \text{ kg} = \underline{10.5 \times 10^{-31} \text{ kg}}$$

The relativistic mass is larger than the rest mass by $\frac{1.42}{9.1} \simeq 0.16$; that is, 16 percent larger.

(b) The speed can be extracted from Eq. (30.4) relating rest mass and relativistic mass. Rearranging we get

$$\left(1 - \frac{v^2}{c^2}\right)^{1/2} = \frac{m_0}{m} = \frac{9.1 \times 10^{-31} \text{ kg}}{10.5 \times 10^{-31} \text{ kg}} = 0.867$$

Squaring gives

$$1 - \frac{v^2}{c^2} = 0.752 \quad \text{so} \quad v^2 = (1 - 0.752)c^2 = 0.248c^2$$

$$v = 0.498c \simeq \underline{0.5c}$$

The positron's speed is about 1.5×10^8 m/s.

PROBLEM: In the process called *pair production,* a gamma ray disappears in the presence of a nearby atom, and in its place an electron and a positron appear. How much energy must the gamma ray have to be able to produce this pair of particles? Assume they are at rest when produced.

Solution: Energy conservation requires that the gamma ray supply an energy equal to the rest mass energy of the two particles.

$$E = 2m_0 c^2$$

where $m_0 = 9.1 \times 10^{-31}$ kg = rest mass of both the electron and the positron.

$$E = 2(9.1 \times 10^{-31} \text{ kg})(3 \times 10^8 \text{ m/s})^2 = \underline{1.64 \times 10^{-13} \text{ J}}$$

If the gamma ray does not have this much energy it cannot spontaneously create an electron-positron pair. If it is more energetic than this, it can create the pair and also endow them and the nearby atom with some kinetic energy.

PROBLEM: The earth receives energy from the sun at the rate of 1.35×10^3 W/m^2. The earth-sun distance is 1.5×10^{11} m. (a) From this, deduce the rate at which the sun is losing mass in the form of radiant energy. (b) What percent of its total mass of 2×10^{30} kg does it lose each year? (1 year = 3.16×10^7 s.)

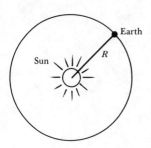

Solution: (a) The sun's energy is radiated out in all directions (Fig. 30.1). The earth receives only a small fraction of that energy. If one square meter, at a distance of $R = 1.5 \times 10^{11}$ m, receives 1.35×10^3 W of power, all other square meters on the surface of a sphere of radius R receive the same amount. The total energy output per unit time from the sun must be

FIG. 30.1

$$\frac{E}{t} = \text{rate per unit area} \times \text{surface area of sphere of radius } R$$

$$\frac{E}{t} = \left(1.35 \times 10^3 \frac{\text{W}}{\text{m}^2} \right) (4\pi R^2)$$

(since the surface area of a sphere is $4\pi R^2$). Therefore,

$$\frac{E}{t} = \left(1.35 \times 10^3 \frac{\text{W}}{\text{m}^2} \right) (4\pi)(1.5 \times 10^{11} \text{ m})^2 = 3.8 \times 10^{26} \text{ W}$$

Each second 3.8×10^{26} joules of energy must be produced. By Einsten's famous relationship, $E = mc^2$, that energy can result from the conversion of mass. Dividing this by t gives

$$\frac{E}{t} = \frac{m}{t} c^2$$

Solving for the rate at which the sun loses mass, we get

$$\frac{m}{t} = \frac{E/t}{c^2} = \frac{3.8 \times 10^{26} \text{ J/s}}{(3 \times 10^8 \text{ m/s})^2} = \underline{4.2 \times 10^9 \text{ kg/s}}$$

The sun is losing mass at the rate of 4.2×10^9 kilograms per second.

(b) In one year, the mass loss of the sun is

$$\left(4.2 \times 10^9 \frac{\text{kg}}{\text{s}} \right) \left(3.16 \times 10^7 \frac{\text{s}}{\text{yr}} \right) = 1.33 \times 10^{17} \text{ kg}$$

The percentage of the total solar mass that this yearly loss represents is

$$\frac{1.33 \times 10^{17} \text{ kg}}{2 \times 10^{30} \text{ kg}} = 6.65 \times 10^{-14} \simeq \underline{6.7 \times 10^{-12} \text{ percent}}$$

The yearly mass loss, which seems such a large amount, is in fact infinitesimal compared to the sun's total mass. We can expect the sun to be around for quite some time! And yet a very small fraction of this infinitesimal loss supplies virtually all the energy necessary to sustain life on earth.

Avoiding Pitfalls

1. Relativistic formulas should be used when objects move with speeds in excess of about one-tenth the speed of light.

2. It is easy to reverse Δt and Δt_0 in the time dilation formula. Remember that the person measuring an interval of time between events occuring *in his own reference frame* (one

moving at the same speed as he is) measures the *proper time* Δt_0, and that this time interval is *shorter* than that measured by any other observer in motion with respect to him.

3. It is easy to reverse L and L_0 in the length contraction formula. Remember that the observer at rest with respect to the object measures its *proper length* L_0, and that this length is *greater* than that measured by any other observer in motion with respect to the object.

4. Pay particular attention to carrying through the mathematical details when using relativistic formulas. With all the squares and square roots, it is so easy to obtain an incorrect result because of a simple algebraic error.

5. Work done on an object increases its kinetic energy. Classically, this meant the object's speed increased. But at relativistic speeds, the increase in kinetic energy manifests itself as an increase in both the speed *and mass* of the object.

6. The equation $E = mc^2$ is used in two quite different situations. One is to calculate the total energy of a body whose relativistic mass is m. The second is to calculate the energy release corresponding to the disappearance of a small amount of mass. You may find it helpful to write $E = \Delta mc^2$ in such cases, as a reminder that usually only a very small fraction of the total mass of the object in question is converted to energy.

7. Up to this point we have conserved energy and calculated a system's gains and losses of energy without visualizing any *substance* to the energy. With the advent of Einstein's equation, $E = mc^2$, we may now envision energy gains and losses as tiny increases and decreases in a system's mass.

Drill Problems
Answers in Appendix

1. A girl decides that it is time to slenderize. She has heard that if she moves fast enough, she will appear thinner to her stationary friends.

   (a) How fast must she move to appear thinner by a factor of 50 percent?
   (b) At this speed, she observes by a watch on her wrist that a time interval of 8 minutes has elapsed. How long would her stationary friends measure this time interval to be?

2. Select the *one best answer* for the following multiple choice questions:

   (a) If you were to travel at speeds close to the speed of light, you would notice that *your:* (1) mass had increased; (2) pulse rate had decreased; (3) shape had undergone a distortion; (4) all of the above; (5) none of the above.
   (b) A person traveling at a relativistic speed past a stationary observer appears to: (1) have a shorter life; (2) expand; (3) be more massive; (4) all of the above; (5) none of the above.

(c) If you were massless and traveling at the speed of light next to a beam of light, the beam would appear to: (1) be at relative rest, traveling beside you at an equal speed c; (2) still travel away from you, but at a speed less than c; (3) travel away from you at speed c.

(d) In relativity theory, the ratio of relativistic mass to rest mass is such that it must equal: (1) zero for low speeds, approaching one for high speeds; (2) zero for low speeds, approaching infinity for high speeds; (3) one for low speeds, approaching infinity for high speeds; (4) one for any speed.

3. (a) What electrical potential difference would be required to give an electron, accelerated from rest, a speed of $0.8c$? (b) At that speed, what is the ratio of the electron's kinetic energy to its rest mass energy?

31

Wave-Particle Duality

Terms

Define or describe briefly what is meant by the following terms. If you have difficulty, refer to the textbook section given in parentheses.

electromagnetic wave (31.1)

electromagnetic spectrum (31.2

photon (31.3)

Planck's constant (31.3)

electron volt (31.3)

work function (31.4)

photoelectric current (31.4)

photomultiplier tube (31.4)

Compton effect (31.4)

wave-particle duality (31.5)

de Broglie wavelength (31.6)

uncertainty principle (31.7)

electron microscope (31.8)

Equation Review

For each equation, be able to state the situation to which it applies, what quantity each symbol represents, and the units for measuring each quantity in some consistent set of units.

$$\lambda = \frac{c}{f} \tag{17.8}$$

$$E = hf \tag{31.1}$$

$$1 \text{ eV} = 1.6 \times 10^{-19} \text{ J} \tag{31.2}$$

$$-hf + \phi + \tfrac{1}{2}mv^2 = 0 \tag{31.3}$$

$$-hf_0 + \phi = 0 \tag{31.4}$$

$$\lambda = \frac{h}{mv} \tag{31.6}$$

$$\Delta x\, \Delta p \geqslant \frac{h}{2\pi} \tag{31.7}$$

$$\Delta E\, \Delta t \geqslant \frac{h}{2\pi} \tag{31.8}$$

Problems with Solutions and Discussion

PROBLEM: A typical laser produces a beam of light of wavelength 634 nm at a power of 2 milliwatts. How many photons pass through the cross-sectional area of the beam each second?

Solution: Each photon has an energy determined by Eq. (31.1):

$$E = hf = \frac{hc}{\lambda}$$

where $\lambda = 634$ nm $= 634 \times 10^{-9}$ m. Therefore the energy per photon is

$$E = \frac{(6.63 \times 10^{-34}\ \text{J} \cdot \text{s})(3 \times 10^8\ \text{m/s})}{634 \times 10^{-9}\ \text{m}} = 3.14 \times 10^{-19}\ \text{J}$$

The power output of the laser is 2 mW $= 2 \times 10^{-3}$ W $= 2 \times 10^{-3}$ J/s. Therefore, the number of photons generated by the laser and passing through the cross-sectional area of the beam each second is

$$\frac{2 \times 10^{-3}\ \text{J/s}}{3.14 \times 10^{-19}\ \text{J/photon}} = \underline{6.4 \times 10^{15}\ \text{photons/s}}$$

PROBLEM: Consider light from a source radiating 100 watts of power at a frequency of 5×10^{14} Hz. (a) Calculate the energy and momentum of each photon of this radiation. (b) How many photons per second are emitted from this source?

Solution: (a) The energy of a photon is given by Eq. (31.1):

$$E = hf = (6.63 \times 10^{-34}\ \text{J} \cdot \text{s})(5 \times 10^{14}\ \text{s}^{-1}) = \underline{3.3 \times 10^{-19}\ \text{J}}$$

The momentum of a photon is given by

$$p = \frac{h}{\lambda}, \quad \text{where} \quad \lambda = \frac{c}{f} = \frac{3 \times 10^8\ \text{m/s}}{5 \times 10^{14}\ \text{s}^{-1}} = 6 \times 10^{-7}\ \text{m} = 600\ \text{nm}$$

$$p = \frac{6.63 \times 10^{-34}\ \text{J} \cdot \text{s}}{6 \times 10^{-7}\ \text{m}} = \underline{1.1 \times 10^{-27}\ \text{kg} \cdot \text{m/s}}$$

(b) The number of photons per second is the energy per second divided by the energy per photon,

$$\frac{\text{photons}}{\text{s}} = \frac{100\ \text{J/s}}{3.3 \times 10^{-19}\ \text{J/photon}} = \underline{3 \times 10^{20}\ \text{photons/s}}$$

PROBLEM: (a) For the photons of the last problem, use Einstein's relationship between mass and energy to determine the *relativistic mass* of each photon. (b) Use the relationship between rest mass and relativistic mass to determine the *rest mass* of one of these photons.

Solution: (a) For a photon, $E = hf$. But according to Einstein, this energy is equivalent to a certain mass, by the relationship $E = mc^2$. Equating these two expressions for E gives

$$mc^2 = hf$$

$$m = \frac{hf}{c^2} = \frac{(6.63 \times 10^{-34} \text{ J} \cdot \text{s})(5 \times 10^{14} \text{ s}^{-1})}{(3 \times 10^8 \text{ m/s})^2} = 3.7 \times 10^{-36} \frac{\text{J} \cdot \text{s}}{\text{m}^2/\text{s}}$$

The unit $\dfrac{\text{J} \cdot \text{s}^2}{\text{m}^2}$ can be analyzed by recalling that $1 \text{ J} = 1 \dfrac{\text{kg} \cdot \text{m}^2}{\text{s}^2}$. Substituting we get

$$\frac{(\text{kg m}^2/\text{s}^2)(\text{s}^2)}{\text{m}^2} = \text{kg}$$

As expected, the mass has units of kg. We find that $m = \underline{3.7 \times 10^{-36} \text{ kg}}$. The energy of each photon is the energy that would appear if 3.7×10^{-36} kg of mass were annihilated.

(b) The rest mass m_0 and the relativistic mass m are related by Eq. (30.4):

$$m = \frac{m_0}{(1 - v^2/c^2)^{1/2}}$$

Solving for m_0 gives

$$m_0 = m \left(1 - \frac{v^2}{c^2} \right)^{1/2}$$

For a photon, $v = c$, so

$$m_0 = m \left(1 - \frac{c^2}{c^2} \right)^{1/2} = (3.7 \times 10^{-36} \text{ kg})(1 - 1)^{1/2} = 0$$

The photon's rest mass is <u>zero</u>, which is the expected result.

PROBLEM: Tungsten has a work function of 4.5 eV. What is the maximum kinetic energy of photoelectrons ejected from a tungsten surface by photons whose frequency is 2.0×10^{15} Hz?

Solution: The incoming photon energy hf is absorbed by an electron in the tungsten. Part of this energy is used to escape from the metal surface, and the rest is the kinetic energy of the free electron. This energy exchange is expressed by Eq. (31.3):

$$hf = \phi + KE \quad \text{where} \quad f = 2 \times 10^{15} \text{ Hz}$$

ϕ, the work function of the metal, is given in electron volts, so it must be converted to joules before substituting into the equation.

$$\phi = 4.5 \text{ eV} \left(\frac{1.6 \times 10^{-19} \text{ J}}{1 \text{ eV}} \right) = 7.2 \times 10^{-19} \text{ J}$$

Solving for KE and putting in numerical values, we get

$$KE = hf - \phi = (6.63 \times 10^{-34} \text{ J} \cdot \text{s})(2.0 \times 10^{15} \text{ s}^{-1}) - 7.2 \times 10^{-19} \text{ J}$$
$$= 13.3 \times 10^{-19} \text{ J} - 7.2 \times 10^{-19} \text{ J} = \underline{6.1 \times 10^{-19} \text{ J}}$$

Expressed in eV, this becomes

$$KE = 6.1 \times 10^{-19} \text{ J} \left(\frac{1 \text{ eV}}{1.6 \times 10^{-19} \text{ J}} \right) = \underline{3.8 \text{ eV}}$$

The ejected photoelectron has an energy of 3.8 eV, or 6.1×10^{-19} J.

Note that we may also carry out the solution in units of eV, if we prefer, by converting the incoming photon's energy into eV:

$$E = hf = 13.3 \times 10^{-19} \text{ J} \left(\frac{1 \text{ eV}}{1.6 \times 10^{-19} \text{ J}} \right) = 8.3 \text{ eV}$$

Substituting this into the photoelectron equation gives

$$KE = hf - \phi = 8.3 \text{ eV} - 4.5 \text{ eV} = \underline{3.8 \text{ eV}}$$

This is in agreement with the original result. Since all terms in the equation are energy terms, we may use whatever energy unit is most convenient, provided we are consistent throughout.

PROBLEM: An *electron volt* is defined as the energy gained by an electron which falls from rest through a potential difference of one volt. (a) How many eV of energy would a *proton* have after falling through the same potential difference? (b) Compare the de Broglie wavelength of the electron and proton.

Solution: (a) The kinetic energy gained by a charged particle falling through a potential difference is equal to its loss in electrical potential energy, or $KE = |\Delta PE_q| = qV$. Since the proton has the same *charge q* as the electron (in magnitude) and falls through one volt, it also has <u>one electron volt</u> of energy. In *mks* units this becomes

$$KE = (1.6 \times 10^{-19} \text{ C})(1 \text{ V}) = 1.6 \times 10^{-19} \text{ J} = 1 \text{ eV, by definition}$$

(b) The electron's de Broglie wavelength is given by Eq. (31.6):

$$\lambda = \frac{h}{mv}$$

We need the momentum mv of the electron; this can be obtained from its kinetic energy. Since this electron is nonrelativistic, $KE = \frac{1}{2}mv^2$, or $mv^2 = 2KE$. Multiplying both sides by m gives

$$m^2v^2 = 2mKE = p^2, \quad \text{or} \quad p = \sqrt{2mKE}$$

Thus,

$$p = \sqrt{2(9.1 \times 10^{-31} \text{ kg})(1.6 \times 10^{-19} \text{ J})} = \sqrt{2.9 \times 10^{-49} \text{ kg}^2 \cdot \text{m}^2/\text{s}^2}$$
$$= 5.4 \times 10^{-25} \text{ kg} \cdot \text{m/s}$$

and

$$\lambda = \frac{h}{p} = \frac{6.63 \times 10^{-34} \text{ J} \cdot \text{s}}{5.4 \times 10^{-25} \text{ kg} \cdot \text{m/s}} = 1.2 \times 10^{-9} \text{ m} = \underline{1.2 \text{ nm}}$$

The proton's mass is 1.67×10^{-27} kg, so its momentum is

$$p = \sqrt{2mKE} = \sqrt{2(1.67 \times 10^{-27} \text{ kg})(1.6 \times 10^{-19} \text{ J})} = 2.31 \times 10^{-23} \text{ kg} \cdot \text{m/s}$$

So for the proton,

$$\lambda = \frac{h}{p} = \frac{6.63 \times 10^{-34} \text{ J} \cdot \text{s}}{2.31 \times 10^{-23} \text{ kg} \cdot \text{m/s}} = 2.9 \times 10^{-11} \text{ m} = \underline{0.029 \text{ nm}}$$

The electron, having a smaller mass and thus a smaller momentum, has a larger de Broglie wavelength; its wavelike properties will be easier to detect than the proton's.

PROBLEM: What is the de Broglie wavelength of an electron after acceleration through a potential difference of 150,000 V in an X-ray tube?

Solution: (You may or may not know by looking at it that 150,000 V will give an electron a relativistic velocity. If you do not, and in other cases when you do not know for sure, it is best to take the advice given in Chapter 30 of the text: when in doubt, assume it is a relativistic problem.) The loss of electrical potential energy, $|\Delta PE_q| = qV$, is transformed into a gain in kinetic energy, $KE = mc^2 - m_0c^2$, or

$$qV = m_0c^2 \left[\frac{1}{\left(1 - \dfrac{v^2}{c^2}\right)^{1/2}} - 1 \right]$$

Here we have expressed the relativistic mass m in terms of the rest mass m_0. From this we can extract the velocity v and proceed with the calculation of the de Broglie wavelength using Eq. (31.6):

$$\lambda = \frac{h}{mv}$$

The only other fine point of the calculation is to recall that m is the relativistic mass in the case of high-speed particles. The speed of the electron is determined first. From the conservation of energy equation above, we find that

$$\frac{1}{\left(1 - \dfrac{v^2}{c^2}\right)^{1/2}} = \frac{qV}{m_0c^2} + 1 = \frac{(1.6 \times 10^{-19} \text{ C})(1.5 \times 10^5 \text{ V})}{(9.1 \times 10^{-31} \text{ kg})(3 \times 10^8 \text{ m/s})^2} + 1$$

$$= 0.293 + 1 = 1.293$$

Inverting gives

$$\left(1 - \frac{v^2}{c^2}\right)^{1/2} = \frac{1}{1.293} = 0.773$$

(Keep this value in mind for the relativistic mass calculation.) Squaring and solving for v, we get

$$\frac{v^2}{c^2} = 1 - (0.773)^2 = 1 - 0.598 = 0.402$$

$$v = \sqrt{0.402}\, c = 0.634c = 1.9 \times 10^8 \text{ m/s}$$

The relativistic mass m is given by

$$m = \frac{m_0}{\left(1 - \dfrac{v^2}{c^2}\right)^{1/2}} = \frac{9.1 \times 10^{-31} \text{ kg}}{0.773} = 1.18 \times 10^{-30} \text{ kg}$$

The de Broglie wavelength is thus

$$\lambda = \frac{h}{mv} = \frac{6.63 \times 10^{-34} \text{ J} \cdot \text{s}}{(1.18 \times 10^{-30} \text{ kg})(1.9 \times 10^8 \text{ m/s})} = \underline{2.96 \times 10^{-12} \text{ m}}$$

Note that if we had used the nonrelativistic kinetic energy, $KE = \frac{1}{2}mv^2$, we would get

$$v = \sqrt{\frac{2KE}{m_0}} = \sqrt{\frac{2qV}{m_0}} = 2.3 \times 10^8 \text{ m/s}$$

The velocity is calculated to be *higher* than it really is because at these speeds, some energy is embodied in the increased mass, leaving a lower velocity than expected classically. Non-relativistically, the de Broglie wavelength would be

$$\lambda = \frac{h}{m_0v} = \frac{6.63 \times 10^{-34} \text{ J} \cdot \text{s}}{(9.1 \times 10^{-31} \text{ kg})(2.3 \times 10^8 \text{ m/s})} = 3.17 \times 10^{-12} \text{ m}$$

The error in the de Broglie wavelength is only 7 percent, whereas the error in the electron velocity is 21 percent. This is because in the momentum mv, the "too-high" velocity is partly compensated for by the "too-low" mass.

PROBLEM: A beam of neutrons is diffracted using a crystal of rock salt which has a "grating space" of 0.281 nm. Determine the speed of the neutrons if the first maximum of the diffraction pattern occurs at 30°. The neutron rest mass is 1.67×10^{-27} kg.

Solution: The diffraction experiment gives information about the neutron's wavelength; then de Broglie's formula, $\lambda = h/mv$, can be used to determine the speed v. Equation (21.1) gives the angular position of maxima for a diffraction grating:

$$\sin \theta = \frac{n\lambda}{d}$$

where d = grating spacing = 0.281 nm = 0.281×10^{-9} m. Solving for λ gives

$$\lambda = \frac{d \sin 30°}{n} = \frac{(0.281 \times 10^{-9}\text{ m})(0.5)}{(1)} = \underline{1.4 \times 10^{-10}\text{ m}}$$

Solving the de Broglie equation for v gives

$$v = \frac{h}{\lambda m} = \frac{6.63 \times 10^{-34}\text{ J} \cdot \text{s}}{(1.4 \times 10^{-10}\text{ m})(1.67 \times 10^{-27}\text{ kg})} = \underline{2.84 \times 10^3\text{ m/s}}$$

The neutrons in this beam have a speed of about 2800 m/s. Since neutrons are uncharged, we cannot use electric or magnetic fields to study their velocity as in the case of charged particles such as electrons, protons, and ions. However, diffraction experiments like this, which make use of their wave nature, can be used to determine their velocity.

PROBLEM: What is the de Broglie wavelength of the sun (mass 2×10^{30} kg) as a result of its motion at a speed of 250 km/s around the galactic center?

Solution: Using the de Broglie relationship gives

$$\lambda = \frac{h}{mv} = \frac{6.63 \times 10^{-34}\text{ J} \cdot \text{s}}{(2 \times 10^{30}\text{ kg})(2.5 \times 10^5\text{ m/s})} = \underline{1.3 \times 10^{-69}\text{ m}}$$

It is not surprising that no one has detected any wavelike behavior on the part of the sun. We have learned that wave characteristics (in particular, diffraction) are most readily observable when the wavefront encounters apertures and obstacles comparable in size to the wavelength. Are we then to understand that if the sun could be observed encountering slits of width 10^{-69} m, solar matter would diffract when passing through the slits? Obviously not, since the fundamental particles making up the sun's matter themselves have dimensions much much larger than the imagined slits. We will *never* be able to observe wavelike behavior on the part of objects whose de Broglie wavelengths are much smaller than the dimensions of the smallest constituent particles of matter. However for objects whose de Broglie wavelength *is* measurable and for which diffraction experiments *can* be devised, de Broglie's relationship has never failed to correctly predict the experimental results.

PROBLEM: (a) What is the momentum of an electron which has a de Broglie wavelength equal to the wavelength of a photon of yellow light (590 nm). (b) If this electron's momentum is uncertain by 10 percent, what is the minimum uncertainty in its position?

Solution: (a) The electron's momentum and de Broglie wavelength are related by Eq. (31.5): $\lambda = h/p$. Solving for p gives

$$p = \frac{h}{\lambda} = \frac{6.63 \times 10^{-34}\text{ J} \cdot \text{s}}{590 \times 10^{-9}\text{ m}} = \underline{1.1 \times 10^{-27}\text{ kg} \cdot \text{m/s}}$$

(b) According to the uncertainty principle, the minimum uncertainty in the electron's position is related to the momentum uncertainty by

$$\Delta x \, \Delta p = \frac{h}{2\pi}$$

Thus,

$$\Delta x = \frac{h}{2\pi \, \Delta p}$$

The momentum uncertainty is not 10 percent, but *10 percent of p*, the electron momentum:

$$\Delta p = 0.1p = (0.1)(1.1 \times 10^{-27}\text{ kg} \cdot \text{m/s}) = 1.1 \times 10^{-28}\text{ kg} \cdot \text{m/s}$$

Putting this in the above equation, we find that

$$\Delta x = \frac{6.63 \times 10^{-34} \text{ J} \cdot \text{s}}{2\pi(1.1 \times 10^{-28} \text{ kg} \cdot \text{m/s})} = 9.6 \times 10^{-7} \text{ m} = \underline{960 \text{ nm}}$$

The electron's uncertainty in position, 960 nm, is greater than its de Broglie wavelength, 590 nm.

PROBLEM: (a) If a photon of red light, wavelength 660 nm, is emitted by an atom in 10^{-9} s, what is the minimum uncertainty in its energy? (b) What percentage of its total energy does this represent?

Solution: (a) The uncertainty principle relating energy and time is given by Eq. (31.8):

$$\Delta E \, \Delta t \geq \frac{h}{2\pi}$$

The minimum energy uncertainty corresponding to $\Delta t = 10^{-9}$ s is thus,

$$\Delta E = \frac{h}{2\pi \, \Delta t} = \frac{6.63 \times 10^{-34} \text{ J} \cdot \text{s}}{2\pi(10^{-9} \text{ s})} = \underline{1.05 \times 10^{-25} \text{ J}}$$

(b) The photon's energy is given by

$$E = hf = \frac{hc}{\lambda} = \frac{(6.63 \times 10^{-34} \text{ J} \cdot \text{s})(3 \times 10^{8} \text{ m/s})}{660 \times 10^{-9} \text{ m}} = 3 \times 10^{-19} \text{ J}$$

Then the fractional uncertainty in its energy is

$$\frac{\Delta E}{E} = \frac{1.05 \times 10^{-25} \text{ J}}{3 \times 10^{-19} \text{ J}} = 3.5 \times 10^{-7} = \underline{3.5 \times 10^{-5} \text{ percent}}$$

Part (b) can be solved without calculating the photon's energy as follows:

$$\frac{\Delta E}{E} = \frac{h/2\pi \, \Delta t}{hc/\lambda} = \frac{\lambda}{2\pi \, \Delta t c} = \frac{660 \times 10^{-9} \text{ m}}{2\pi(10^{-9} \text{ s})(3 \times 10^{8} \text{ m/s})}$$

$$= 3.5 \times 10^{-7} = \underline{3.5 \times 10^{-5}} \text{ percent}$$

This is in agreement with the original result.

PROBLEM: Somewhere between the earth and the sun it is estimated that there are 10^{23} photons crossing an area of 1 m^2 each second. Assuming an average wavelength of 500 nm for these photons (green light), determine the force of this radiation on (a) a relatively large spherical interplanetary dust particle with a radius of 3×10^{-6} m, and (b) a small spherical dust particle whose radius is one-tenth as great. (c) Compare the accelerations of these two particles (density 3000 kg/m^3) as a result of the force, and determine the distance each travels in an hour if starting from rest. Assume that the dust particles absorb all the radiation they receive.

Solution: In being absorbed, photons transfer momentum to the dust particles, resulting in a force called the *radiation force*. The impulse-momentum equation relates the change in momentum to the resulting force:

$$\vec{F} \, \Delta t = m\vec{v} - m\vec{v}_0 = \Delta \vec{p}$$

$$\text{or} \quad \vec{F} = \frac{\Delta \vec{p}}{\Delta t}$$

The photon momentum is given by $p = h/\lambda$. For the green photons above, $\lambda = 500$ nm $= 500 \times 10^{-9}$ m, so each photon has a momentum of

$$p = \frac{6.63 \times 10^{-34} \text{ J} \cdot \text{s}}{500 \times 10^{-9} \text{ m}} = 1.33 \times 10^{-27} \text{ kg} \cdot \text{m/s}$$

230

(a) The cross-sectional area of the large dust particle is πr^2, where $r = 3 \times 10^{-6}$ m. The number of photons per second it intercepts is

$$10^{23}\,\frac{\text{photons}}{\text{s}\cdot\text{m}^2}\,(\pi r^2) = 10^{23}\,\frac{\text{photons}}{\text{s}\cdot\text{m}^2}\,\pi(3 \times 10^{-6}\text{ m})^2 = 2.83 \times 10^{12}\,\frac{\text{photons}}{\text{s}}$$

Since each photon transfers 1.33×10^{27} kg·m/s, the total momentum change is

$$\frac{\Delta p}{\Delta t} = \left(2.83 \times 10^{12}\,\frac{\text{photons}}{\text{s}}\right)\left(1.33 \times 10^{-27}\,\frac{\text{kg}\cdot\text{m/s}}{\text{photon}}\right) = 3.76 \times 10^{-15}\text{ N}$$

But by the impulse-momentum equation, this is just the *force* exerted on the dust particle by the photons striking it. The large dust particle experiences a radiation force of 3.76 × 10^{-15} N.

(b) The small dust particle has a radius only one-tenth as great, so its area is $\frac{1}{100}$ as great. It intercepts a factor of 100 less photons and experiences a radiation force which is smaller by a factor of 100, or

$$F' = \frac{3.76 \times 10^{-15}\text{ N}}{100} = \underline{3.76 \times 10^{-17}\text{ N}}$$

(c) The acceleration of the large particle, by Newton's second law, is

$$a = \frac{F}{m}$$

where $m = \rho V = (3000\text{ kg/m}^3)(\tfrac{4}{3}\pi r^3) = (3000\text{ kg/m}^3)(\tfrac{4}{3}\pi)(3 \times 10^{-6}\text{ m})^3 = 3.39 \times 10^{-13}$ kg.

$$a = \frac{F}{m} = \frac{3.76 \times 10^{-15}\text{ N}}{3.39 \times 10^{-13}\text{ kg}} = 1.1 \times 10^{-2}\text{ m/s}^2 = \underline{1.1\text{ cm/s}^2}$$

For the small particle,

$$a' = \frac{F'}{m'} \quad \text{where } m' = \rho V' = \rho(\tfrac{4}{3}\pi r'^3)$$

Since the radius is one-tenth as great, the mass, which varies as r^3, will be $\dfrac{1}{(10)^3} = \frac{1}{1000}$ as great,

$$m' = \frac{3.39 \times 10^{-13}\text{ kg}}{1000} = 3.39 \times 10^{-16}\text{ kg}$$

Therefore,

$$a' = \frac{F'}{m'} = \frac{3.76 \times 10^{-17}\text{ N}}{3.39 \times 10^{-16}\text{ kg}} = 1.1 \times 10^{-1}\text{ m/s}^2 = \underline{11\text{ cm/s}^2}$$

The small dust particle experiences 10 times the acceleration of the large particle. If both start from rest, in one hour the large particle travels a distance

$$x = v_0 t + \tfrac{1}{2}at^2 = 0 + \tfrac{1}{2}(1.1 \times 10^{-2}\text{ m/s}^2)(3600\text{ s})^2 = 7.1 \times 10^4\text{ m} = \underline{71\text{ km}}$$

The small particle with an acceleration 10 times as great will travel 10 times as far, or 710 km.

Discussion: The problem we have solved is a comet tail problem. A comet is thought to be a frozen mass of ice and dust. When a comet approaches the sun, the ice melts, freeing dust particles. These particles experience an acceleration which is inversely proportional to their radius, and they stream out from the comet's head in the antisolar direction, forming the comet's spectacular tail. (We have assumed that the particles absorbed all the sunlight they received; actually, they reflect some of the sunlight, increasing the momentum transfer and enabling the tail to be visible by this reflected light.)

Avoiding Pitfalls

1. An *electron volt* is not a unit of charge or of potential difference; it is a unit of *energy*, equal to 1.6×10^{-19} J.

2. The electron volt is a hybrid unit, not a proper *mks* unit. Energies expressed in electron volts must be converted to joules before substituting into equations in which other quantities are expressed in *mks* units.

3. The de Broglie wavelength equation is valid for both nonrelativistic and relativistic cases, provided m is understood to be the relativistic mass (which is essentially equal to the rest mass for speeds that are small compared to c).

4. In applying the uncertainty principle, Δx and Δp must be expressed in units of length and momentum respectively. They cannot be unitless percentages or fractions; they must be percentages or fractions *of* the particular quantity. The same considerations apply to ΔE and Δt in the energy-time form of the uncertainty principle.

Drill Problems
Answers in Appendix

1. A very weak radio signal that can barely be detected by a conventional radio has an intensity of about 10^{-14} W/m². If its frequency is 1 megahertz, how many photons cross a square meter of surface area each second?

2. Photons with an energy of 5 eV eject electrons from a metal surface whose work function is 3 eV.

   (a) What is the maximum speed of the ejected electrons?
   (b) What is the wavelength of the incoming photons?

3. Photon A has twice the energy of photon B.

   (a) What is the ratio of their wavelengths?

   (b) What is the ratio of their momenta?

4. (a) What is the de Broglie wavelength of a nonrelativistic electron moving with a speed of 10^5 m/s? (b) If the electron's momentum is known to one part in one hundred, what is the uncertainty in the electron's position?

32

The Bohr Model of the Atom

Terms

Define or describe briefly what is meant by the following terms. If you have difficulty, refer to the textbook section given in parentheses.

"plum-pudding" model (32.1)

alpha particle (32.1)

nuclear model (32.1)

line spectrum (32.2)

Balmer's equation (32.2)

Rydberg constant (32.2)

Bohr model (32.3)

atomic number (32.3)

energy-level diagram (32.3)

quantum number (32.3)

ground state (32.3)

excited state (32.3)

Equation Review

For each equation, be able to state the situation to which it applies, what each quantity represents, and the units for measuring each quantity in some consistent set of units.

$$\frac{1}{\lambda} = R \left(\frac{1}{1^2} - \frac{1}{n^2} \right) \quad n = 2, 3, \ldots \tag{32.1}$$

$$\frac{1}{\lambda} = R \left(\frac{1}{2^2} - \frac{1}{n^2} \right) \quad n = 3, 4, \ldots \tag{32.2}$$

$$\frac{1}{\lambda} = R \left(\frac{1}{3^2} - \frac{1}{n^2} \right) \quad n = 4, 5, \ldots \tag{32.3}$$

$$E_i - E_f = hf \tag{32.4}$$

$$E = -\frac{kZe^2}{2r} \tag{32.7}$$

$$mv_n r_n = \frac{nh}{2\pi} \quad \text{for } n = 1, 2, 3, \ldots \tag{32.8}$$

$$r_n = \frac{(0.53 \times 10^{-10} \text{ m})n^2}{Z} \tag{32.9}$$

$$E_n = -\frac{2\pi^2 k^2 e^4 m}{h^2} \frac{Z^2}{n^2} \tag{32.10}$$

$$E_n = -\frac{(13.6 \text{ eV})Z^2}{n^2} \tag{32.11}$$

$$\frac{1}{\lambda} = RZ^2 \left(\frac{1}{n_f^2} - \frac{1}{n_i^2} \right) \tag{32.12}$$

Problems with Solutions and Discussions

PROBLEM: (a) To get a feel for the size of the nucleus, determine how many gold nuclei could fit into the volume occupied by the gold atom. Assume that the radius of a gold atom is 1.5×10^{-10} m and the radius of a gold nucleus is 6×10^{-14} m. (b) Do the same type of calculation to determine how many suns could fit into the volume of a solar system consisting only of the sun and earth. The sun's radius is about 7×10^8 m and the radius of the earth's orbit is 1.5×10^{11} m.

Solution: (a) The volume of the nucleus, assumed to be spherical, is $V_n = \frac{4}{3}\pi r_n^3$. The volume of the atom, also assumed to be spherical, is $V_a = \frac{4}{3}\pi r_a^3$. To get the number of nuclear volumes in the atom's volume, we divide the latter by the former:

$$\frac{V_a}{V_n} = \frac{\frac{4}{3}\pi r_a^3}{\frac{4}{3}\pi r_n^3} = \left(\frac{r_a}{r_n} \right)^3 = \left(\frac{1.5 \times 10^{-10} \text{ m}}{6 \times 10^{-14} \text{ m}} \right)^3 = (2.5 \times 10^3)^3 = \underline{1.56 \times 10^{10}}$$

Over *15 billion* gold nuclei could fit in the volume of a gold atom. Or the gold nucleus is less than one-billionth the size of the atom. We can now understand the comment that the atom is mostly empty space. The fact that ordinary matter seems so solid and impenetrable and the fact that enormous pressures are required to make one atom intrude into the empty space of another atom are indications of how strong the electric force is and how basic the force is to the structure of matter as we know it.

(b) A similar calculation for the sun (*s*) and the limited solar system (*ss*) gives

$$\frac{V_{ss}}{V_s} = \frac{\frac{4}{3}\pi r_{ss}^3}{\frac{4}{3}\pi r_s^3} = \left(\frac{r_{ss}}{r_s} \right)^3 = \left(\frac{1.5 \times 10^{11} \text{ m}}{7 \times 10^8 \text{ m}} \right)^3 = (2.14 \times 10^2)^3 = 9.8 \times 10^6$$

About 10 million suns could fit into the volume of a sphere with a radius equal to the radius of the earth's orbit. There is a greater proportion of empty space in an atom than in this limited solar system. (To get a comparable proportion, we could extend our solar system to include Saturn's orbit.) The occasional reference to an atom as being similar to a tiny solar system is quite apt, at least with regard to relative sizes of constituents.

PROBLEM: In the emission spectrum of the hydrogen atom, what are the longest and shortest wavelengths present in the Balmer series of spectral lines?

Solution: The Balmer lines are those representing transitions which end in the $n = 2$ state; their wavelengths are given by Eq. (32.12):

$$\frac{1}{\lambda} = RZ^2 \left(\frac{1}{n_f^2} - \frac{1}{n_i^2} \right)$$

where $Z = 1$ for the hydrogen atom and $n_f = 2$ for the Balmer series.

$$\frac{1}{\lambda} = R\left(\frac{1}{2^2} - \frac{1}{n_i^2}\right) \quad n_i = 3, 4, 5, \ldots$$

The longest wavelength (lowest energy) transition ending in the $n = 2$ state is the transition from $n = 3$ to $n = 2$:

$$\frac{1}{\lambda} = R\left(\frac{1}{2^2} - \frac{1}{3^2}\right) = 1.097 \times 10^7 \text{ m}^{-1} (\frac{1}{4} - \frac{1}{9}) = 1.524 \times 10^6 \text{ m}^{-1}$$

Inverting we get

$$\lambda = 6.563 \times 10^{-7} \text{ m} = \underline{656 \text{ nm}}$$

This wavelength is in the red portion of the visible spectrum.

The longest wavelength of the series (sometimes called the *series limit*) is obtained by setting $n_i = \infty$. This is the wavelength of the photon produced when a free electron at rest falls to the $n = 2$ orbit of the hydrogen atom.

$$\frac{1}{\lambda} = R\left(\frac{1}{2^2} - \frac{1}{\infty^2}\right) = \frac{R}{4} = \frac{1.097 \times 10^7 \text{ m}^{-1}}{4} = 2.74 \times 10^6 \text{ m}^{-1}$$

$$\lambda = 3.65 \times 10^{-7} \text{ m} = \underline{365 \text{ nm}}$$

This photon lies in the near ultraviolet. All of the Balmer series of spectral lines lie between 364 nm and 656 nm, and no lines from any other spectral series of hydrogen lie within this range.

PROBLEM: Consider a beryllium atom ($Z = 4$) with three electrons removed. Determine the energy of the first five orbits for this ion and sketch an energy-level diagram.

Solution: Equation (32.11) gives the energies of the various levels in a hydrogenlike atom. For $Z = 4$, we have

$$E_n = -\frac{(13.6 \text{ eV})Z^2}{n^2} = -\frac{(13.6 \text{ eV})(16)}{n^2} = -\frac{218 \text{ eV}}{n^2}$$

The energies are calculated below and sketched (not to scale).

Energy	Quantum number
0	$n = \infty$
-8.7 eV	$n = 5$
-13.6 eV	$n = 4$
-24.2 eV	$n = 3$
-54.5 eV	$n = 2$
-218 eV	$n = 1$

$$E_1 = -\frac{218 \text{ eV}}{1^2} = \underline{-218 \text{ eV}}$$

$$E_2 = -\frac{218 \text{ eV}}{2^2} = -\frac{218 \text{ eV}}{4} = \underline{-54.5 \text{ eV}}$$

$$E_3 = -\frac{218 \text{ eV}}{3^2} = -\frac{218 \text{ eV}}{9} = \underline{-24.2 \text{ eV}}$$

$$E_4 = -\frac{218 \text{ eV}}{4^2} = -\frac{218 \text{ eV}}{16} = \underline{-13.6 \text{ eV}}$$

$$E_5 = -\frac{218 \text{ eV}}{5^2} = -\frac{218 \text{ eV}}{25} = \underline{-8.7 \text{ eV}}$$

PROBLEM: For the beryllium ion above, determine (a) the radius of the $n = 1$ orbit, and (b) the wavelength and frequency of the photon emitted in an electron transition from $n = 4$ to $n = 2$.

Solution: (a) The radius of a hydrogenlike atom or ion is given by Eq. (32.9):

$$r_n = \frac{(0.53 \times 10^{-10} \text{ m})n^2}{Z}$$

For $Z = 4$, $n = 1$:

$$r_n = \frac{(0.53 \times 10^{-10} \text{ m})}{4} = \underline{1.3 \times 10^{-11} \text{ m}}$$

The radius of the $n = 1$ orbit of the beryllium ion is ¼ that of the hydrogen atom, because the electron is attracted by 4 times as much positive charge in the beryllium nucleus as in the hydrogen nucleus.

(b) The photon energy is equal to the *difference in energy* between the initial and final states,

$$E_{\text{photon}} = \Delta E = E_4 - E_2 = (-13.6 \text{ eV}) - (-54.5 \text{ eV}) = 40.9 \text{ eV}$$

Converting to *mks* units we get

$$E_{\text{photon}} = 40.9 \text{ eV} \left(\frac{1.6 \times 10^{-19} \text{ J}}{1 \text{ eV}} \right) = 6.54 \times 10^{-18} \text{ J}$$

For a photon, $E = hf$, so

$$f = \frac{E}{h} = \frac{6.54 \times 10^{-18} \text{ J}}{6.63 \times 10^{-34} \text{ J} \cdot \text{s}} = \underline{9.86 \times 10^{15} \text{ Hz}}$$

Its wavelength is given by

$$\lambda = \frac{c}{f} = \frac{3 \times 10^8 \text{ m/s}}{9.86 \times 10^{15} \text{ s}^{-1}} = 3.04 \times 10^{-8} \text{ m} = 30 \text{ nm}$$

Referring to Fig. 31.3 in the textbook, we see that this photon is in the ultraviolet region of the electromagnetic spectrum.

PROBLEM: What is the frequency of revolution of an electron in the $n = 3$ orbit of a He^+ ($Z = 2$) ion?

Solution: The frequency of revolution f is the reciprocal of the period T, the time taken for one revolution. This time is related to the electron's speed by a simple constant-speed equation,

$$v = \frac{\text{distance}}{\text{time}} = \frac{2\pi r}{T}$$

where $2\pi r$ is the circumference of the orbit. Solving for T gives $T = \dfrac{2\pi r}{v}$, so

$$f = \frac{1}{T} = \frac{v}{2\pi r}$$

v and r are not independent; they are related by Bohr's angular momentum condition:

$$mvr = \frac{nh}{2\pi}$$

Solving for v,

$$v = \frac{nh}{2\pi mr}$$

and substituting into the frequency equation, we get

$$f = \left(\frac{nh}{2\pi mr} \right) \frac{1}{2\pi r} = \frac{nh}{4\pi^2 mr^2}$$

Now we need only the radius of the $n = 3$ orbit to complete the calculation, using Eq. (32.9):

$$r_n = \frac{(0.53 \times 10^{-10} \text{ m}) n^2}{Z}$$

where $Z = 2$ and $n = 3$,

$$r_3 = \frac{(0.53 \times 10^{-10} \text{ m})(3)^2}{2} = 2.38 \times 10^{-10} \text{ m}$$

Putting this into the frequency equation above gives

$$f = \frac{(3)(6.63 \times 10^{-34} \text{ J}\cdot\text{s})}{4\pi^2(9.1 \times 10^{-31} \text{ kg})(2.38 \times 10^{-10} \text{ m})^2} = \underline{9.8 \times 10^{14} \text{ Hz}}$$

The frequency of the electron in this orbit is almost 10^{15} Hz.

PROBLEM: Suppose the charge on the electron and on the proton in the hydrogen atom were both doubled. How would this change the energy of the photon emitted in the $n = 2$ to $n = 1$ transition?

Solution: The energy of an emitted photon is equal to the difference in energy between the initial and final states of the atom,

$$E_{\text{photon}} = \Delta E = E_i - E_f = E_2 - E_1$$

These energies are given by Eq. (32.10):

$$E_n = -\left[\frac{2\pi^2 k^2 e^4 m}{h^2}\right] \frac{Z^2}{n^2}$$

E_2 and E_1 differ from each other only in their quantum numbers n, so

$$E_{\text{photon}} = \left[\frac{2\pi^2 k^2 e^4 m}{h^2}\right] Z^2 \left(\frac{1}{n_1^2} - \frac{1}{n_2^2}\right)$$

How does this expression change if the proton and electron charge suddenly double? In the expression for E_{photon}, this charge e appears raised to the fourth power. If we replace e by $2e$, we have $(2e)^4 = 16e^4$; the expression for E is larger by a factor of 16. All other quantities in the expression are unchanged. (Recall that Z is the *number* of protons; there is still only one.) Thus the photon from the doubled hydrogen atom is 16 times more energetic than the corresponding photon in the normal hydrogen atom. Note that this result is independent of the particular transition we looked at; *any* photon from the doubled atom would be 16 times more energetic than the corresponding photon from the normal atom.

PROBLEM: Assume that Bohr's condition on the angular momentum of an orbiting electron also applies to the angular momentum of an orbiting planet. What is the quantum number of the earth's orbit around the sun? (The earth's mass is 5.98×10^{24} kg and the radius of its orbit is 1.5×10^{11} m.)

Solution: Bohr's condition on angular momentum was that it must be an integral multiple of $h/2\pi$. This is stated by Eq. 32.8:

$$mvr = \frac{nh}{2\pi}$$

where n, the quantum number, is the number of these units of $h/2\pi$ in the object's angular momentum. Solving for n, we get

$$n = \frac{2\pi mvr}{h}$$

where m and r are given above. We need to find v. Assuming that the earth moves with constant speed in a circular orbit, we find that

$$v = \frac{\text{distance}}{\text{time}} = \frac{2\pi r}{T}$$

where $T = 1$ year $= 3.16 \times 10^7$ s. Substituting this expression for v gives

$$n = \frac{2\pi mr}{h}\left(\frac{2\pi r}{T}\right) = \frac{4\pi^2 r^2 m}{hT}$$

Then substituting in numerical values, we get

$$n = \frac{4\pi^2(1.5 \times 10^{11} \text{ m})^2(5.98 \times 10^{24} \text{ kg})}{(6.63 \times 10^{-34} \text{ J} \cdot \text{s})(3.16 \times 10^7 \text{ s})} = \underline{2.5 \times 10^{74}}$$

When n gets this large, the spacing between two adjacent allowed orbits is so tiny that we could never expect to be able to determine whether or not the earth's orbital angular momentum follows Bohr's condition or not. For macroscopic objects we are safe in acting as if there are no restrictions on the value that the angular momentum may assume. Any radius and its corresponding velocity are permitted.

PROBLEM: What is the orbital angular momentum of an electron in the $n = 5$ orbit of a single-electron atom?

Solution: According to Bohr's theory, the quantum number n, in addition to numbering the orbits, indicates how many units of $h/2\pi$ make up the orbital angular momentum of an electron. By Eq. (32.8):

$$mvr = \text{angular momentum} = \frac{nh}{2\pi}$$

For $n = 5$,

$$\text{angular momentum} = 5\frac{h}{2\pi} = \frac{5(6.63 \times 10^{-34} \text{ J} \cdot \text{s})}{2\pi} = \underline{5.3 \times 10^{-34} \text{ J} \cdot \text{s}}$$

This result is independent of the atomic number of the atom or ion. An electron in the $n = 5$ orbit of hydrogen has exactly the same angular momentum as an electron in the $n = 5$ orbit of a He^+ or Li^{2+} ion, according to the Bohr model. Since mvr is constant for these electrons, v and r for the electron in each ion must vary such that their product is constant. Since r varies as $1/Z$, v must be directly proportional to Z.

PROBLEM: Bohr's condition on angular momentum states that it can occur only in units of $h/2\pi$. Show that the units of $h/2\pi$ are proper units for angular momentum.

Solution: First we must recall what the proper units for angular momentum are. Angular momentum, mvr, has *mks* units of $(\text{kg})\left(\dfrac{\text{m}}{\text{s}}\right)(\text{m}) = \dfrac{\text{kg} \cdot \text{m}^2}{\text{s}}$. Now, $\dfrac{h}{2\pi}$ has the units of h which are $\text{J} \cdot \text{s}$. By definition a joule may be expressed in these units:

$$1 \text{ J} = 1 \text{ N} \cdot \text{m} = 1 \frac{\text{kg} \cdot \text{m}^2}{\text{s}^2}$$

So

$$\text{J} \cdot \text{s} = \left(\frac{\text{kg} \cdot \text{m}^2}{\text{s}^2}\right)\text{s} = \frac{\text{kg} \cdot \text{m}^2}{\text{s}}$$

The units of $h/2\pi$ are therefore the units of angular momentum. Because of its place in the Bohr theory of the atom, the quantity $h/2\pi$ is often thought of as a natural unit of angular momentum. In atomic nuclear physics, the angular momenta of the various particles—nuclei, photons, etc.—are traditionally expressed in units of $h/2\pi$.

PROBLEM: Assume that a large number of hydrogen atoms are initially in the $n = 4$ state. The atoms then proceed to make transitions to lower energy states. Make an energy-level diagram (not necessarily to scale) and to show the possible routes by which an atom may return to ground state. How many distinct spectral lines (lines of different wavelength) will be emitted?

Solution: The four possible routes an atom may take in returning to ground state are shown in Fig. 32.1. Each transition involving a photon of a different wavelength is given a different letter. Note that some parts of the various routes are identical; for example,

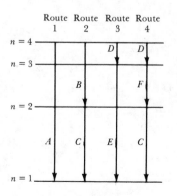

FIG. 32.1

photon D is emitted in Route 3 and in Route 4. Counting up the possibilities, we see there are <u>six distinct transitions</u>, labeled A through F, giving spectral lines of different wavelength.

Avoiding Pitfalls

1. Recalling the order of the size of the atom and of its nucleus can serve as a check on the results of many problems. The dimensions of an atom are of the order of 10^{-10} m; those of a nucleus, 10^{-14} m.

2. Atoms of a particular element absorb the same series of spectral lines that they emit.

3. When using Balmer's equation and other similar equations giving the wavelengths of a particular series of spectral lines, do not forget the final inversion of $1/\lambda$ to obtain λ.

4. An atom emits (or absorbs) a photon whose energy is equal to the energy *difference* between the initial and final states of the atom.

5. In Bohr's theory an electron's orbital angular momentum can only take on values that are integral multiples of $h/2\pi$.

6. The Bohr condition on an electron's orbital angular momentum is equivalent to the statement that the circumference of an allowed orbit contains an integral number of electron wavelengths.

7. The quantum number n of an atomic state determines the energy, angular momentum, orbital radius, and number of de Broglie wavelengths fitting into the orbit circumference for an electron in that state.

8. When electrons are bound to atoms, their total energy is negative. Thus energy level diagrams range from zero at the top (electron separated from atom, $PE_q = KE = 0$) to large negative energies (electron bound to atom in one of many excited states). The *ground state* is the lowest energy state (largest negative energy).

Drill Problems
Answers in Appendix

1. In the Bohr model of the atom, state how the following properties of an electron depend on the quantum number n:

   (a) radius of orbit;
   (b) speed;
   (c) frequency;
   (d) energy;
   (e) angular momentum.

2.   What is the wavelength and energy of the photon emitted when the electron in a Li^{2+} ion falls from the $n = 9$ to the $n = 4$ state?

3.   (a) What is the radius of the $n = 5$ orbit of a He^+ ion? (b) What is the de Broglie wavelength of an electron in this orbit? (c) What is the angular momentum of this electron?

33

Quantum Mechanics and Atoms

Terms

Define or describe briefly what is meant by the following terms. If you have difficulty, refer to the textbook section given in parentheses.

quantum mechanics (33.1)

Schrödinger's equation (33.1)

wave functions (33.1)

principal quantum number (33.2)

angular momentum quantum number (33.2)

magnetic quantum number (33.2)

spin magnetic quantum number (33.2)

Pauli exclusion principle (33.3)

subshell (33.3)

ground state (33.3)

electron configuration (33.3)

periodic table (33.4)

excitation (33.5)

fluorescence (33.5)

phosphorescence (33.5)

metastable state (33.5)

spontaneous emission (33.6)

stimulated emission (33.6)

laser (33.6)

population inversion (33.6)

forbidden transition (33.6)

coherent light (33.6)

Bremsstrahlung radiation (33.7)

characteristic lines (33.7)

Equation Review

For each equation, be able to state the situation to which it applies, what quantity each symbol represents, and the units for measuring each quantity in some consistent set of units.

$$n = 1, 2, 3, 4, \ldots \tag{33.1}$$

$$l = 0, 1, 2, \ldots, (n - 1) \tag{33.2}$$

$$m_l = 0, \pm 1, \pm 2, \ldots, \pm l \tag{33.3}$$

$$m_s = \pm \tfrac{1}{2} \tag{33.4}$$

$$hf_{\max} = eV \tag{33.5}$$

Problems with Solutions and Discussion

PROBLEM: For each of the wave functions whose quantum numbers $|n\ l\ m_l\ m_s\rangle$ are listed, state why the symbol *cannot* represent an allowed wave function. (a)$|\ 0\ 1\ 1\ \tfrac{1}{2}\rangle$; (b) $|1\ 0\ 0\ 0\rangle$; (c) $|2\ 2\ 1\ -\tfrac{1}{2}\rangle$; (d) $|2\ -1\ -1\ \tfrac{1}{2}\rangle$; (e) $|2\ 1\ 2\ \tfrac{1}{2}\rangle$.

Solution: (a) The symbol indicates that $n = 0$; this is not an allowed value for n. The smallest value n can have is 1, which represents an electron in the smallest possible orbit around the nucleus. Higher values of n (for example, $n = 2, 3, \ldots$) would also be satisfactory.

(b) If $n = 1$, l must be zero (l can range from 0 to $n - 1$), and m_l must be zero (m_l can assume values from zero up to $\pm l$). So the first three quantum numbers are possible. But m_s can assume one of only two values, $+\tfrac{1}{2}$ and $-\tfrac{1}{2}$. Zero is not a possible value for m_s.

(c) If $n = 2$, then l can be 0 or 1; its maximum value is $n - 1$, or 1 in this case. It can never assume a value equal to n.

(d) l ranges from 0 to $n - 1$. Since n is always a positive number, l is always a positive number or zero; it cannot be a negative number.

(e) The values of m_l can range from 0 to $\pm l$; it can never exceed l in absolute value. Thus, for $l = 1$, m_l can be 0 or 1 or -1, but it can never be 2 (or -2).

PROBLEM: Using arrows, represent schematically the possible orientations of an $l = 3$ state in a magnetic field.

Solution: The quantum number representing the orientation of the orbit in a magnetic field is m_l. Now m_l can assume values from 0 to $\pm l$, so for $l = 3$, m_l can be $-3, -2, -1, 0, 1, 2, 3$. Note that there are $2l + 1$ possible values of m_l, so for $l = 3$, there are $2l + 1 = 2(3) + 1 = 7$ values that m_l may assume.

The arrows representing these seven orientations are shown in Fig. 33.1. These arrows are perpendicular to the plane of the electron orbit as in Fig. 33.2 of the textbook. It appears that at the atomic level, not only are energy and angular momentum restricted

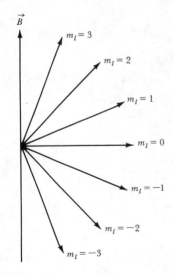

FIG. 33.1

243

to certain values, but even the orientation of the orbit in space can assume only certain values. This idea is sometimes referred to as *space quantization*.

PROBLEM: List the $|nlm_lm_s\rangle$ states available to an electron in the 2*p* subshell.

Solution: The 2*p* subshell has $n = 2$. The letter *p* specifies the value of *l*; by referring to Table 33.2 in the textbook, we see that a *p* state has $l = 1$. So the first two quantum numbers of each state will be $|2\ 1\quad\rangle$. The third quantum number m_l can go from 0 to ± 1; it can be 0 or -1 or 1. Each of these m_l states can have an m_s of $+\frac{1}{2}$ or $-\frac{1}{2}$. So the states are:

$$|2\ 1\ -1\ \tfrac{1}{2}\rangle \qquad |2\ 1\ -1\ -\tfrac{1}{2}\rangle$$
$$|2\ 1\ 0\ \tfrac{1}{2}\rangle \qquad |2\ 1\ 0\ -\tfrac{1}{2}\rangle$$
$$|2\ 1\ 1\ \tfrac{1}{2}\rangle \qquad |2\ 1\ 1\ -\tfrac{1}{2}\rangle$$

There are 6 states available in this *p* subshell (or any *p* subshell, for that matter). Thus a *p* subshell can hold no more than six electrons.

PROBLEM: How many electrons can an *f* subshell hold?

Solution: Referring to Table 33.2 in the textbook, we see that an *f* subshell has $l = 3$. This means that m_l can have values of $-3, -2, -1, 0, 1, 2, 3$; *seven* values in all. But for each value of m_l, m_s can take on *two* values, $\frac{1}{2}$ or $-\frac{1}{2}$. Therefore the *f* subshell consists of 7 (the values of m_l) $\times$ 2 (the values of m_s for each m_l) = 14 states, and can hold <u>14 electrons</u> without duplicating quantum numbers.

PROBLEM: Write the quantum numbers for each electron in a magnesium atom ($Z = 12$).

Solution: The magnesium atom whose $Z = 12$ has 12 protons and 12 electrons. Referring to Table 33.3 in the textbook, we see that the first 2 electrons will fill the 1*s* subshell, the next 2 will fill the 2*s* subshell, the next 6 will fill the 2*p* subshell. We have accounted for $2 + 2 + 6 = 10$ electrons; the last 2 will go into the 3*s* subshell, filling it. The quantum numbers for these electrons will be:

$$\left.\begin{array}{l}|1\ 0\ 0\ \tfrac{1}{2}\rangle \\ |1\ 0\ 0\ -\tfrac{1}{2}\rangle\end{array}\right\} \quad 1s \text{ subshell } (n = 1,\ l = 0)$$

$$\left.\begin{array}{l}|2\ 0\ 0\ \tfrac{1}{2}\rangle \\ |2\ 0\ 0\ -\tfrac{1}{2}\rangle\end{array}\right\} \quad 2s \text{ subshell } (n = 2,\ l = 0)$$

$$\left.\begin{array}{l}|2\ 1\ -1\ \tfrac{1}{2}\rangle \\ |2\ 1\ -1\ -\tfrac{1}{2}\rangle \\ |2\ 1\ 0\ \tfrac{1}{2}\rangle \\ |2\ 1\ 0\ -\tfrac{1}{2}\rangle \\ |2\ 1\ 1\ \tfrac{1}{2}\rangle \\ |2\ 1\ 1\ -\tfrac{1}{2}\rangle\end{array}\right\} \quad 2p \text{ subshell } (n = 2,\ l = 1)$$

$$\left.\begin{array}{l}|3\ 0\ 0\ \tfrac{1}{2}\rangle \\ |3\ 0\ 0\ -\tfrac{1}{2}\rangle\end{array}\right\} \quad 3s \text{ subshell } (n = 3,\ l = 0)$$

PROBLEM: Write the electron configuration for (a) magnesium ($Z = 12$), (b) sulfur ($Z = 16$), (c) copper ($Z = 29$), which has one electron in the 4*s* subshell, and (d) bromine ($Z = 35$).

Solution: (a) Referring to the solution of the previous problem, we may immediately write the electron configuration of magnesium: $\underline{1s^2 2s^2 2p^6 3s^2}$.

(b) Sulfur has 16 electrons. We determine its configuration by putting electrons into the lowest energy subshells possible, in the order given by Table 33.3, until the shell is filled, continuing until we use up all the electrons: 2 electrons into $1s$, 2 electrons into $2s$, 6 electrons into $2p$, 2 electrons into $3s$. 12 electrons are now placed, 4 remain; they go into the $3p$ subshell. Thus the configuration is $1s^2 2s^2 2p^6 3s^2 3p^4$. Notice that the superscripts, representing numbers of electrons, must add up to Z, the total number of electrons in the neutral atom.

(c) Filling shells in the order shown in Table 33.3, we have: $1s^2 2s^2 2p^6 3s^2 3p^6$. At this point we have used up 18 of the 29 electrons and would expect the next 2 to go in the $4s$ subshell. However, with the additional information that copper has only one electron in the $4s$ subshell, we must put the other in the next subshell, which is $3d$. So the electron configuration of copper must be $1s^2 2s^2 2p^6 3s^2 3p^6 3d^{10} 4s^1$. (The order of the last two subshells may be written as shown or reversed.) Copper thus has a filled $3d$ subshell. We would expect the next element in the periodic table ($Z = 30$, zinc) to have both the $3d$ and $4s$ subshells filled, and those following zinc would have electrons in the $4p$ subshell.

(d) Bromine ($Z = 35$) has 35 electrons. Filling shells in the order shown in Table 33.3, we have $1s^2 2s^2 2p^6 3s^2 3p^6 4s^2 3d^{10} 4p^5$. Bromine is one electron short of having a closed $4p$ subshell; it will readily accept another electron, making a Br^- ion whose electron configuration is extremely stable, consisting entirely of completely filled subshells.

PROBLEM: A helium-neon laser emits light of wavelength 632.8 nm, as an electron in the $5s$ subshell of neon falls to the $3p$ subshell. If the energy of the $5s$ state is 20.66 eV above the ground state, what is the energy of the $3p$ state?

Solution: The photon energy is equal to the *difference* in energy between the $5s$ and $3p$ levels.

$$\Delta E = E_{photon}$$

$$E_1 - E_2 = \frac{hc}{\lambda} = \frac{(6.63 \times 10^{-34} \text{ J}\cdot\text{s})(3 \times 10^8 \text{ m/s})}{632.8 \times 10^{-9} \text{ m}} \left(\frac{1 \text{ eV}}{1.6 \times 10^{-19} \text{ J}} \right)$$

$$20.66 \text{ eV} - E_2 = 1.96 \text{ eV}$$

$$E_2 = 20.66 - 1.96 = \underline{18.70 \text{ eV}}$$

The $3p$ state of neon is 18.7 eV above the ground state.

PROBLEM: Lasers can be used to drill holes in metal. An intensity of 5×10^7 W/cm^2 is required to produce a hole which is not sealed by the molten metal. Consider a laser pulse which delivers 50 J of energy, focused onto an area of 10^{-2} cm^2. (a) What is the duration of the pulse which will achieve the required intensity? (b) If 10% of the laser energy is used to melt the metal, how many pulses from this laser are required to melt 200 milligrams of iron ($L_f = 270$ J/g)?

Solution: (a) Intensity is power per unit area as can be seen by its units of W/cm^2. Power is energy per unit time, so

$$\text{Intensity} = \frac{P}{A} = \frac{E}{tA} = 5 \times 10^7 \text{ W/cm}^2$$

Solving for t gives

$$t = \frac{E}{A(5 \times 10^7 \text{ J/s}\cdot\text{cm}^2)}$$

where the unit for power (watt) has been expressed as J/s. Putting in numerical values, we get

$$t = \frac{50 \text{ J}}{(10^{-2} \text{ cm}^2)(5 \times 10^7 \text{ J/s}\cdot\text{cm}^2)} = 1 \times 10^{-4} \text{ s}$$

The duration of the pulse must be 0.1 ms to achieve the necessary intensity.

(b) 10 percent of the 50-J pulse, or $(0.1)(50 \text{ J}) = 5 \text{ J}$, is effective in melting the iron. To melt 0.2 g of iron will require a heat energy of

$$Q = mL_f = (0.2 \text{ g})(270 \text{ J/g}) = 54 \text{ J}$$

At 5 J per pulse, the number of pulses needed to deliver 54 J of energy is

$$\frac{54 \text{ J}}{5 \text{ J/pulse}} = 10.8 \approx \underline{11 \text{ pulses}}$$

The total time of laser action for the 11 pulses is $11(1 \times 10^{-4} \text{ s}) = 1.1 \text{ ms}$.

PROBLEM: What minimum voltage is required for an X-ray tube to produce X-rays of wavelength 0.02 nm?

Solution: An X-ray tube produces a maximum-frequency photon whose energy is related to the voltage of the tube by Eq. (33.5):

$$hf_{max} = eV$$

An electron accelerated by falling through a voltage V gains a kinetic energy equal to its loss of electrical potential energy, or $KE = eV$. If the electron is brought to rest in a single collision in the anode, all its kinetic energy appears as the energy of a radiated photon. Thus Eq. (33.5) is just a conservation of energy equation. Recalling that $f = c/\lambda$, and that the maximum frequency corresponds to the minimum wavelength, we may write

$$\frac{hc}{\lambda_{min}} = eV$$

Solving for V gives

$$V = \frac{hc}{\lambda_{min} e} = \frac{(6.63 \times 10^{-34} \text{ J} \cdot \text{s})(3 \times 10^8 \text{ m/s})}{(0.02 \times 10^{-9} \text{ m})(1.6 \times 10^{-19} \text{ C})} = 6.2 \times 10^4 \text{ V} = \underline{62,000 \text{ V}}$$

The tube must have a voltage of at least 62,000 V to produce these X-rays.

PROBLEM: If the voltage of an X-ray tube is doubled, by what factor will the minimum wavelength of the X-rays that it produces change?

Solution: Solving $\dfrac{hc}{\lambda_{min}} = eV$ for λ_{min} gives

$$\lambda_{min} = \frac{hc}{eV}$$

Thus if V doubles, λ_{min} will be half as great, or smaller by a factor of two. For the X-ray tube of the previous problem, increasing the voltage from 62,000 V to 124,000 V would cause the minimum wavelength produced by the tube to decrease from 0.02 nm to 0.01 nm.

Avoiding Pitfalls

1. The quantum numbers n and l cannot assume negative values; the quantum numbers m_l and m_s can be negative.

2. The quantum numbers n and m_s can never be zero; the quantum numbers l and m_l can have a value of zero.

3. It is traditional to use letters to designate the value of the orbital angular momentum quantum number l (see Table 33.2). But l may never equal n; its maximum value is $n - 1$.

4. No two electrons in an atom can have the same set of quantum numbers (that is, be in the same state).

5. In building up an atom, it is not necessary to fill all electron states in a given shell of quantum number n before electrons go to states of higher n. For example, the $4s$ subshell

is usually filled before electrons go into the $3d$ subshell. Table 33.3 is useful for determining the order in which the subshells are filled.

6. Laser action requires the presence of a metastable state, so that electrons in that state do not immediately make spontaneous transitions to the ground state but wait for the transition to be stimulated by a photon of the appropriate wavelength.

7. An X-ray tube produces a continuous energy distribution of photons because the accelerated electron may lose any fraction of its kinetic energy in a collision in the anode, resulting in emission of a photon whose energy is equal to the electron energy loss. The maximum energy photon results when the electron loses *all* its kinetic energy in one collision.

Drill Problems
Answers in Appendix

1. Select the one best answer for the following multiple choice questions:

   (a) How many electrons can a d subshell hold?
   (1) 2; (2) 6; (3) 8; (4) 10; (5) 14; (6)18.

   (b) The probability of finding an electron at a given point:
   (1) cannot be stated with precision;
   (2) is proportional to the value of the wave function ψ at that point;
   (3) is proportional to the square of the wave function ψ at that point;
   (4) is the same as the probability of finding it at any other point.

   (c) Lithium has atomic number $Z = 3$. Which of the following wave functions, indicated by listing their quantum numbers $|nlm_lm_s\rangle$, might represent the outermost electron in a lithium atom in its ground state?
   (1) $|1\ 0\ 0\ ½\rangle$; (2) $|1\ 0\ ½\ -½\rangle$; (3) $|1\ 1\ 0\ ½\rangle$; (4) $|2\ 0\ 0\ ½\rangle$;
   (5) $|2\ 0\ 1\ -½\rangle$; (6) $|2\ 1\ 2\ ½\rangle$.

   (d) In the operation of a laser, which one of the following is a necessary condition?
   (1) destructive interference; (2) inversion; (3) diffraction; (4) refraction.

   (e) There is a definite short wavelength limit to the continuous X-ray spectrum produced by a certain X-ray tube. This limit:
   (1) depends on the voltage across the tube only and is independent of the anode material;
   (2) depends on both the voltage across the tube and the material of the anode;
   (3) depends on the material of the anode only and is independent of the voltage across the tube;
   (4) depends on the energy levels of the anode atoms;
   (5) depends on other factors not mentioned in the above statements.

2. Determine the electron configuration of Xe ($Z = 54$).

3. An X-ray spectrum produced by a tube with a copper anode shows a characteristic line of wavelength 0.153 nm, caused by the transition of a copper electron from the $n = 2$ level to replace an electron knocked out of the $n = 1$ level.

    (a) Determine the difference in energy between these two levels in the copper atom.

    (b) Estimate the X-ray tube voltage required to have a spectrum which shows the 0.153-nm characteristic line.

    (c) What are the wavelength and frequency of the most energetic X-rays produced by this tube if the operating voltage is in fact 20,000 V?

The Nucleus and Radioactivity

34

Terms

Define or describe briefly what is meant by the following terms. If you have difficulty, refer to the textbook section given in parentheses.

neutron (34.1)

nucleons (34.1)

neutron number (34.1)

atomic number (34.1)

mass number (34.1)

isotope (34.1)

natural abundance (34.1)

strong nuclear force (34.2)

mass defect (34.3)

atomic mass unit (34.3)

binding energy (34.3)

nuclear reaction (34.4)

Q of a nuclear reaction (34.4)

radioactive decay (34.5)

alpha particle (34.5)

beta particle (34.5)

alpha decay (34.5)

beta decay (34.5)

neutrino (34.5)

gamma decay (34.5)

gamma ray photon (34.5)

half-life (34.6)

activity (34.6)

curie (34.6)

becquerel (34.6)

decay constant (34.6)

carbon dating (34.7)

decay series (34.8)

ionizing radiation (34.9)

exposure (34.9)

roentgen (34.9)

rad (34.9)

relative biological effectiveness (34.9)

effective dose (34.9)

rem (34.9)

cosmic rays (34.10)

genetic damage (34.10)

somatic damage (34.10)

Geiger counter (34.11)

scintillation detector (34.11)

semiconductor diode detector (34.11)

cloud chamber (34.11)

bubble chamber (34.11)

Equation Review

For each equation, be able to state the situation to which it applies, what quantity each symbol represents, and the units for measuring each quantity in some consistent set of units.

$$R = (1.2 \times 10^{-15} \text{ m})A^{1/3}$$

$$\Delta m_X = [Zm_H + (A - Z)m_n] - m_X \tag{34.1}$$

$$BE = \Delta mc^2 \tag{34.2}$$

$$1\text{ u} = 931.5\text{ MeV}/c^2 \tag{34.3}$$

$$BE/\text{nucleon} = \frac{BE}{A} \tag{34.4}$$

$$Q = (\Sigma m_{\text{reactants}} - \Sigma m_{\text{products}})c^2 \tag{34.6}$$

$$\text{Decay rate} = \frac{\Delta N}{\Delta t} \tag{34.7}$$

$$\frac{\Delta N}{\Delta t} = -\lambda N \tag{34.8}$$

$$\lambda = \frac{0.693}{T} \tag{34.9}$$

$$N = N_0 e^{-\lambda t} \tag{34.10}$$

$$1\text{ rad} = 10^{-2}\text{ J/kg} \tag{34.12}$$

$$\text{Effective dose (in rem)} = \text{dose (in rad)} \times \text{RBE} \tag{34.13}$$

Problems with Solutions and Discussion

PROBLEM: (a) Determine the radius of a tungsten-184 nucleus, $^{184}_{74}\text{W}$. (b) What nucleus has half the radius of a tungsten-184 nucleus?

Solution: (a) The nuclear radius is related to the number of nucleons A by

$$R = (1.2 \times 10^{-15}\text{ m})A^{1/3}$$

In a tungsten-184 nucleus, $A = 184$, so its radius is

$$R = (1.2 \times 10^{-15}\text{ m})(184)^{1/3} = (1.2 \times 10^{-15}\text{ m})(5.69) = \underline{6.83 \times 10^{-15}\text{ m}}$$

(b) A nucleus with radius R' which is one half R satisfies the same relationship:

$$R' = (1.2 \times 10^{-15}\text{ m})A'^{1/3}$$

$$\text{where } R' = \frac{R}{2} = \frac{6.83 \times 10^{-15}\text{ m}}{2} = 3.415 \times 10^{-15}\text{ m}$$

$$3.415 \times 10^{-15}\text{ m} = (1.2 \times 10^{-15}\text{ m})A'^{1/3}$$

$$A'^{1/3} = 2.846$$

Cubing we get

$$A' = 23$$

A <u>sodium-23 nucleus</u> has half the radius of a tungsten-184 nucleus.

PROBLEM: Determine the chemical symbols for the nuclei which have the following properties: (a) $Z = 5$, $N = 6$; (b) $Z = 6$, $N = 5$; (c) $Z = 11$, $A = 24$; (d) $Z = 36$, $A = 84$; (e) $N = 19$, $A = 35$; (f) $N = 16$, $A = 32$. (g) Which of these are isotopes of the same element?

Solution: The nuclear symbol is written in the form $^A_Z X$, where X is the chemical symbol of the element. Z is the number of protons, called the atomic number, and it determines the element. When Z is known, the element can be identified from Appendix C in the textbook or a periodic table; its chemical symbol also appears there. N is the number of neutrons and A, the mass number, is $Z + N$, the total number of nucleons.

(a) $Z = 5$ and $A = Z + N = 5 + 6 = 11$. Referring to Appendix C, the element with atomic number 5 is boron; chemical symbol, B. The symbol for this nucleus is $^{11}_5\text{B}$.

(b) $Z = 6$ and $A = Z + N = 6 + 5 = 11$. Carbon has atomic number 6; the appropriate symbol for this nucleus is $^{11}_6\text{C}$. Although this nucleus and $^{11}_5\text{B}$ have the same mass number A, they are not isotopes of the same element since they do not have the same number of protons.

(c) $Z = 11$ is the atomic number for sodium; chemical symbol, Na. Since $A = 24$, the symbol for this nucleus is $^{24}_{11}\text{Na}$.

(d) Looking up $Z = 36$ in Appendix C of the textbook, we find that krypton (Kr) has this atomic number. The symbol is $^{84}_{36}\text{Kr}$.

(e) Z is not given but it can be determined from $A = Z + N$. Therefore, $Z = A - N = 35 - 19 = 16$. Sulfur (S) has atomic number 16, so the required symbol is $^{35}_{16}\text{S}$.

(f) Again, A is determined from $Z = A - N = 32 - 16 = 16$. This is also sulfur: $^{32}_{16}\text{S}$.

(g) $^{32}_{16}\text{S}$ and $^{35}_{16}\text{S}$ are isotopes of the same element. They both have 16 protons but one has three more neutrons than the other.

PROBLEM: Which nucleus is more stable, $^{15}_8\text{O}$ (15.003065 u) or $^{15}_7\text{N}$ (15.000109 u)?

Solution: That nucleus which has more binding energy per nucleon will be the more stable. The binding energy of a nucleus is the energy corresponding to the *mass defect,* the difference between the mass of an atom's constituent particles and its atomic mass. It is calculated from Eq. (34.1):

$$\Delta m_X = \underbrace{[Zm_{\text{H}} + (A - Z)m_n]}_{\text{mass of constituent particles}} - \underbrace{m_X}_{\text{atomic mass}}$$

The term Zm_{H} is the mass of the protons and electrons, since m_{H} is the mass of a hydrogen atom (one proton and one electron), and the term $(A - Z)m_n = Nm_n$ is the mass of the N neutrons.

For oxygen-15: $Z = 8$, $(A - Z) = N = 15 - 8 = 7$, m_0 is given above, and m_{H} and m_n are obtained from Table 34.2 of the textbook. Putting in these values gives

$$\Delta m_0 = Zm_{\text{H}} + Nm_n - m_0 = 8(1.007825 \text{ u}) + 7(1.008665 \text{ u}) - 15.003065 \text{ u}$$

$$= 8.06260 \text{ u} + 7.060655 \text{ u} - 15.003065 \text{ u} = 15.123255 \text{ u} = 0.12019$$

Using Eq. (34.3), 1 u = 931.5 MeV/c^2, as a conversion factor, we get

$$\Delta m = 0.12019 \text{ u} \left(\frac{931.5 \text{ MeV}/c^2}{1 \text{ u}} \right) = 112 \text{ MeV}/c^2$$

Einstein's equation gives the corresponding binding energy:

$$BE = \Delta mc^2 = (112 \text{ MeV}/c^2)c^2 = 112 \text{ MeV}$$

This is the total binding energy of the nucleus, which contains 15 nucleons. The binding energy per nucleon is

$$\frac{BE}{\text{nucleon}} = \frac{112 \text{ MeV}}{15 \text{ nucleons}} = 7.47 \frac{\text{MeV}}{\text{nucleon}}$$

The calculation of the binding energy per nucleon for the $^{15}_7\text{N}$ nucleus can be carried out in the same way. It can also be set up as a simple arithmetic problem as follows:

$^{15}_7\text{N}$:			
$7 \text{ H} =$	$7(1.007825 \text{ u})$;	7.054775 u	protons + electrons
$+ 8 \text{ n} =$	$8(1.008665 \text{ u})$;	$+ \ 8.069320 \text{ u}$	neutrons
		15.124095 u	total mass of constituents
		$- \ 15.000109 \text{ u}$	mass of atom
$\Delta m =$		0.123986 u	mass defect

252

Since 1 u of mass corresponds to 931.5 MeV of energy,

$$BE = (0.123986)(931.5) = 115.5 \text{ MeV}$$

The binding energy per nucleon in this case is

$$\frac{BE}{\text{nucleon}} = \frac{115.5 \text{ MeV}}{15 \text{ nucleons}} = 7.70 \frac{\text{MeV}}{\text{nucleon}}$$

^{15}N has a greater binding energy per nucleon than ^{15}O. We therefore conclude that $\underline{^{15}\text{N is}}$ $\underline{\text{a more stable nucleus than } ^{15}\text{O}}$. (In fact, Appendix C reveals that ^{15}O is radioactive with a half-life of 122 s.)

We might have guessed the correct answer to this problem by observing the relative masses of the two atoms. Since neutrons have greater mass than protons and $^{15}_{7}$N has more neutrons than $^{15}_{8}$O, we would expect the $^{15}_{7}$N nucleus to be the more massive of the two. However, it is less massive, indicating a greater mass defect and a greater binding energy per nucleon. If the two nuclei being compared do *not* have the same total number of nucleons, no useful information can be deduced by looking at their atomic masses, and the entire calculation must be carried out.

PROBLEM: Determine the missing nucleus in each of the following nuclear reactions:
(a) Silicon-28 is bombarded with alpha particles, causing the ejection of a proton:

$$^{28}_{14}\text{Si} + {}^{4}_{2}\text{He} \rightarrow {}^{1}_{1}\text{H} + \underline{\quad \overset{?}{\quad} \quad}$$

(b) In a neutron detector, a boron-10 nucleus absorbs a neutron and an alpha particle is emitted:

$$^{10}_{5}\text{B} + {}^{1}_{0}n \rightarrow {}^{4}_{2}\text{He} + \underline{\quad \overset{?}{\quad} \quad}$$

(c) Thorium-228 spontaneously undergoes alpha decay:

$$^{228}_{90}\text{Th} \rightarrow {}^{4}_{2}\text{He} + \underline{\quad \overset{?}{\quad} \quad}$$

(d) Iodine-131 emits a negative beta particle:

$$^{131}_{53}\text{I} \rightarrow {}^{0}_{-1}e + \underline{\quad \overset{?}{\quad} \quad}$$

(e) Nitrogen-13 undergoes positron decay:

$$^{13}_{7}\text{N} \rightarrow {}^{0}_{1}e + \underline{\quad \overset{?}{\quad} \quad}$$

(f) Uranium-235 undergoes neutron-induced fission, which produces 3 neutrons and 2 fission fragments:

$$^{235}_{92}\text{U} + {}^{1}_{0}n \rightarrow \underline{\quad \overset{?}{\quad} \quad} + {}^{89}_{35}\text{Br} + 3 {}^{1}_{0}n$$

Solution: (a) The total Z entering the reaction is $14 + 2 = 16$; therefore the Z of the missing nucleus must be 15, to make a total Z of 16 after the reaction. Referring to Appendix C, the atom with $Z = 15$ is phosphorus, P. Similarly for A: entering the reaction are $28 + 4 = 32$ nucleons; only if the unknown mass number is 31 will there still be 32 nucleons after the reaction. The missing nucleus is $\underline{^{31}_{15}\text{P}}$.

(b) To conserve both Z and A in the reaction, the missing nucleus must have $Z = 3$ and $A = 7$. It must therefore be $\underline{^{7}_{3}\text{Li}}$.

(c) Since this decay occurs spontaneously, only one reactant appears on the left side of the equation. No other particle is necessary to make this reaction occur. Alpha decay reduces Z by 2 and A by 4; the remaining daughter nucleus is $\underline{^{224}_{88}\text{Ra}}$.

(d) A negatively charged electron is ejected from the nucleus, so one neutron has become a proton; Z increases by 1 but the *number* of nucleons A has not changed. The daughter nucleus is $\underline{^{131}_{54}\text{Xe}}$.

(e) The ejection of a positron from the nucleus means a proton has become a neutron; Z decreases by 1 while A is unchanged. The daughter nucleus is $\underline{^{13}_{6}\text{C}}$.

(f) On the left we have $Z = 92$; on the right $Z = 35 + ? = 92$. The missing atomic number is $92 - 35 = 57$, which identifies lanthanum, according to Appendix C. The mass numbers on the left add up to 236; on the right, $236 = ? + 89 + 3(1)$. The missing mass number is $236 - 92 = 144$. The other fission fragment is $^{144}_{57}$La.

PROBLEM: (a) Determine whether or not radioactive sulfur-35, $^{35}_{16}$S, is an alpha emitter. (b) What is the Q for alpha emission?

Solution: (a) We are asked whether a spontaneous alpha decay of ^{35}S is possible. This reaction is represented by:

$$^{35}_{16}\text{S} \xrightarrow{?} {}^{4}_{2}\text{He} + {}^{31}_{14}\text{Si}$$

The daughter nucleus is determined by requiring the sum of the superscripts and subscripts on both sides of the reaction to be equal. The reaction will occur spontaneously if the decay products (alpha and silicon-31) are less massive than the parent. Referring to Appendix C of the textbook for the required masses gives:

$$\begin{array}{ll} {}^{35}\text{S: } 34.969033 \text{ u} & {}^{4}\text{He: } 4.002603 \text{ u} \\ & {}^{31}\text{Si: } \underline{30.975364 \text{ u}} \\ & \phantom{{}^{31}\text{Si: }} 34.977967 \text{ u} \end{array}$$

We see that a sulfur-35 nucleus does not have enough mass to form the necessary decay products of an alpha decay; this reaction will not occur spontaneously.

(b) The Q-value of a reaction is the energy corresponding to the mass difference between the reactants and the product nuclei:

$$\Delta m = 34.969033 - 34.977967 = -0.008934 \text{ u}$$

So the Q value is $(-0.008934)(931.5) = \underline{-8.3 \text{ MeV}}$. A negative Q implies that the reaction cannot take place unless the nucleus in question somehow receives at least that much excess energy. The reaction will not occur spontaneously.

PROBLEM: A radioactive sample has a decay constant of 5.615×10^{-7} s^{-1}. (a) What is its half-life? (b) What might this sample be? (c) How much will be left after 43 days?

Solution: (a) Half-life and decay constant are related by Eq. (34.9):

$$\lambda = \frac{0.693}{T}$$

Solving for T gives

$$T = \frac{0.693}{\lambda} = \frac{0.693}{5.615 \times 10^{-7} \text{ s}^{-1}} = 1.234 \times 10^{6} \text{ s}$$

Converting to days, we get

$$T = 1.234 \times 10^{6} \text{ s} \left(\frac{1 \text{ day}}{8.64 \times 10^{4} \text{ s}} \right) = \underline{14.3 \text{ days}}$$

(b) The half-lives of selected isotopes are listed in column (7) of Appendix C in the textbook. Scanning down this column, we find that this sample may be phosphorus-31, which has a half-life of 14.28 days.

(c) During each half-life the number of radioactive nuclei N remaining decreases by a factor of two; thus after n half-lives, the number decreases by 2^{n}. After 43 days, $43/14.3 \simeq 3$ half-lives have elapsed. So the original N_0 nuclei have decreased to

$$N = \frac{N_0}{2^3} = \frac{N_0}{8} = \underline{0.125 \ N_0}$$

About 12½ percent of the original sample is left after 43 days.

PROBLEM: The half-life of tritium, ^3_1H, is 12.33 yr. (a) Determine the activity of 0.15 milligrams of this substance. Express your result in curies. (b) What percent of the original ^3_1H nuclei will remain after 79 years?

Solution: (a) The activity or decay rate is the number of decays per second that occur, $\Delta N/\Delta t$. This depends on the number of radioactive nuclei N present and the decay constant λ. According to Eq. (34.8):

$$\frac{\Delta N}{\Delta t} = -\lambda N$$

The decay constant can be obtained from the half-life T by Eq. (34.9):

$$\lambda = \frac{0.693}{T} = \frac{0.693}{(12.33 \text{ yr})(3.16 \times 10^7 \text{ s/yr})} = 1.78 \times 10^{-9} \text{ s}^{-1}$$

To get $\Delta N/\Delta t$, we need the number of atoms N in one milligram of ^3_1H. The atomic mass of tritium is 3, so one mole of tritium has a mass of 3 grams. Recall also that 1 mole of any substance contains Avogadro's number of atoms, or 6.02×10^{23} atoms. The mass of our sample is $1.5 \times 10^{-4} \text{ g}/3 \text{ g} = 5 \times 10^{-5}$ of a mole, so it contains

$$5 \times 10^{-5} \text{ mole} \left(\frac{6.02 \times 10^{23} \text{ atoms}}{1 \text{ mole}} \right) = 3 \times 10^{19} \text{ atoms}$$

The activity of the sample is therefore

$$\frac{\Delta N}{\Delta t} = \lambda N = (1.78 \times 10^{-9} \text{ s}^{-1})(3 \times 10^{19} \text{ atoms}) = \underline{5.34 \times 10^{10} \text{ decays/s}}$$

(We have omitted the negative sign; it simply indicates that as t increases, N decreases.) Expressing this result in curies, we get

$$5.34 \times 10^{10} \text{ decays/s} \left(\frac{1 \text{ Ci}}{3.7 \times 10^{10} \text{ decays/s}} \right) = \underline{1.4 \text{ Ci}}$$

This is a very large activity.

(b) After 79 years, $79/12.33 = 6.4$ half-lives have elapsed. Even though the elapsed time is not an integral number of half-lives, we may use the $1/2^n$ method, provided we have a calculator with a y^x function key.

$$N = \frac{N_0}{2^n} = \frac{N_0}{2^{6.4}} = \frac{N_0}{84.4} = \underline{0.012 N_0}$$

About 1 percent of the original sample is left after 79 years.

An alternate method of solution uses Eq. (34.10): $N = N_0 e^{-\lambda t}$. If we substitute λ in units of s^{-1}, t must be in seconds so that the exponent of e will be unitless.

$$t = 79 \text{ yr}(3.16 \times 10^7 \text{ s/yr}) = 2.49 \times 10^9 \text{ s}$$

Putting in these numerical values we get

$$N = N_0 e^{-\lambda t} = N_0 e^{-(1.78 \times 10^{-9} \text{s}^{-1})(2.49 \times 10^9 \text{s})} = N_0 e^{-4.43} = \underline{0.012 \, N_0 \text{ or } 1.2 \text{ percent of } N_0}$$

This result is in agreement with that obtained previously.

PROBLEM: A metastable state of technetium-99 which decays by emitting a gamma ray of energy 140 keV has wide use in nuclear medicine. Suppose 10^{11} of these photons are absorbed by 6 kg of your body's mass. (a) What dose of radiation in rad have you received? (b) What effective dose in rem have you received? (c) How many 8-MeV alpha particles would constitute the same effective dose, when absorbed by the same body mass?

Solution: (a) The dose is the energy received per unit mass. Each gamma ray has an energy of

$$(140 \times 10^3 \text{ eV})(1.6 \times 10^{-19} \text{ J/eV}) = 2.24 \times 10^{-14} \text{ J}$$

The energy per unit mass received by the body will be

$$\frac{E}{m} = \frac{(\text{No. of photons})(\text{energy per photon})}{m}$$

$$= \frac{(10^{11})(2.24 \times 10^{-14} \text{ J})}{6 \text{ kg}} = 3.7 \times 10^{-4} \text{ J/kg}$$

Using Eq. (34.12) to convert to rads gives

$$3.7 \times 10^{-4} \text{ J/kg} \left(\frac{1 \text{ rad}}{10^{-2} \text{ J/kg}} \right) = \underline{3.7 \times 10^{-2} \text{ rad}} = \underline{37 \text{ mrad}}$$

(b) The effective dose in rem is obtained by multiplying the dose in rad by the relative biological effectiveness (RBE) of the particular type of radiation. Referring to Table 34.4, gamma rays have an RBE of 1.0; therefore the dose in rad and the effective dose in rem are numerically equal. Your effective dose is 37 mrem, or $\underline{3.7 \times 10^{-2} \text{ rem}}$.

(c) Alpha particles have an RBE of 10, so the same effective dose (3.7×10^{-2} rem) is received by absorbing a small number of rad. Solving Eq. (34.13) for dose (in rad), we find that

$$\text{Dose (in rad)} = \frac{\text{effective dose (in rem)}}{\text{RBE}} = \frac{3.7 \times 10^{-2} \text{ rem}}{10} = 3.7 \times 10^{-3} \text{ rad}$$

The energy received per mass is thus

$$3.7 \times 10^{-3} \text{ rad} \left(\frac{10^{-2} \text{ J/kg}}{1 \text{ rad}} \right) = 3.7 \times 10^{-5} \text{ J/kg}$$

The total mass of 6 kg receives

$$(3.7 \times 10^{-5} \text{ J/kg})(6 \text{ kg}) = 2.2 \times 10^{-4} \text{ J}$$

Now each alpha particle delivers 8 MeV of energy. Converting this to joules, we get

$$(8 \times 10^{6} \text{ eV})(1.6 \times 10^{-19} \text{ J/eV}) = 1.28 \times 10^{-12} \text{ J}$$

So the number of alpha particles required is

$$\text{No. of } \alpha = \frac{\text{total energy received}}{\text{energy per } \alpha} = \frac{2.2 \times 10^{-4} \text{ J}}{1.28 \times 10^{-12} \text{ J}} = \underline{1.7 \times 10^{8}}$$

The number of alpha particles we have calculated is considerably smaller than the number of gamma rays (10^{11}) because the alpha particles have higher energy than the gamma rays and they do more damage per unit distance than gamma rays of equal energy. This last calculation, though interesting, is not of physical consequence, because in actual fact, it would be very difficult to distribute the dose of alpha radiation uniformly through 6 kg of body mass. The range of penetration of an alpha particle is very short. Several centimeters of air or a sheet of paper—or the body's layer of skin—is sufficient to stop most alpha particles. They would expend all their energy in the outer layer of the skin and the vast majority of the 6-kg body mass would receive no dose at all.

Avoiding Pitfalls

1. It is the atomic number Z which determines the element to which a nucleus belongs. Z can be located on a periodic table or an isotope chart to verify the element's chemical symbol.

2. The atomic mass unit u is not an *mks* unit, nor is the energy unit of electron volts, eV. A useful relationship to memorize for easy conversions of mass to energy and vice versa, without having to convert either of these hybrid units back into *mks* units, is that the rest mass energy of 1 u of mass is 931.5 MeV.

3. Since the mass defect of a stable atom is quite small, it is necessary to carry 5 or 6 significant figures in the atomic masses used to calculate it, in order to obtain the binding energy to 2 or 3 significant figures.

4. Nuclear reactions conserve charge (total Z does not change), number of nucleons (total A does not change), and energy (provided mass is included as a form of energy).

5. A nuclear reaction which occurs spontaneously (for example, α, β, or γ decay of naturally radioactive elements) has only one nucleus on the left side of the reaction equation. A reaction which is artificially induced will have two reactants on the left side of the equation.

6. In using equations relating decay constant λ, half-life T, and elapsed time t, any convenient time units may be used provided they are consistent throughout. For example, in using $N = N_0 e^{-\lambda t}$, if λ is in yr^{-1}, t must be in yr to insure that the exponent is unitless.

Drill Problems
Answers in Appendix

1. Write equations representing the following nuclear transformations:

    (a) Bismuth-214 undergoes alpha decay;
    (b) Neutrons are produced when beryllium-9 is bombarded with alpha particles;
    (c) In a plutonium breeder reactor, uranium-238 captures a neutron to form uranium 239. This is followed by two negative beta decays. Write the equations for these three nuclear transformations.

2. How much energy is required to remove a proton from an iron-56 (55.934939 u) to form manganese-55 (54.938046 u)?

3. Suppose a laboratory is supplied with radioactive iodine-131 once a month. The half-life of $^{131}_{53}\text{I}$ is about 8 days.

    (a) How much iodine-131 should be bought on June 1, if an experiment requiring 3 micrograms is to be performed on June 24?
    (b) What is the sample's activity on June 24? (The number of atoms in 3 μg of $^{131}_{53}\text{I}$ is 1.38×10^{16} atoms—you may wish to confirm this by a calculation.)

35

Fission, Fusion, and Elementary Particles

Terms

Define or describe briefly what is meant by the following terms. If you have difficulty, refer to the textbook section given in parentheses.

fission (35.1)

nuclear power plant (35.2)

enriched uranium (35.2)

reactor core (35.2)

moderator (35.2)

deuterium (35.2)

chain reaction (35.2)

control rods (35.2)

breeder reactor (35.3)

fusion (35.4)

proton-proton cycle (35.4)

elementary particles (35.5)

quarks (35.5, 35.7)

antiparticle (35.5)

hadrons (35.6)

leptons (35.6)

baryons (35.6)

conservation of baryon number (35.6)

strange particles (35.6)

conservation of strangeness (35.6)

quantum chromodynamics (35.7)

mesons (35.7)

gluons (35.7)

Problems with Solutions and Discussion

PROBLEM: Estimate the amount of energy that would be released if a nucleus of mass number $A = 200$ could be made to undergo fission into 4 equal pieces, each with $A = 50$.

Solution: Referring to the average binding energy per nucleon curve given in Fig. 35.2 of the textbook:

$$\text{For } A = 200, \quad BE/\text{nucleon} \simeq 7.9 \text{ MeV}$$
$$\text{For } A = 50, \quad BE/\text{nucleon} \simeq 8.8 \text{ MeV}$$

(The higher binding energy per nucleon signifies a greater mass defect for the smaller nuclei, or more missing mass. This excess missing mass is converted to energy during fission.) The difference is $(8.8 - 7.9) = 0.9$ MeV/nucleon. So for 200 nucleons participating in the fission, the energy release is about

$$(0.9 \text{ MeV/nucleon})(200 \text{ nucleons}) = \underline{180 \text{ MeV}}$$

PROBLEM: Consider a neutron-induced fission of a uranium-235 nucleus, in which 2 neutrons are ejected:

$$\underset{(235.0439 \text{ u})}{{}^{235}_{92}\text{U}} \quad + \quad \underset{(1.0087 \text{ u})}{{}^{1}_{0}n} \quad \rightarrow \quad \underset{(95.9326 \text{ u})}{{}^{96}_{37}\text{Rb}} \quad + \quad \underline{\quad ? \quad} \quad + \quad 2{}^{1}_{0}n \quad + \quad 178.4 \text{ MeV}$$

Determine the identity and the mass of the missing fission fragment.

Solution: Conserving Z gives $92 + 0 = 37 + \underline{\quad ? \quad} + 0$; the missing isotope has $Z = 92 - 37 = 55$. Referring to Appendix C, we find that cesium (Cs) has an atomic number of 55. Conserving A gives $235 + 1 = 96 + ? + 2$; the missing isotope has $A = 236 - 98 = 138$. The identity of the fragment is ${}^{138}_{55}\text{Cs}$.

The equation relating the various masses is the defining equation for the Q value of a reaction:

$$Q = [\Sigma m_{\text{reactants}} - \Sigma m_{\text{products}}]c^2$$

where Q is the energy released in the reaction, in this case 178.4 MeV, and is the energy equivalent to the mass difference between the reactants and products.

$$\Delta m\cancel{c^2} = [\Sigma m_{\text{reactants}} - \Sigma m_{\text{products}}]\cancel{c^2}$$

Since a mass of 1 u has a rest energy of 931.5 MeV, the energy of 178.4 MeV is equivalent to a mass of

$$\frac{178.4 \text{ MeV}}{931.5 \text{ MeV}} (1 \text{ u}) = 0.1915 \text{ u} = \Delta m$$

The mass of the reactants is

$$\Sigma m_{\text{reactants}} = m_{\text{U}} + m_n = 235.0439 \text{ u} + 1.0087 \text{ u} = 236.0526 \text{ u}$$

and the mass of the products is

$$\Sigma m_{\text{products}} = m_{\text{Rb}} + m_{\text{Cs}} + 2m_n = 95.9326 + m_{\text{Cs}} + 2(1.0087) = 97.9500 \text{ u} + m_{\text{Cs}}$$

Solving the mass equation above for $\Sigma m_{products}$ and putting in numerical values gives

$$\Sigma m_{products} = \Sigma m_{reactants} - \Delta m$$

$$95.9500 \text{ u} + m_{Cs} = 236.0526 \text{ u} - 0.1915 \text{ u}$$

$$m_{Cs} = 235.8611 - 97.9500 = \underline{137.9111 \text{ u}}$$

The mass of $^{138}_{55}Cs$ is 137.9111 u.

PROBLEM: Determine the power output of a nuclear reactor if 2 kg of $^{235}_{92}U$ is fissioned in 10 days. Assume an energy release of 200 MeV/fission.

Solution: We first determine the number of uranium-235 atoms in 2 kg. Since 1 mole has a mass of 235 grams, 2 kg or 2000 g contains $^{2000}/_{235}$ = 8.51 moles. At 6.02×10^{23} atoms/mole, we have

$$\left(6.02 \times 10^{23} \frac{\text{atoms}}{\text{mole}} \right) (8.51 \text{ moles}) = 5.1 \times 10^{24} \text{ atoms}$$

Thus if 2 kg of $^{235}_{92}U$ are used up, 5.1×10^{24} fissions have occurred, releasing an energy of

$$(5.1 \times 10^{24} \text{ fissions}) \left(200 \frac{\text{MeV}}{\text{fission}} \right) = 1 \times 10^{27} \text{ MeV}$$

Converting to joules gives

$$(1 \times 10^{27} \text{ MeV}) \left(\frac{10^6 \text{ eV}}{1 \text{ MeV}} \right) \left(\frac{1.6 \times 10^{-19} \text{ J}}{1 \text{ eV}} \right) = 1.6 \times 10^{14} \text{ J}$$

This energy is released over a period of 10 days. Converting to seconds, we get

$$t = 10 \text{ days}(8.64 \times 10^4 \text{ s/day}) = 8.64 \times 10^5 \text{ s}$$

Since the power output is the energy released per unit time, then

$$P = \frac{E}{t} = \frac{1.6 \times 10^{14} \text{ J}}{8.64 \times 10^5 \text{ s}} = 1.85 \times 10^8 \text{ J/s} = \underline{185 \text{ MW}}$$

The reactor's power output is 185 megawatts.

PROBLEM: It is thought that if a star has created enough helium and is sufficiently hot, three alpha particles can have a triple collision and form carbon. What is the Q value of this fusion reaction?

Solution:

$$^4_2He + ^4_2He + ^4_2He \rightarrow ^{12}_6C + Q$$

The mass difference is

$$\Delta m = \Sigma m_{reactants} - \Sigma m_{products} = 3m_{He} - m_C$$

where the atomic mass of carbon-12 is defined as 12.000. . . . u and the mass of ^4_2He is 4.002603 u.

$$\Delta m = 3(4.002603 \text{ u}) - 12.000000 = 12.007809 \text{ u} - 12.000000 \text{ u} = 0.007809 \text{ u}$$

The corresponding energy is obtained using $E = mc^2$ and the conversion factor to express the result in MeV:

$$Q = \Delta mc^2 = (0.007809 \text{ u})c^2 \left(\frac{931.5 \text{ MeV}/c^2}{1 \text{ u}} \right) = \underline{7.27 \text{ MeV}}$$

This fusion reaction releases 7.3 MeV of energy.

PROBLEM: Compare the efficiency of (a) fission and (b) fusion by calculating the percent of the total mass that is converted to energy in each case. Assume the energy release is 200 MeV per uranium fission and about 24 MeV for the fusion of two deuterons into an alpha particle.

Solution: (a) 200 MeV of energy corresponds to the conversion of a certain amount of mass, Δm. Since the energy content of 1 u of mass is 931.5 MeV, the mass corresponding to 200 MeV can be determined by a simple proportion:

$$\frac{200 \text{ MeV}}{931.5 \text{ MeV}} = \frac{\Delta m}{1 \text{ u}}$$

$$\Delta m = 0.2147 \text{ u}$$

The total mass of a uranium-235 nucleus is about 235 u, so the fraction of the total mass which is converted is

$$\frac{\Delta m}{m} = \frac{0.2147 \text{ u}}{235 \text{ u}} = 0.0009 \simeq \underline{0.001}$$

About one-tenth of one percent, or 0.1 percent, of the total mass is converted to energy in a fission reaction.

(b) The calculation is similar for fusion: an energy release of 24 MeV corresponds to a mass loss of

$$\Delta m = 1 \text{ u} \left(\frac{24 \text{ MeV}}{931.5 \text{ MeV}} \right) = 0.02576 \text{ u}$$

The total mass participating in the fusion is the mass of the two deuterons, or ~ 4 u. So

$$\frac{\Delta m}{m} = \frac{0.02576 \text{ u}}{4 \text{ u}} = 0.0064 \simeq \underline{0.6 \text{ percent}}$$

The percent of total mass which is converted to energy is more than six times greater for fusion (0.6 percent) than for fission (0.1 percent).

PROBLEM: Determine whether the laws of conservation of charge, baryon number, and strangeness are violated in any of the following reactions involving elementary particles:

(a) $p + n \rightarrow \Sigma^0 + K^0 + \pi^+$
(b) $\eta^0 \rightarrow \pi^+ + \pi^+ + \pi^-$
(c) $K^+ \rightarrow \pi^+ + \pi^+ + \pi^-$
(d) $\pi^+ + n \rightarrow K^+ + K^0$
(e) $\pi^- + p \rightarrow \Lambda^0 + K^0$
(f) $\pi^- + p \rightarrow \Sigma^0 + \eta^0$

Solution: For each reaction we will look at charge (q), baryon number (B), and strangeness (S) to see if each is conserved.

(a) q: $1 + 0 \rightarrow 0 + 0 + 1$; Conserved. (Charge is expressed in units of e, the proton charge.)
 B: $1 + 1 \rightarrow 1 + 0 + 0$; Not conserved.
 S: $0 + 0 \rightarrow -1 + 1 + 0$; Conserved.
This reaction violates conservation of baryon number.

(b) q: $0 = +1 + 1 - 1$; Not conserved.
 B: $0 = 0 + 0 + 0$; Conserved.
 S: $0 = 0 + 0 + 0$; Conserved.
This reaction violates conservation of charge.

(c) q: $1 = +1 + 1 - 1$; Conserved.
 B: $0 = 0 + 0 + 0$; Conserved.
 S: $1 = 0 + 0 + 0$; Not conserved.
This reaction violates conservation of strangeness.

(d) q: $1 + 0 = 1 + 0$;   Conserved.
  B: $0 + 1 = 0 + 0$;   Not conserved.
  S: $0 + 0 = 1 + 1$;   Not conserved.
  This reaction violates conservation of baryon number and strangeness.
(e) q: $-1 + 1 = 0 + 0$;   Conserved.
  B: $0 + 1 = 1 + 0$;   Conserved.
  S: $0 + 0 = -1 + 1$;   Conserved.
  This reaction does not violate any conservation laws; it is an observed reaction.
(f) q: $-1 + 1 = 0 + 0$;   Conserved.
  B: $0 + 1 = 1 + 0$;   Conserved.
  S: $0 + 0 = -1 + 0$;   Not conserved.
  This reaction violates conservation of strangeness.

PROBLEM: By applying the conservation laws, deduce what the missing particle in the following reaction might be: $p + p \rightarrow \pi^+ + n + \Lambda^0 + \underline{\quad ? \quad}$

Solution: q: $1 + 1 = 1 + 0 + 0 + \underline{\quad ? \quad}$; to conserve charge, q must be $+1$ for the unknown particle.
  B: $1 + 1 = 0 + 1 + 1 + \underline{\quad ? \quad}$; to conserve baryon number, the unknown particle must have a baryon number of 0.
  S: $0 + 0 = 0 + 0 - 1 + \underline{\quad ? \quad}$; the missing S must be $+1$ to conserve strangeness. The particle has $q = +1$, $B = 0$, $S = +1$. Referring to Table 35.2, the particle possessing these properties is the $\underline{K^+ \text{ meson}}$.

PROBLEM: Table 35.2 shows that a neutral pion decays into two photons: $\pi^0 \rightarrow \gamma + \gamma$. Find the energy of the photons. Why must they be equal?

Solution: $\underline{\text{Momentum conservation}}$ requires the two photons to be equal. Assuming the pion to be at rest, it has zero momentum before the decay. The only way the momentum can still be zero after the decay is if the two photons have equal momentum and travel in opposite directions.
  The energy of the photons must equal the rest energy of the pion in order to conserve energy.

$$2E_\gamma = m_\pi c^2 = 135 \text{ MeV (from Table 35.2)}$$
$$E_\gamma = \frac{135 \text{ MeV}}{2} = \underline{67.5 \text{ MeV}}$$

Each photon has an energy of 67.5 MeV and they travel in opposite directions.

PROBLEM: Confirm that the charge, baryon number, and strangeness of the Σ^- equals the sum of the number for its constituent quarks (dds).

Solution: The properties of the three Σ^- quarks are obtained from Table 35.3 and are listed and summed below:

	d		d		s		
$q =$	$(-\tfrac{1}{3}e)$	$+$	$(-\tfrac{1}{3}e)$	$+$	$(-\tfrac{1}{3}e)$	$=$	$-e$
$B =$	$\tfrac{1}{3}$	$+$	$\tfrac{1}{3}$	$+$	$\tfrac{1}{3}$	$=$	1
$S =$	0	$+$	0	$+$	(-1)	$=$	-1

Referring to Table 35.2, we find that the Σ^- has a baryon number of 1, a strangeness of -1, and a charge of $-e$, in agreement with the sums of the individual quark properties.

PROBLEM: Determine the charge, baryon number, and strangeness of the Δ^- hadron which is composed of the quark triplet (ddd).

Solution:

$$
\begin{array}{ccccccc}
 & d & & d & & d & \\
q = & (-\tfrac{1}{3}\,e) + & & (-\tfrac{1}{3}\,e) + & & (-\tfrac{1}{3}\,e) = & -e \\
B = & \tfrac{1}{3} & + & \tfrac{1}{3} & + & \tfrac{1}{3} & = & 1 \\
S = & 0 & + & 0 & + & 0 & = & 0
\end{array}
$$

The Δ^- hadron has $q = -e$, $B = 1$, and $S = 0$.

Avoiding Pitfalls

1. In checking to see if elementary particle reactions obey conservation laws, recall that an antiparticle has the same mass but the *opposite electric charge* and the *opposite strangeness* as the particle itself.

2. In elementary particle physics it is traditional to give rest mass energies (in MeV) rather than atomic masses (in u). But mass can always be easily obtained from rest mass energy if you recall that the energy content of 1 u is 931.5 MeV.

Drill Problems
Answers in Appendix

1. Relatively slow protons (a few keV or less) can cause fission of a lithium-7 nucleus into two alpha particles. What is the energy release for this reaction?

2. The first hydrogen bomb was exploded in November 1952 on an island in the Pacific. It had an energy release equivalent to about 5 million tons of TNT (and it completely destroyed the island). For how long could this energy operate a 1000-MW power plant? (1 ton of TNT releases 4.2×10^9 J of energy.)

3. The following reactions do not occur. State what conservation law is violated in each case.

(a) $p + \bar{n} \rightarrow n + \bar{p}$;

(b) $\pi^- + p \rightarrow \Sigma^0 + \eta^0$;

(c) $p + n \rightarrow \Sigma^0 + K^0 + \pi^+$

(d) $\pi^- + p \rightarrow \pi^0 + \Lambda^0$;

(e) $p + \bar{p} \rightarrow \Sigma^+ + K^0 + \pi^-$.

4. A neutron and an antineutron with negligible kinetic energy annihilate each other, producing two photons. What is the energy of each photon?

Appendix Answers
to Drill Problems

Chapter 1

1.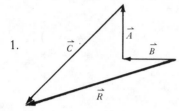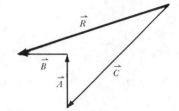

2. (a) 30 miles, north; (b) 8 miles, west; (c) 10 miles, 53° east of north; (d) 10 miles, 37° south of east

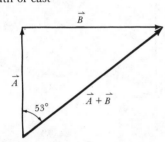 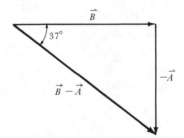

3. $A_x = A \cos 30° = 10 \text{ N} (.87) = 8.7 \text{ N}$
   $A_y = A \sin 30° = 10 \text{ N} (.5) = 5 \text{ N}$
   $B_x = -B \cos 20° = -10 \text{ N} (.94) = -9.4 \text{ N}$
   $B_y = B \sin 20° = 10 \text{ N} (.34) = 3.4 \text{ N}$

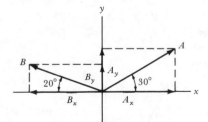

$R_x = 8.7 \text{ N} - 9.4 \text{ N} = -0.7 \text{ N}$
$R_y = 5 \text{ N} + 3.4 \text{ N} = 8.4 \text{ N}$
$R = \sqrt{(-0.7 \text{ N})^2 + (8.4 \text{ N})^2} = \underline{8.43 \text{ N}}$
$\tan \theta = |8.43/-0.7| = 12; \underline{\theta = 85°}$

4. $\tan\theta = {}^{20}\!/_{15} = 1.33; \theta = \underline{53°}$

$R = \sqrt{(15\ \text{m})^2 + (20\ \text{m})^2} = \sqrt{625\ \text{m}^2} = \underline{25\ \text{m}}$

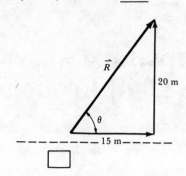

Chapter 2

1. $\Sigma F_y = 0: T_1 + T_2 - 30 - 150 = 0$
$T_1 + T_2 = 180$
$\Sigma\tau_A = 0: -30\ \text{lb}(3\ \text{ft}) - 150\ \text{lb}(5\ \text{ft}) + T_2(6\ \text{ft}) = 0$

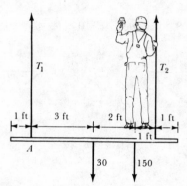

$T_2 = \dfrac{840\ \text{lb} \cdot \text{ft}}{6\ \text{ft}} = \underline{140\ \text{lb}}$

$T_1 = 180 - 140 = \underline{40\ \text{lb}}$

2. $T = 1000\ \text{N}; f = 800\ \text{N}; C = \underline{400\ \text{N}}$

3. $\Sigma F_x = 0: F - T\cos 60° = 0$
$F = T(0.5)$
$\Sigma F_y = 0: T\sin 60° - mg = 0$

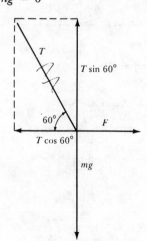

266

$$T(0.87) = (25)(9.8)$$
$$T = \frac{245}{0.87} = 280 \text{ N}$$
$$F = (280 \text{ N})(0.5) = \underline{140 \text{ N}}$$

Chapter 3

1. The proper *mks* units for acceleration are m/s².

$$\left(\frac{8 \text{ hands}}{\text{moment}^2}\right)\left(\frac{20 \text{ cm}}{1 \text{ hand}}\right)\left(\frac{1 \text{ m}}{100 \text{ cm}}\right)\left(\frac{1 \text{ moment}}{4 \text{ s}}\right)\left(\frac{1 \text{ moment}}{4 \text{ s}}\right) = \underline{\frac{1}{10}\frac{\text{m}}{\text{s}^2}}$$

2. Average speed:

$$\bar{v} = \frac{\text{distance}}{\text{time}} = \frac{80 + 40}{1.5} = \frac{120}{1.5} = \underline{80 \text{ km/hr}}$$

Average velocity:

$$\bar{\bar{v}} = \frac{\text{displacement}}{\text{time}}$$

The two displacements are added as vectors to get the resultant $\vec{d}$.

$$\bar{\bar{v}} = \frac{\vec{d}}{t} = \frac{40 \text{ km}}{1.5 \text{ hr}} = \underline{27 \text{ km/hr, west}}$$

3. (a) Find the stopping distance and compare to 50 m. Because acceleration changes, the motion must be broken into two legs.

$$\text{Leg 1: } x - x_0 = v_0 t + \tfrac{1}{2}at^2 = (20)(.7) = 14 \text{ m}$$

$$\text{Leg 2: Final } v = 0; \ x_0 = 14 \text{ m}$$

Solving $2a(x - x_0) = v^2 - v_0^2$ for x:

$$x = \frac{-v_0^2}{2a} + x_0 = \frac{-(20)^2}{2(-5)} + 14 = 40 + 14 = 54 \text{ m}$$

Car stops in 54 m, but roadblock is 50 m away; so car will hit roadblock.
   (b) Find final v for Leg 2 where $(x - x_0) = 50 - 14 = 36 \text{ m}$

$$v^2 - v_0^2 = 2a(x - x_0)$$
$$v^2 = 2a(x - x_0) + v_0^2 = 2(-5)(36) + (20)^2 = -360 + 400 = 40$$
$$v = \underline{6.3 \text{ m/s}}$$

   (c) Minimum deceleration would just stop car in 36 m; $v_0 = 20$ m/s; $v = 0$. Rewriting the above equation:

$$a = \frac{v^2 - v_0^2}{2(x - x_0)} = \frac{0 - (20)^2}{2(36)} = \underline{-5.6 \text{ m/s}^2}$$

4. (a) In the x direction:

$$x = v_{x_0}t, \text{ so } v_{x_0} = \frac{x}{t} = \frac{100}{3} = 33 \text{ ft/s}$$

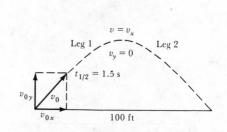

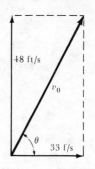

In the y direction, consider first half of trip: $v_y = 0$ at top, time for half the trip = 1.5 s because the motion is symmetric.

$$v_y = v_{y0} + at, \text{ so } v_{y0} = v_y - at$$
$$v_{y0} = -(-32)(1.5) = 48 \text{ ft/s}$$

v_o is now constructed from its components:

$$v_o = \sqrt{(33)^2 + (48)^2} = \underline{58 \text{ ft/s}}$$
$$\tan \theta = \frac{48}{33} = 1.44, \text{ so } \theta = \underline{55°}$$

(b) Maximum height can be obtained from Leg 1, taking $v_y = 0$ (top) and $t = 1.5$ s:

$$y = \frac{v_{y0} + v_y}{2}(t) = \left(\frac{48 + 0}{2}\right)(1.5) = \underline{36 \text{ ft}}$$

Chapter 4

1. $a = \dfrac{F}{m}$, where $m = \dfrac{w}{g} = \dfrac{160}{32} = 5$ slugs, and $F = 20{,}000$ lb

$$a = \frac{20{,}000}{5} = \underline{4000 \text{ ft/s}^2}$$

In g's, $a = 4000/32 = \underline{125g}$

2.   (a) Each body is separately isolated:

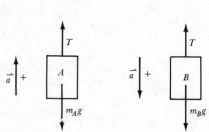

Block A: $T - m_A g = m_A a$

$T - (2)(9.8) = 2a$

$T = 2a + 19.6$ (1)

Block B: $m_B g - T = m_B a$

$(5)(9.8) - T = 5a$

Substituting T from (1): $49 - (2a + 19.6) = 5a$; $a = \underline{4.2 \text{ m/s}^2}$

(b) From (1): $T = 2(4.2) + 19.6 = \underline{28 \text{ N}}$

(c) $v^2 - \cancel{v_0^2} = 2a(x - x_0)$

$v^2 = 2(4.2)(10) = 84$; $v = \underline{9.2 \text{ m/s}}$

3.   Sled will slide down if $w \cos 60° > F_s(\text{max})$:

$$w \cos 60° = (60)(0.5) = 30 \text{ lb}$$
$$F_s(\text{max}) = \mu_s F_c = \mu_s w \sin 60° = (0.2)(60)(0.866) = 10.4 \text{ lb}$$

$w \cos 60°$ *is* greater than $F_s(\text{max})$; <u>sled will slide back down</u>. From $\Sigma \vec{F} = m\vec{a}$,

$$\vec{a} = \frac{\Sigma \vec{F}}{m} = \frac{w \cos 60° - F_k}{m}$$

where

$$F_k = \mu_k \, w \sin 60° = (0.1)(60)(0.866) = 5.2 \text{ N}$$

(Sled is now in motion; *kinetic* coefficient is used.)

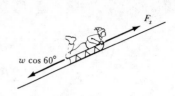
w cos 60°

$$a = \frac{30 - 5.2}{1.87} = 13 \text{ ft/s}^2$$

$$v^2 - v_0^2 = 2a(x - x_0) = 2(13)(5.9) = 153 \text{ ft}^2/\text{s}^2$$

$$v = \underline{12.4 \text{ ft/s}}$$

Chapter 5

1. $\vec{F}t = m\vec{v} - m\vec{v}_0$ so $\vec{F} = \dfrac{m\vec{v} - m\vec{v}_0}{t}$

$$F = \frac{(0.15)(-40) - (0.15)(30)}{3 \times 10^{-3}} = \frac{-10.5}{3 \times 10^{-3}} = \underline{-3500 \text{ N}}$$

$\vec{v}_0 = 30$ m/s

$v = -40$ m/s

2. Treating x and y axes individually,

$$x \text{ direction: } \vec{p}_{\text{before}} = \vec{p}_{\text{after}}$$
$$0 = m\vec{v}_x + M\vec{V}_x$$

where $m = 100$ kg; $M = 20{,}000 - 100 = 19{,}900$ kg

$$\vec{V}_x = \frac{-m\vec{v}_x}{M} = \frac{-mv \cos 30°}{M} = \frac{-(100)(200)(0.866)}{19{,}900} = \underline{-0.87 \text{ m/s}}$$

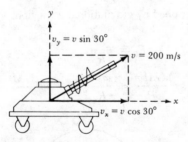

$v_y = v \sin 30°$

$v = 200$ m/s

$v_x = v \cos 30°$

The cannon recoils with a speed of 0.87 m/s or about 2 mi/hr, in the $-x$ direction. In the y direction the cannon is in contact with the earth; the earth recoils to absorb the y momentum, but its mass is so large that its recoil velocity is unmeasurably small.

3. $\vec{p}_{\text{before}} = \vec{p}_{\text{after}}$
    $M\vec{V}_0 = M\vec{V} + m\vec{v}$
    $(6)(3) = (6)V + (2)(6)$
    $V = \dfrac{18 - 12}{6} = \underline{1 \text{ m/s in the forward direction}}$

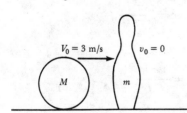

$V_0 = 3$ m/s $v_0 = 0$

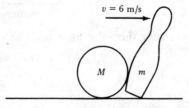

$v = 6$ m/s

Chapter 6

1. The second hand makes 1 revolution every 60 seconds.

    (a) $v = \dfrac{\text{distance}}{\text{time}} = \dfrac{\text{one circumference}}{60\ \text{s}} = \dfrac{2\pi r}{60} = \dfrac{2\pi(0.1)}{60} = \underline{0.01\ \text{m/s}}$

    (b) $\omega = \dfrac{v}{r} = \dfrac{0.01\ \text{m/s}}{0.1\ \text{m}} = \underline{0.1\ \text{rad/s}}$

    or $\omega = \dfrac{\theta}{t} = \dfrac{2\pi\ \text{rad}}{60\ \text{s}} = \underline{0.1\ \text{rad/s}}$

    (c) $a_c = \dfrac{v^2}{r} = \dfrac{(0.01\ \text{m/s})^2}{0.1\ \text{m}} = \underline{0.001\ \text{m/s}^2}$

    or $a_c = \omega^2 r = (0.1\ \text{rad/s})^2(0.1\ \text{m}) = \underline{0.001\ \text{m/s}^2}$

2. The earth radius R is *not* the radius of the circular motion r.

    $$r = R\cos 40° = (6.38 \times 10^6\ \text{m})(0.766) = 4.9 \times 10^6\ \text{m}$$
    $$F = ma_c = mr\omega^2$$
    $$\text{where } \omega = \frac{1\ \text{rev}}{24\ \text{hr}}\left(\frac{2\pi\ \text{rad}}{1\ \text{rev}}\right)\left(\frac{1\ \text{hr}}{3600\ \text{s}}\right) = 7.3 \times 10^{-5}\ \text{rad/s}$$
    $$F = (60\ \text{kg})(4.9 \times 10^6\ \text{m})(7.3 \times 10^{-5}\ \text{rad/s})^2 = \underline{1.6\ \text{N}}$$

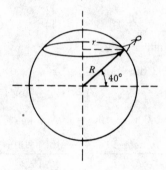

3. Centripetal force is supplied by gravitational attraction:

    $$F_G = \frac{mv^2}{R}$$
    $$\frac{GMm}{R^2} = \frac{mv^2}{R}, \text{ so } v^2 = \frac{GM}{R}$$

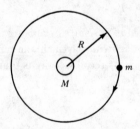

Now

$$v = \frac{\text{distance}}{\text{time}} = \frac{2\pi R}{T}, \text{ so } v^2 = \frac{4\pi^2 R^2}{T^2}$$

Substituting this value for v, and solving for M:

$$\frac{GM}{R} = \frac{4\pi^2 R^2}{T^2}, \text{ or } M = \frac{4\pi^2 R^3}{GT^2}$$

Alternate solution:

$$F = mR\omega^2, \text{ so } \frac{GMm}{R^2} = mR\omega^2$$

Now $\omega = \dfrac{\theta}{t}$, and when $t = T$ (1 rev), $\theta = 2\pi$ (in radians). So $\omega = \dfrac{2\pi}{T}$. Then,

$$\frac{GM}{R^2} = R\left(\frac{4\pi^2}{T^2}\right)$$

which can be solved for M giving:

$$M = \frac{4\pi^2 R^3}{GT^2}$$

Chapter 7

1. (a) $\omega_1 = 400\,\dfrac{\text{rev}}{\text{min}}\left(\dfrac{2\pi\text{ rad}}{1\text{ rev}}\right)\left(\dfrac{1\text{ min}}{60\text{ s}}\right) = 41.9$ rad/s

    $\omega_2 = 500\,\dfrac{\text{rev}}{\text{min}}\left(\dfrac{2\pi\text{ rad}}{1\text{ rev}}\right)\left(\dfrac{1\text{ min}}{60\text{ s}}\right) = 52.4$ rad/s

    $\alpha = \dfrac{\omega_2 - \omega_1}{t} = \dfrac{52.4 - 41.9}{2} = \underline{5.25\text{ rad/s}^2}$

    (b) $\omega^2 - \omega_0^2 = 2\alpha(\theta - \theta_0)$ where $\omega_0 = 0$, $\omega = 52.4$ rad/s, and $\alpha = 5.25$ rad/s^2

    $\theta - \theta_0 = \dfrac{\omega^2 - \cancel{\omega_0^2}^{\,0}}{2(5.25)} = \dfrac{(52.4)^2}{10.5} = 261\text{ rad}\left(\dfrac{1\text{ rev}}{2\pi\text{ rad}}\right) = \underline{42\text{ rev}}$

2. $\Sigma\tau = I\alpha$, where $\Sigma\tau = \tau_{\text{man}} = Fl = F(0.8\text{ m})$, since force is applied 0.8 m from axis.

    $\alpha = \dfrac{\cancel{\theta^2}^{\,0} - \omega_0^2}{2(\theta - \theta_0)}$ where $\theta - \theta_0 = \dfrac{1}{8}\text{ rev}\left(\dfrac{2\pi\text{ rad}}{1\text{ rev}}\right) = 0.78$ rad

    and $\omega_0 = 5\,\dfrac{\text{turns}}{\text{min}}\left(\dfrac{2\pi\text{ rad}}{1\text{ turn}}\right)\left(\dfrac{1\text{ min}}{60\text{ s}}\right) = 0.52$ rad/s

    $\alpha = \dfrac{-(0.52)^2}{2(0.78)} = 0.17$ rad/s^2

    $F(0.8) = 200(0.17)$, so $F = \dfrac{34}{0.8} = \underline{42\text{ N}}$

3. (a) $I = I_{\text{rod}} + 2mr^2 = \tfrac{1}{12}ml^2 + 2m\left(\dfrac{l}{2}\right)^2 = \tfrac{1}{12}ml^2 + m\dfrac{l^2}{2}$

    $= \tfrac{7}{12}ml^2 = \tfrac{7}{12}(0.2\text{ kg})(0.8\text{ m})^2 = \underline{0.075\text{ kg}\cdot\text{m}^2}$

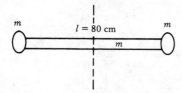

    (b) $L = I\omega$, where $\omega = 2\,\dfrac{\text{rev}}{\text{s}}\left(\dfrac{2\pi\text{ rad}}{1\text{ rev}}\right) = 12.6$ rad/s

    $L = (0.075\text{ kg}\cdot\text{m}^2)(12.6\text{ rad/s}) = \underline{0.94\text{ kg}\cdot\text{m}^2/\text{s}}$

4. $L_{\text{before}} = L_{\text{after}}$

    $I_A\omega_A = (I_A + I_B)\omega_f,$

    and $I_B = 3I_A$, so $I_A\omega_A = (I_A + 3I_A)\omega_f$

    $\omega_f = \dfrac{\cancel{I_A}\omega_A}{4\cancel{I_A}} = \dfrac{\omega_A}{4} = \dfrac{100\text{ rpm}}{4} = \underline{25\text{ rpm}}$

Chapter 8

1. $KE_t = KE_A$, so $\frac{1}{2}m_t v_t^2 = \frac{1}{2}m_A v_A^2$. It is given that $m_t = 4m_A$. Substituting:

$$\frac{1}{2}(4m_A)v_t^2 = \frac{1}{2}m_A v_A^2, \text{ or } v_t^2 = \frac{v_A^2}{4}; \text{ so } v_t = \frac{v_A}{2}$$

The truck's <u>speed is half</u> the Audi's speed. For momenta:

$$p_A = m_A v_A; \text{ and } p_t = m_t v_t = (4m_A)\left(\frac{v_A}{2}\right) = 2m_A v_A$$

The truck's <u>momentum is twice</u> the Audi's momentum.

2. By energy conservation: $E_{\text{bottom}} = E_{\text{top}}$. Choosing $y = 0$ at bottom of hill, we get

$$(KE + KE_r)_{\text{bottom}} = \underline{(PE_g)_{\text{top}}}$$

$\frac{1}{2}mv^2 + \frac{1}{2}I\omega^2 = mgh$, where I for sphere $= \frac{2}{5}mR^2$ and $\omega = \dfrac{v}{R}$

$$\frac{1}{2}mv^2 + \frac{1}{2}(\tfrac{2}{5}mR^2)\left(\frac{v^2}{R^2}\right) = mgh$$

$$\frac{1}{2}v^2 + \frac{1}{5}v^2 = gh; \text{ or } \tfrac{7}{10}v^2 = gh; \text{ so } v = \underline{\sqrt{\tfrac{10}{7}\,gh}}$$

3. (a) $PE_s = \frac{1}{2}kx^2$, where $k = 400$ N/m and $x = 10$ cm $= 0.1$ m
   $PE_s = \frac{1}{2}(400 \text{ N/m})(0.1 \text{ m})^2 = \underline{2 \text{ J}}$

   (b) Energy is conserved from moment after bullet impact to moment of maximum compression of spring: KE(of block + bullet after impact) $\Rightarrow PE_s$ (of spring) + work done against friction.

$\frac{1}{2}(m + M)v^2 = 2$ J $+ F_k x$, where $x = 10$ cm $= 0.1$ m

and $F_k = \mu_k F_c = \mu_k(m + M)g = (0.41)(1 \text{ kg})(9.8 \text{ m/s}^2) = 4$ N

$\frac{1}{2}(1 \text{ kg})v^2 = 2$ J $+ (4 \text{ N})x = 2 + (4 \text{ N})(0.1 \text{ m})$

$v^2 = \dfrac{2(2 \text{ J} + 0.4 \text{ J})}{1 \text{ kg}} = 4.8 \text{ m}^2/\text{s}^2$; so $v = \underline{2.2 \text{ m/s}}$

   (c) Momentum is conserved for the bullet-block collision:

$$p_{\text{before}} = p_{\text{after}}$$
$$mv_0 = (M + m)v$$

where $v =$ speed of bullet + block right after collision $= 2.2$ m/s from (b).

$$v_0 = \frac{(M + m)v}{m} = \frac{(1 \text{ kg})(2.2 \text{ m/s})}{(0.02 \text{ kg})} = \underline{110 \text{ m/s}}$$

Chapter 9

1. $T_C = (\tfrac{5}{9})(T_F - 32) = (\tfrac{5}{9})(38 - 32) = (\tfrac{5}{9})(6) = \underline{3.3 \text{ °C}}$ or, by the proportions method:

$$\frac{38 - 32}{180} = \frac{T_C - 0}{100}$$

$$T_C = \frac{100}{180}(6) = \underline{3.3 \text{ °C}}$$

$T_K = T_C + 273 = 3 + 273 = \underline{276 \text{ K}}$

2. $Q_{\text{steam}} + Q_{\text{ice}} = 0$

$$\underbrace{-mL_v + mc\,\Delta T}_{\substack{\text{steam condenses and} \\ \text{cools from } 100°C}} + \underbrace{mL_f + mc\,\Delta T}_{\substack{\text{ice melts and} \\ \text{warms from } 0°C}} = 0$$

(The first term must be given a negative sign to indicate the loss of heat; the second term will be negative because ΔT is negative.)

$$-(0.1 \text{ kg})(2.26 \times 10^6 \text{ J/kg}) + (0.1 \text{ kg})(4180 \text{ J/kg} \cdot \text{C}^\circ)(T - 100^\circ\text{C})$$
$$+(0.5 \text{ kg})(3.35 \times 10^5 \text{ J/kg}) + (0.5 \text{ kg})(4180 \text{ J/kg} \cdot \text{C}^\circ)(T - 0^\circ\text{C}) = 0$$
$$-226{,}000 \text{ J} + (418 \text{ J/C}^\circ)T - 41{,}800 \text{ J} + 167{,}500 \text{ J} + (2090 \text{ J/C}^\circ) = 0$$
$$(2508 \text{ J/C}^\circ)T = 98{,}300 \text{ J}; \text{ so } T = \underline{39^\circ\text{C}}$$

3. 20 percent energy from eggs $\Rightarrow$ gravitational potential energy. Let x = number of eggs. Then:

$$(0.20)(x \text{ eggs})(\text{energy/egg}) = mg(y - y_0)$$

where $\text{energy/egg} = 80 \dfrac{\text{kcal}}{\text{egg}} \left(\dfrac{4180 \text{ J}}{1 \text{ kcal}} \right) = 334{,}400 \text{ J/egg}$

$$(0.2)(x)(334{,}400 \text{ J/egg}) = (700 \text{ N})(1000 \text{ m})$$
$$x = \underline{10.5 \text{ eggs}}$$

He should eat $\underline{11 \text{ eggs}}$.

Chapter 10

1. (a) $\dfrac{\Delta Q}{\Delta t} = \dfrac{KA(T_2 - T_1)}{L}$,

where A = area of 2 sides + 2 ends + top and bottom;
$= 2(0.6 \text{ m} \times 0.35 \text{ m}) + 2(0.35 \text{ m} \times 0.4 \text{ m}) + 2(0.6 \text{ m} \times 0.4 \text{ m})$
$= 0.42 \text{ m}^2 + 0.28 \text{ m}^2 + 0.48 \text{ m}^2 = 1.18 \text{ m}^2$

From Table 10.1, K for styrofoam $= 0.016 \dfrac{\text{W}}{\text{m} \cdot \text{C}^\circ}$

$T_2 - T_1 = 30 \text{ C}^\circ$; L = thickness $= 0.01$ m;
$\dfrac{\Delta Q}{\Delta t} = \dfrac{(0.016)(1.18)(30)}{0.01} = 57 \text{ J/s or } \underline{57 \text{ watts}} \text{ flowing in}$

(b) $\dfrac{\Delta Q}{\Delta t} = \dfrac{\Delta m}{\Delta t} L_f$; so $\dfrac{\Delta m}{\Delta t} = \dfrac{\Delta Q}{\Delta t L_f} = \dfrac{57}{335{,}000} = 1.7 \times 10^{-4} \dfrac{\text{kg}}{\text{s}} \left(\dfrac{3600 \text{ s}}{1 \text{ hr}} \right)$
$= \underline{0.61 \text{ kg/hr}}$

2. Object 1: $R_1 = \epsilon \sigma A T_1^4$ $\dfrac{R_2}{R_1} = \dfrac{\epsilon \sigma A T_2^4}{\epsilon \sigma A T_1^4} = \dfrac{T_2^4}{T_1^4}$
Object 2: $R_2 = \epsilon \sigma A T_2^4$

$T_2 = 900 + 273 = 1173 \text{ K}$ $\dfrac{R_2}{R_1} = \left(\dfrac{1173 \text{ K}}{573 \text{ K}} \right)^4 = (2.05)^4 = 17.7$
$T_1 = 300 + 273 = 573 \text{ K}$

The temperatures used *must* be absolute; an answer of $(3)^4 = 81$, based on *Celsius* temperatures would be *incorrect*.

3. $\dfrac{\Delta E_{\text{system}}}{\Delta t} = -100 \text{ W}$

$H_{cv} = hA(T_2 - T_1) = (5)(1.4)(80 - 40)$
$= 280 \text{ W (positive, heat is } gained \text{ by person)}$
$H_r = \epsilon \sigma A(T_2^4 - T_1^4) = (0.98)(5.67 \times 10^{-8})(1.4)(353^4 - 313^4)$
$= 461 \text{ W (positive)}$

$H_{cv} + H_r + H_e = \dfrac{\Delta E}{\Delta t}$; so $280 + 461 + H_e = -100$
$H_e = -100 - 280 - 461 = -841 \text{ W}$
$H_e = -\dfrac{\Delta m}{\Delta t} L_e$; so $\dfrac{\Delta m}{\Delta t} = \dfrac{-H_e}{L_e} = \dfrac{-(-841)}{2.4 \times 10^6} = 3.5 \times 10^{-4} \text{ kg/s}$
$= \underline{0.35 \text{ gm/s}}$

Chapter 11

1. For the hot container: $\Delta S = \dfrac{\Delta Q}{T} = \dfrac{-2000\ \text{J}}{(273 + 100)\text{K}} = \dfrac{-2000\ \text{J}}{373\ \text{K}} = -5.36\ \text{J/K}$

   For the cold container: $\Delta S = \dfrac{\Delta Q}{T} = \dfrac{+2000\ \text{J}}{(273 + 0)\text{K}} = +7.33\ \text{J/K}$

   Total entropy change $= -5.36\ \text{J/K} + 7.33\ \text{J/K}$

   $\qquad\qquad\qquad\qquad = \underline{+1.97\ \text{J/K}}$, an <u>increase</u> in entropy.

2. First find equilibrium temperature: $Q_{ice} + Q_{water} = 0$

   $$(mL_v + cm\,\Delta T)_{ice} + (cm\,\Delta T)_{water} = 0$$

   $$(0.02)(333{,}000) + (4180)(0.02)(T - 0) + (4180)(0.43)(T - 10) = 0$$

   $$T = \underline{6°C}$$

   For ice melting: $T = 0°C = 273\ \text{K}$;

   so $S = \dfrac{\Delta Q}{T} = \dfrac{mL_f}{T} = \dfrac{(0.02)(333{,}000)}{273} = 24.4\ \text{J/K}$

   For ice-water warming: $\Delta T = 6\ \text{C}°$; so $\overline{T} = 3°C = 276\ \text{K}$

   $$S \simeq \dfrac{\Delta Q}{\overline{T}} = \dfrac{cm\,\Delta T}{\overline{T}} = \dfrac{(4180)(0.02)(6)}{276} = 1.8\ \text{J/K}$$

   For water cooling: $\Delta T = -4\ \text{C}°$, so $\overline{T} = 8°C = 281\ \text{K}$

   $$S \simeq \dfrac{\Delta Q}{\overline{T}} = \dfrac{cm\,\Delta T}{\overline{T}} = \dfrac{(4180)(0.43)(-4)}{281} = 25.6\ \text{J/K}$$

   Total entropy change $\simeq +24.4 + 1.8 - 25.6 = \underline{0.6\ \text{J/K}}$, an <u>increase</u> in entropy

3. (a) First stage: $e_1 = 1 - \dfrac{T_{cold}}{T_{hot}} = 1 - \dfrac{350\ \text{K}}{500\ \text{K}} = 1 - 0.7 = 0.3$

   $e_1 = \dfrac{W_1}{Q_{hot}}$, where $Q_{hot} = 1000\ \text{J}$; so $W_1 = e_1 Q_{hot} = (0.3)(1000\ \text{J}) = 300\ \text{J}$

   Second stage: $e_2 = 1 - \dfrac{T_{cold}}{T_{hot}}$, where T_{hot} for stage 2 $= T_{cold}$ for stage 1 $= 350\ \text{K}$

   $$e_2 = 1 - \dfrac{200\ \text{K}}{350\ \text{K}} = 1 - 0.57 = 0.43$$

   $W_2 = e_2 Q_{hot}$, where Q_{hot} for stage 2 $= Q_{cold}$ for stage 1

   $\qquad\qquad = 1000\ \text{J (heat in)} - 300\ \text{J (work out)} = 700\ \text{J}$;

   $$W_2 = (0.43)(700\ \text{J}) = 300\ \text{J}$$

   Total work done per 1000 J of heat $= W_1 + W_2 = 300\ \text{J} + 300\ \text{J} = 600\ \text{J}$

   Overall efficiency of engine = total work out/heat in $= 600\ \text{J}/1000\ \text{J} = \underline{0.6}$ or <u>60 percent</u>

   (b) Single engine: $e = 1 - \dfrac{T_{cold}}{T_{hot}} = 1 - \dfrac{200\ \text{K}}{500\ \text{K}} = 1 - 0.4 = \underline{0.6}$, <u>the same</u> as for two-stage engine

Chapter 12

1. $P_1 V_1 = P_2 V_2$ (at constant T). In this case, $V_2 = 3V_1$, so $P_1 V_1 = P_2(3V_1)$; or $P_2 = P_1/3$. Then, $\dfrac{P_2}{T_2} = \dfrac{P_3}{T_3}$ (at constant V). Here $P_2 = P_1/3$ and $P_3 = P_1$.

$$\dfrac{P_1/3}{T_2} = \dfrac{P_1}{T_3}; \text{ so } \tfrac{1}{3}T_3 = T_2; \text{ or } T_3 = 3T_2$$

Therefore, T has increased by a factor of 3.

2. (a) Molecular mass of CCl_4 = 12(C) + 4 × 35(Cl) = 12 + 140 = 152. One mole of CCl_4 has a mass of <u>152 grams</u>.

(b) $PV = nRT$, so $V = \dfrac{nRT}{P} = \dfrac{(1)(8.314)(273)}{1.013 \times 10^5} = \underline{0.0224 \ m^3}$ or 22.4 liters

Note that this result is independent of the type of gas; one mole of *any* ideal gas occupies 22.4 liters at standard conditions.

3. (a) $\dfrac{P_1 V_1}{T_1} = \dfrac{(3P_1)(\frac{1}{2}V_1)}{T_2}$; or $T_2 = \frac{3}{2}T_1 = \underline{450\ K}$

(b)

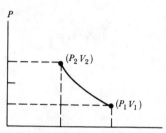

(c) $Q + W = \Delta E = \frac{3}{2}nR\,\Delta T$; where $\Delta T = 450 - 300 = 150$ K
$Q + 1000 \text{ J} = \frac{3}{2}(1)(8.314)(150) = 1870 \text{ J}$
$Q = 1870 - 1000 = \underline{870 \text{ J}}$, positive so <u>heat flowed in</u>.

Chapter 13

1. ρ_{blood} = 1050 kg/m³, from Table 13.1. For a 60-kg person, m_{blood} = (0.07)(60) = 4.2 kg.

$$V = \frac{m}{\rho} = \frac{4.2}{1050} = \underline{4 \times 10^{-3} \ m^3}$$

Since 10^{-3} m³ = 1 liter, this is 4 liters or a little more than 4 quarts. One kg weighs 2.2 lb, therefore 4.2 kg of blood weighs (4.2)(2.2) = <u>9.2 lb</u>

2. $P_1 - P_2 = \rho g(y_2 - y_1)$ becomes $P - P_a (= P_{gauge}) = \rho g h$

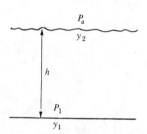

(a) For h = 3 m:
P_{gauge} = (1000 kg/m³)(9.8 m/s²)(3 m) = <u>29,400 N/m²</u>

(b) $h = \dfrac{P_{gauge}}{\rho g} = \dfrac{39,200 \text{ N/m}^2}{(1000 \text{ kg/m}^3)(9.8 \text{ m/s}^2)} = 4$ m

3. Buoyant force = difference between real weight w and apparent weight w':

$$B = w - w' = 82.3 - 74.5 = 7.8 \text{ N}$$

By Archimedes' principle: $B = (mg)_{displaced\ water} = (\rho V g)_{displaced\ water}$

So $V = \dfrac{B}{\rho g} = \dfrac{7.8}{(1000)(9.8)} = 7.96 \times 10^{-4} \ m^3$

$$\rho = \frac{m}{V}, \text{ where } m = \frac{w}{g} = \frac{82.3}{9.8} = 8.4 \text{ kg}$$

$$\rho = \frac{8.4}{7.96 \times 10^{-4} \ m^3} = 10,550 \text{ kg/m}^3$$

From Table 13.1, the metal is <u>silver</u>.

Chapter 14

1. (a) Pressure term: $P = \dfrac{N}{m^2} \times \dfrac{m}{m} = \dfrac{N \cdot m}{m^3} = \dfrac{J}{m^3}$

Speed term: $\frac{1}{2}\rho v^2 = \left(\dfrac{kg}{m^3}\right)\left(\dfrac{m^2}{s^2}\right) = \left(\dfrac{kg \cdot m^2}{s^2}\right)\left(\dfrac{1}{m^3}\right) = \dfrac{J}{m^3}$

Elevation term: $\rho g(y_2 - y_1) = \left(\dfrac{kg}{m^3}\right)\left(\dfrac{m}{s^2}\right)(m) = \left(\dfrac{kg \cdot m^2}{s^2}\right)\left(\dfrac{1}{m^3}\right) = \dfrac{J}{m^3}$

(b) $Re = \dfrac{2\bar{v}r\rho}{\eta} = \dfrac{(m/s)(m)(kg/m^3)}{N \cdot s/m^2} = \dfrac{kg \cdot m/s}{N \cdot s} = \dfrac{kg \cdot m/s}{(kg \cdot m/s^2)s} = \dfrac{\cancel{s}}{\cancel{s}}$; $\underline{Re\ is}$ $\underline{dimensionless.}$

2. Continuity equation: $v_1 A_1 = v_2 A_2$, where position 1 is at the outlet and position 2 is at the top surface of the water. Thus, $v_2 = \dfrac{A_1}{A_2} v_1$.

Bernoulli's equation: $\frac{1}{2}\rho v_1^2 + \rho g y_1 = \frac{1}{2}\rho v_2^2 + \rho g y_2$, (where $P_1 = P_2$);

$$\frac{1}{2}(v_1^2 - v_2^2) = g(y_2 - y_1)$$

Substituting v_2 from above:

$$\frac{1}{2}v_1^2\left(1 - \frac{A_1^2}{A_2^2}\right) = g(y_2 - y_1)$$

$$A_1 = \pi r^2 = \pi(4.5\ \text{cm})^2 = 64\ \text{cm}^2;\ A_2 = 600\ \text{cm}^2;\ \frac{A_1^2}{A_2^2} = \frac{64^2}{600^2} = 0.011$$

$$\frac{1}{2}v_1^2(1 - 0.011) = (9.8)(0.3)$$

$$v_1^2 = 2(9.8)(0.3)(0.989) = 5.82;\ v_1 = \underline{2.4\ \text{m/s}}$$

$$v_2 = {}^{64}\!/_{600}(2.4) = \underline{0.26\ \text{m/s}}$$

3. $Q = \dfrac{\pi r^4}{8\eta l}(P_2 - P_1)$; so $P_2 - P_1 = \dfrac{8Q\eta l}{\pi r^4}$

$$= \frac{8(1 \times 10^{-7}\ \text{m}^3/\text{s})(20\ \text{N} \cdot \text{s/m}^2)(0.03\ \text{m})}{\pi(10^{-3}\ \text{m})^4}$$

$P_2 - P_1 = \underline{1.53 \times 10^5\ \text{N/m}^2}$

$Re = \dfrac{2\bar{v}r\rho}{\eta}$, where $\bar{v} = \dfrac{Q}{A} = \dfrac{1\ \text{cm}^3/10\ \text{s}}{(0.1\ \text{cm})^2} = 3.18\ \text{cm/s} = 0.0318\ \text{m/s}$

$Re = \dfrac{2(0.0318)(0.001)(1200)}{20} = 0.0038$. Since $Re < 2000$, the flow is $\underline{\text{laminar.}}$

Were it not, Poiseuille's equation above for pressure difference would not be applicable. It would probably be difficult to induce turbulence into the flow of molasses.

Chapter 15

1. $\dfrac{F}{A} = Y\dfrac{\Delta L}{L}$; ΔL_{max} occurs when $\dfrac{F}{A}$ = ultimate tensile strength. Ultimate tensile strength $\simeq 10^8\ \text{N/m}^2$; $Y \simeq 10^{10}\ \text{N/m}^2$; for hair, from Table 15.1

$$\Delta L_{max} = \left(\frac{F}{A}\right)\frac{L}{Y} = (10^8\ \text{N/m}^2)\left(\frac{0.5\ \text{m}}{10^{10}\ \text{N/m}^2}\right)$$

$$= 5 \times 10^{-3}\ \text{m} = \underline{5\ \text{mm}}$$

2. Each cable has the same tensile force: $F = mg = 100\ \text{kg}\ (9.8\ \text{m/s}^2) = 980\ \text{N}$, and the same area: $A = 10^{-2}\ \text{cm}^2 = 10^{-6}\ \text{m}^2$; $\dfrac{F}{A} = 980\ \text{N}/10^{-6}\ \text{m}^2 = 9.8 \times 10^8\ \text{N/m}^2$ for both.

Steel: $L = 2\ \text{m}$; $Y = 20 \times 10^{10}\ \text{N/m}^2$;

$$\Delta L = \left(\frac{F}{A}\right)\frac{L}{Y} = (9.8 \times 10^8\ \text{N/m}^2)\left(\frac{2\ \text{m}}{20 \times 10^{10}\ \text{N/m}^2}\right)$$

$$= 9.8 \times 10^{-3}\ \text{m} = 0.98\ \text{cm}$$

Aluminum: $L = 4\ \text{m}$; $Y = 7.1 \times 10^{10}\ \text{N/m}^2$;

$$\Delta L = \left(\frac{F}{A}\right)\frac{L}{Y} = (9.8 \times 10^8 \text{ N/m}^2)\left(\frac{4 \text{ m}}{7.1 \times 10^{10} \text{ N/m}^2}\right)$$

$$= 5.52 \times 10^{-2} \text{ m} = 5.52 \text{ cm}$$

Total elongation $= 0.98 \text{ cm} + 5.52 \text{ cm} = \underline{6.5 \text{ cm}}$

3. $V = \frac{4}{3}\pi r^3 = \frac{4}{3}\pi(0.2 \text{ m})^3 = 0.0335 \text{ m}^3$;
    $B = 16 \times 10^{10} \text{ N/m}^2$ for steel, from Table 15.3
    $V' = \frac{4}{3}\pi(0.2 - 0.002 \text{ m})^3 = \frac{4}{3}\pi(0.198 \text{ m})^3 = 0.0325 \text{ m}^3$
    $\Delta V = V' - V = 0.0325 \text{ m}^3 - 0.0335 \text{ m}^3 = -0.001 \text{ m}^3$

$$\Delta P = -B\left(\frac{\Delta V}{V}\right) = -(16 \times 10^{10} \text{ N/m}^2)\left(\frac{-.001 \text{ m}^3}{0.0335 \text{ m}^3}\right)$$

$$= \underline{4.78 \times 10^9 \text{ N/m}^2} = \underline{47,300 \text{ atm}}$$

Chapter 16

1. $f = \dfrac{1}{2\pi}\sqrt{\dfrac{k}{m}}$, where $k = \dfrac{mg}{x} = \dfrac{(2 \text{ kg})(9.8 \text{ m/s}^2)}{0.1 \text{ m}} = 196 \text{ N/m}$

$$f = \frac{1}{2\pi}\sqrt{\frac{196 \text{ N/m}}{2 \text{ kg}}} = \frac{9.9}{2\pi} = \underline{1.58 \text{ Hz}}$$

(b) $\frac{1}{2}mv_{max}^2 = \frac{1}{2}kx^2$, where $A = 15 \text{ cm} = 0.15 \text{ m}$

$$v_{max} = \sqrt{\frac{kA^2}{m}} = \sqrt{\frac{(196)(0.15)^2}{2}} = \underline{1.48 \text{ m/s}}$$

(c) $E = \frac{1}{2}kA^2 = \frac{1}{2}(196)(0.15)^2 = 2.2 \text{ J}$
    $E = \frac{1}{2}mv^2 + \frac{1}{2}kx^2$, where $v = 1 \text{ m/s}$
    $2.2 \text{ J} = \frac{1}{2}(2 \text{ kg})(1 \text{ m/s})^2 + \frac{1}{2}(196 \text{ N/m})x^2$

$$x^2 = \frac{2.2 - 1}{98} = 0.0122; \text{ so } x = \pm 0.11 \text{ m} = \underline{\pm 11 \text{ cm}}$$

2. (a) Since $f = \dfrac{1}{2\pi}\sqrt{\dfrac{g}{L}}$, $\tau = 2\pi\sqrt{\dfrac{L}{g}}$

    If m doubles, τ is unchanged, as τ does not depend on m.

(b) If L doubles, $\tau' = 2\pi\sqrt{\dfrac{2L}{g}} = \sqrt{2} \cdot 2\pi\sqrt{\dfrac{L}{g}}$

    $\tau' = 1.4\tau$; the period increases by $\sqrt{2}$, or 1.4.

3. Comparing $x = A \sin 2\pi ft$ with $x = 4 \sin 5\pi t$, gives $A = \underline{4 \text{ cm}}$

$$2\pi f = 5\pi; \text{ so } f = \frac{5}{2} = 2.5 \text{ Hz}; \text{ therefore } \tau = \frac{1}{f} = \frac{1}{2.5} = \underline{0.4 \text{ s}}$$

$$\frac{1}{2}mv_{max}^2 = \frac{1}{2}kA^2; \text{ so } v_{max} = \sqrt{\frac{k}{m}}\,A$$

The ratio $\dfrac{k}{m}$ can be obtained from the frequency formula: $\dfrac{k}{m} = (2\pi f)^2 = 247 \text{ s}^{-2}$

$$v_{max} = \sqrt{247 \text{ s}^{-2}}\,(4 \text{ cm}) = \underline{63 \text{ cm/s}}$$

Maximum acceleration occurs at $x = A$: $F = ma = -kx$; so $|a_{max}| = \dfrac{k}{m}A$

$$a_{max} = (247 \text{ s}^{-2})(4 \text{ cm}) = \underline{988 \text{ cm/s}^2}$$

Chapter 17

1. (a) Speed of compressional wave in solid, $v = \sqrt{\dfrac{B + (4/3)S}{\rho}}$ from Table 17.1.

For glass, $v = \sqrt{\dfrac{(3.7 \times 10^{10} \text{ N/m}^2) + (4/3)(2.3 \times 10^{10} \text{ N/m}^2)}{2800 \text{ kg/m}^3}} = 4900$ m/s

Distance across screen: $x = 12 \text{ in}\left(\dfrac{1 \text{ ft}}{12 \text{ in}}\right)\left(\dfrac{1 \text{ m}}{3.28 \text{ ft}}\right) = 0.305 \text{ m}$

$$t = \frac{x}{v} = \frac{0.305 \text{ m}}{4900 \text{ m/s}} = \underline{6.2 \times 10^{-5} \text{ s}} = \underline{62 \text{ microseconds}}$$

(b) $\lambda = \dfrac{v}{f} = \dfrac{4900 \text{ m/s}}{4 \times 10^6 \text{ s}^{-1}} = \underline{1.2 \times 10^{-3} \text{ m}} = \underline{1.2 \text{ mm}}$

(c) Horizontally: $x = \frac{1}{2}(12 \text{ in}) = \frac{1}{2}(0.305 \text{ m}) = 0.152 \text{ m}$

$$t_H = \frac{x}{v} = \frac{0.152 \text{ m}}{4900 \text{ m/s}} = 3.1 \times 10^{-5} \text{ s} = 31 \text{ } \mu s$$

Vertically: $x = \frac{1}{2}(9 \text{ in}) = 4.5 \text{ in}\left(\dfrac{1 \text{ ft}}{12 \text{ in}}\right)\left(\dfrac{1 \text{ m}}{3.28 \text{ ft}}\right) = 0.114 \text{ m}$

$$t_V = \frac{x}{v} = \frac{0.114 \text{ m}}{4900 \text{ m/s}} = 2.3 \times 10^{-5} \text{ s} = 23 \text{ } \mu s$$

$$t = 31 - 23 = \underline{8 \text{ } \mu s}$$

The delay between the return of the vertical and horizontal waves allows the instrument to determine where the finger touched the screen.

2.

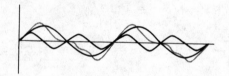

3. As train approaches, $f' = 480$ Hz, $v_0 = 0$, $v = 340$ m/s, v_s is positive, train is approaching observer:

$$f' = f\left(\frac{v + v_0}{v - v_s}\right); \text{ so } 480 = f\left(\frac{340 + 0}{340 - v_s}\right)$$

$$490(340 - v_s) = 340 f \tag{1}$$

As train recedes, $f' = 420$ Hz, $v_0 = 0$, v_s is negative, train is going away from observer:

$$420 = f\left(\frac{340 + 0}{340 - (-v_s)}\right)$$

$$420(340 + v_s) = 340 f \tag{2}$$

Equating the values for $340 f$ in (1) and (2) gives

$$480(340 - v_s) = 420(340 + v_s); v_s = \underline{22.7 \text{ m/s}}$$

Substituting v_s into (1) gives $480(340 - 22.7) = 340 f; f = \underline{448 \text{ Hz}}$

Chapter 18

1. Number of loops = 5; <u>fifth harmonic</u>. $f_5 = 5f_1 = 5(410 \text{ Hz}) = \underline{2050 \text{ Hz}}$.

To get v, $f_1 = \dfrac{v}{2L}$, so $v = 2Lf_1 = 2(0.6 \text{ m})(410 \text{ m/s}) = \underline{492 \text{ m/s}}$

2. (a) $\dfrac{125}{250} = \dfrac{1}{2} = \dfrac{n}{n+1}$; therefore the consecutive harmonics must be f_1 and f_2 for an open pipe; all other consecutive harmonics have larger ratios. For an open pipe with $f_1 = 125$ Hz:

$$f_1 = \dfrac{v}{2L}, \text{ so } L = \dfrac{v}{2f_1} = \dfrac{340 \text{ m/s}}{2(125 \text{ Hz})} = \underline{1.36 \text{ m}}$$

(b) For a closed pipe, $f_1 = \dfrac{v}{4L}$, so $L = \dfrac{v}{4f_1}$

Male voice: $L = \dfrac{340 \text{ m/s}}{4(125 \text{ Hz})} = \underline{0.68 \text{ m}}$; female voice: $L = \dfrac{340 \text{ m/s}}{4(250 \text{ Hz})} = \underline{0.34 \text{ m}}$

3. (a) $\beta_2 - \beta_1 = 10 \log \dfrac{I_2}{I_1}$, where $\beta_1 = 60$ dB and $I_2 = 25I_1$

$\beta_2 - 60 = 10 \log 25 = 10(1.4) = 14$

$\beta_2 = 74$ dB, an increase of $\underline{14 \text{ dB}}$

(b) To get another 14 dB increase, $88 - 74 = 10 \log \dfrac{I_3}{25I_1}$

$\dfrac{14}{10} = 1.4 = \log \dfrac{I_3}{25I_1}$; taking antilogs, $25 = \dfrac{I_3}{25I_1}$;

$$I_3 = 625I_1 \text{ ; } \underline{600 \text{ more singers are needed.}}$$

Chapter 19

1. (a) $n = \dfrac{c}{v} = \dfrac{3.00 \times 10^8 \text{ m/s}}{2.292 \times 10^8 \text{ m/s}} = \underline{1.309}$

(b) $n_a \sin 60° = n_i \sin \theta_i$; $\sin \theta_i = \dfrac{(1)(0.866)}{1.309} = 0.662$; $\theta_i = \underline{41.5°}$

(c) $n_i \sin \theta_c = n_a \sin 90°$

$\sin \theta_c = \dfrac{n_a}{n_i} = \dfrac{1}{1.309} = 0.764$; $\theta_c = \underline{49.8°}$

2. First boundary:

$$n_1 \sin \theta_1 = n_2 \sin \theta_2 \tag{1}$$

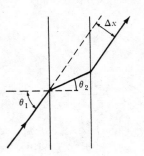

Second boundary: angle of incidence $= \theta_2$ (opposite interior angles between two parallel normals)

$$n_2 \sin \theta_2 = n_1 \sin \theta_3 \tag{2}$$

Comparing (1) and (2): $n_1 \sin \theta_1 = n_1 \sin \theta_3$; so $\theta_3 = \theta_1$; the direction is __unchanged,__ although the ray is displaced from its original path by an amount Δx. This result is independent of n_1 and n_2 and can be extended to a series of adjacent slabs of different n. As long as each slab has parallel sides and the ray eventually emerges into the original medium, its *direction* in that medium will be unaltered.

3. $$n_a = 1; \, n_g = 1.5.$$

First boundary: $n_a \sin 45° = n_g \sin \theta_1$

$$\sin \theta_1 = \frac{(1)(0.707)}{1.5} = 0.471; \ \theta_1 = 28°$$

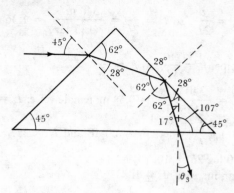

Boundary two: Angle of incidence in glass = 62° (see triangles in diagram); $n_g \sin 62° = n_a \sin \theta_2$

$$\sin \theta_2 = \frac{(1.5)(0.883)}{1} = 1.33$$

Sin greater than $1 \Rightarrow$ total internal reflection at 62°. Boundary 3: Angle of incidence in glass = $107 - 90 = 17°$ (see geometry in sketch)

$$n_g \sin 17° = n_a \sin \theta_3; \ \sin \theta_3 = \frac{(1.5)(0.292)}{1} = 0.438; \ \theta_3 = \underline{26°}.$$

4. Red: $n_R = \dfrac{c}{v_R}$ so $v_R = \dfrac{c}{n_R} = \dfrac{3.00 \times 10^8 \text{ m/s}}{1.514} = 1.982 \times 10^8 \text{ m/s}$

Violet: $v_V = \dfrac{c}{n_V} = \dfrac{3.00 \times 10^8 \text{ m/s}}{1.532} = 1.958 \times 10^8 \text{ m/s}$

$\Delta v = v_R - v_V = (1.982 - 1.958) \times 10^8 \text{ m/s} = 0.024 \times 10^8 \text{ m/s}$

$= \underline{2.4 \times 10^6 \text{ m/s}}$; red light travels faster.

Chapter 20

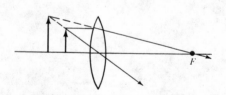

1. Converging lens: $\dfrac{1}{s} = \dfrac{1}{f} - \dfrac{1}{s'} = \dfrac{1}{20} - \dfrac{1}{(-10)} = \dfrac{3}{20}$

$s' = \frac{20}{3} = 6.67 \text{ cm}$

$m = -\dfrac{s'}{s} = -\dfrac{(-10)}{6.67} = \underline{1.5}$ (upright, enlarged)

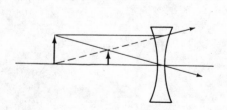

Diverging lens: $\dfrac{1}{s} = \dfrac{1}{f} - \dfrac{1}{s'} = \dfrac{1}{(-20)} - \dfrac{1}{(-10)} = \dfrac{1}{20}$

$s = 20 \text{ cm}$

$m = -\dfrac{s'}{s} = -\dfrac{(-10)}{20} = \underline{\tfrac{1}{2}}$ (upright, reduced)

First image: $\dfrac{1}{s'_1} = \dfrac{1}{f_1} - \dfrac{1}{s_1} = \dfrac{1}{15} - \dfrac{1}{60} = \dfrac{4-1}{60} = \dfrac{3}{60}$; $s'_1 = \frac{60}{3} = 20 \text{ cm}$; $m_1 = -\frac{20}{60} = -\frac{1}{3}$

Second image: $s_2 = d - s_1' = 10 - 20 = -10 \text{ cm}$;

$$\dfrac{1}{s'_2} = \dfrac{1}{f_2} - \dfrac{1}{s_2} = \dfrac{1}{(-20)} - \dfrac{1}{(-10)} = \dfrac{-1+2}{20} = \dfrac{1}{20};$$

$s'_2 = \underline{20 \text{ cm}}$, final image position, <u>real</u> image.

$$m_2 = -\frac{s'_2}{s_2} = -\frac{20}{(-10)} = 2;$$

Total $m = (-\frac{1}{3})(2) = -\frac{2}{3}$, final image is <u>reduced, inverted.</u>

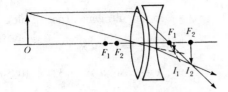

3. (a) $m = -2$ (inverted, twice as large);

$$-2 = -\frac{s'}{s}, \text{ so } s' = 2s$$

$$\frac{1}{10} = \frac{1}{s} + \frac{1}{2s}$$

Multiplying by $2s$: $\frac{2s}{10} = 1 + 2$; $s = \underline{15 \text{ cm}}$

$$s' = 2s = 30 \text{ cm (real image)}.$$

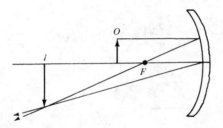

(b) $m = 2$ (upright, twice as large).

$$2 = -\frac{s'}{s}; \text{ so } s' = -2s$$

$$\frac{1}{10} = \frac{1}{s} + \frac{1}{(-2s)}$$

Multiplying by $2s$: $\frac{2s}{10} = 2 - 1$; $s = \underline{5 \text{ cm}}$

$$s' = -2s = -10 \text{ cm (virtual image)}.$$

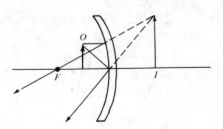

Chapter 21

1. Since $d \ll D$, small angle approximation holds.

Green: $y_{3G} = \frac{n\lambda D}{d} = \frac{(3)(500 \times 10^{-9} \text{ m})(2 \text{ m})}{0.5 \times 10^{-3} \text{ m}} = 6 \times 10^{-3} \text{ m} = 6 \text{ mm}$

Red: $y_{3R} = \frac{n\lambda D}{d} = \frac{(3)(650 \times 10^{-9} \text{ m})(2 \text{ m})}{0.5 \times 10^{-3} \text{ m}} = 7.8 \times 10^{-3} \text{ m} = 7.8 \text{ mm}$

$\Delta y = 7.8 - 6 = \underline{1.8 \text{ mm}}$

2. For $m = 1$: $\lambda = \dfrac{2tn}{m} = \dfrac{(2)(350 \text{ nm})(1.35)}{1} = \dfrac{945 \text{ nm}}{1} = 945$ nm; infrared (not visible).

For $m = 2$: $\lambda = \dfrac{945 \text{ nm}}{2} = \underline{473 \text{ nm}}$; blue-green.

For $m = 3$: $\lambda = \dfrac{945 \text{ nm}}{3} = 315$ nm; ultraviolet (not visible).

3. (a) $d = \frac{1}{3000}$ cm $= 3.33 \times 10^{-4}$ cm $= 3.33 \times 10^{-6}$ m

For $n = 1$: $\sin \theta = \dfrac{\lambda}{d}$; so $\lambda = d \sin \theta = (3.33 \times 10^{-6}$ m$) \sin 11° = 6.35 \times 10^{-7}$ m $= \underline{635 \text{ nm}}$

(b) For $n = 2$: $\sin \theta = \dfrac{2\lambda}{d} = \dfrac{2(6.35 \times 10^{-7} \text{ m})}{3.33 \times 10^{-6} \text{ m}} = 0.381$; $\theta = \underline{22.4°}$

(c) Maximum $\theta = 90°$: $\sin 90° = n_{\max} \dfrac{\lambda}{d}$

$n_{\max} = \sin 90° \dfrac{d}{\lambda} = (1) \left(\dfrac{3.33 \times 10^{-6} \text{ m}}{6.35 \times 10^{-7} \text{ m}} \right) = 5.2$; $\underline{5 \text{ orders}}$ will be seen.

Chapter 22

1. Before: $F_1 = \dfrac{kq_1q_2}{r^2} = 4$ N

After: $F_2 = \dfrac{k(3q_1)(3q_2)}{(2r)^2} = \dfrac{9}{4} \dfrac{kq_1q_2}{r^2} = \frac{9}{4}F_1 = \frac{9}{4}(4 \text{ N}) = \underline{9 \text{ N}}$

2. (a)

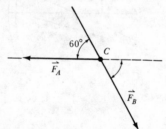

Angle of triangle $= 60°$.

F_A (attraction) $= \dfrac{kq_Aq_C}{r^2} = \dfrac{(9 \times 10^9)(2 \times 10^{-6})^2}{(0.06 \text{ m})^2} = 10$ N/C

F_B(repulsion) $= 10$ N/C (same q's, same r)

Adding graphically as vectors, resultant $\vec{F}$ is the side of an equilateral triangle; therefore:

Total $\vec{F} = \underline{10 \text{ N/C, } 60°}$ below horizontal as shown.

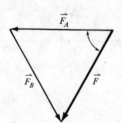

(b) $K_{\text{water}} = 80$; so

$$\vec{F}_{\text{water}} = \dfrac{\vec{F}_{\text{air}}}{K} = \dfrac{10}{80} = \underline{0.125 \text{ N/C}} \text{ in the same direction.}$$

3. Each E has the same magnitude because each charge is equidistant from the center.

$E_A = E_B = E_C = E_D = E = \dfrac{F}{q'} = \dfrac{kqq'}{r^2q'} = \dfrac{kq}{r^2}$; where $r^2 = (0.5 \text{ m})^2 + (0.5 \text{ m})^2 = 0.5 \text{ m}^2$

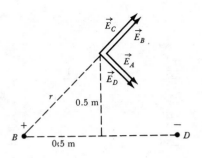

$$E = \frac{(9 \times 10^9)(2 \times 10^{-6})}{0.5} = 36 \text{ N/C}$$

To add $\vec{E}$'s as vectors, resolve them into x and y components. The y components add to zero. The x components give

$$\vec{E} = 4 \, E \cos 45° = 4(36 \text{ N/C})(0.707) = \underline{102 \text{ N/C}} \text{ in the } +x \text{ direction}$$

Chapter 23

1.
$$W = \Delta PE_q = \sum_{i=1}^{2} \frac{kq_i q'}{K}\left(\frac{1}{r_i} - \frac{1}{r_{i_o}}\right)$$

$$= kq_1 q'\left(\frac{1}{r_1} - \frac{1}{r_{1_0}}\right) + kq_2 q'\left(\frac{1}{r_2} - \frac{1}{r_{2_0}}\right) \text{ (where } K = 1)$$

$$W = (9 \times 10^9)(60 \times 10^{-6})(10 \times 10^{-6})\left(\frac{1}{0.6} - \frac{1}{0.85}\right)$$

$$+ (9 \times 10^9)(-20 \times 10^{-6})(10 \times 10^{-6})\left(\frac{1}{0.85} - \frac{1}{0.6}\right)$$

where r_{1_0} = diagonal of square = $\sqrt{(0.6)^2 + (0.6)^2} = 0.85$ m

$$W = (5.4)(1.667 - 1.176) - (1.8)(1.176 - 1.667) = (5.4)(0.491) - (1.8)(-0.491)$$

$$W = 2.65 + 0.88 = \underline{3.53 \text{ J}}$$

2.   (a) $\vec{E}$ points <u>to the left</u> (direction a *positive* charge would move).

 (b) $\Delta PE_q = qV = \frac{1}{2}mv^2$; where $V = \frac{6}{10}$ of total V between plates, since electron has traversed $\frac{6}{10}$ of total distance; $V = \frac{6}{10}(12 \text{ V}) = 7.2$ V

$$v = \sqrt{\frac{2qV}{m}} = \sqrt{\frac{2(1.6 \times 10^{-19})(7.2 \text{ V})}{9.1 \times 10^{-31} \text{ kg}}} = \underline{1.6 \times 10^6 \text{ m/s}}$$

3.   (a) $E = \dfrac{V}{d}$; so $V = Ed = (120 \text{ V/m})(2 \text{ m}) = \underline{240 \text{ V}}$

 (b) Total $\Delta PE = -\Delta PE_g - \Delta PE_q = -mgh - q\,\Delta V$
 $\Delta PE = -(1.67 \times 10^{-27} \text{ kg})(9.8 \text{ m/s}^2)(2 \text{ m}) - (1.6 \times 10^{-19} \text{ C})(240 \text{ V})$
 $= -3.27 \times 10^{-26} \text{ J} - 3.84 \times 10^{-17} \text{ J} = \underline{-3.84 \times 10^{-17} \text{ J}}$

Chapter 24

1.   (a) $I = \dfrac{\Delta q}{\Delta t}$; where $\Delta t = 1 \text{ yr}\left(\dfrac{3.16 \times 10^7 \text{ s}}{1 \text{ yr}}\right) = 3.16 \times 10^7 \text{ s}$

$$I = \frac{0.36 \text{ C}}{3.16 \times 10^7 \text{ s}} = \underline{1.2 \times 10^{-8} \text{ A}} = \underline{12 \text{ nA}}$$

(b) $R = \dfrac{V}{I} = \dfrac{1.55\ \text{V}}{1.1 \times 10^{-8}\ \text{A}} = \underline{1.4 \times 10^{8}\ \Omega} = \underline{140\ \text{M}\Omega}$

2. (a) $R = \rho\dfrac{L}{A}$; where $A = 2\ \text{mm}^2 \left(\dfrac{1\ \text{m}}{10^3\ \text{mm}}\right)^2 = 2 \times 10^{-6}\ \text{m}^2$; $L = 6\ \text{m}$

$\rho = 1.7 \times 10^{-8}\ \Omega \cdot \text{m}$ from Table 24.1

$R = \dfrac{(1.7 \times 10^{-8}\ \Omega \cdot \text{m})(6\ \text{m})}{2 \times 10^{-6}\ \text{m}^2} = \underline{5.1 \times 10^{-2}\ \Omega}$

(b) $I = \dfrac{V}{R} = \dfrac{6\ \text{V}}{5.1 \times 10^{-2}\ \Omega} = \underline{118\ \text{A}}$

(c) $E = \dfrac{V}{d} = \dfrac{6\ \text{V}}{6\ \text{m}} = \underline{1\ \text{V/m}}$

(d) $P = IV = (118\ \text{A})(6\ \text{V}) = \underline{708\ \text{W}}$

3. (a) $I = \dfrac{V}{R} = \dfrac{120\ \text{V}}{10\ \Omega} = \underline{12\ \text{A}}$

(b) $P = IV = (12\ \text{A})(120\ \text{V}) = \underline{1440\ \text{W}}$

(c) $E = Pt$; expressing P in kW: $P = 1440\ \text{W} = 1.44\ \text{kW}$, and t in hours:
$t = 16\ \text{hr/day} \times 28\ \text{days} = 448\ \text{hrs in Feb.}$
$E = (1.44\ \text{kW})(448\ \text{hr}) = 645\ \text{kW} \cdot \text{h}$. At 7¢ per kW·h, $645 \times \$0.07 = \underline{\$45.15}$

Chapter 25

1.

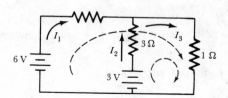

Point A: $I_1 + I_2 - I_3 = 0$ **(1)**

Right Loop: $3 - 3I_2 - I_3 = 0$ **(2)**

Outside Loop: $6 - 2I_1 - I_3 = 0$ **(3)**

From (1), $I_1 = I_3 - I_2$. Substituting into (3) gives
$6 - 2(I_3 - I_2) - I_3 = 0$

$6 - 3I_3 + 2I_2 = 0$ **(3a)**

$\underline{-9 + 3I_3 + 9I_2 = 0}$ [Eq. (2) $\times -3$; then add to Eq. (3a)]
$-3 \qquad\qquad + 11I_2 = 0$

$I_2 = \tfrac{3}{11}\ \text{A}$

Put I_2 into Eq. (2): $3 - 3\left(\tfrac{3}{11}\right) - I_3 = 0$

$I_3 = 3 - \tfrac{9}{11} = \tfrac{24}{11}\ \text{A}$

From Eq. (1), $I_1 = I_3 - I_2 = \tfrac{24}{11} - \tfrac{3}{11} = \underline{1\tfrac{10}{11}\ \text{A}}$

2. (a) Series: $6\ \Omega + 4\ \Omega = 10\ \Omega$
Parallel: $\tfrac{1}{10} + \tfrac{1}{10} = \tfrac{2}{10} = \tfrac{1}{5}$; $R_{eq} = 5\ \Omega$
Series: $5\ \Omega + 5\ \Omega + 2\ \Omega = \underline{12\ \Omega}$

(b) $I_{total} = \dfrac{V}{R_{eq}} = \dfrac{24\ \text{V}}{12\ \Omega} = \underline{2\ \text{A}} = \text{current through 5-}\Omega\text{ resistor.}$

(c) Since parallel branches have equal equivalent R (10 Ω), the current divides evenly, 1 A through each branch. For 4-Ω resistor, $V = IR = (1\ \text{A})(4\ \Omega) = \underline{4\ \text{V}}$

284

3. (a) $P = VI$, so we need I. $R_{eq} = 2 + 4 + 6 = 12\ \Omega$; $I = \dfrac{V}{R_{eq}} = \dfrac{36\ V}{12\ \Omega} = 3\ A$

$P = VI = (36\ V)(3\ A) = \underline{108\ watts}$

(b) 2-Ω resistor: $P = I^2R = (3\ A)^2(2\ \Omega) = \underline{18\ W}$
4-Ω resistor: $P = I^2R = (3\ A)^2(4\ \Omega) = \underline{36\ W}$
6-Ω resistor: $P = I^2R = (3\ A)^2(6\ \Omega) = \underline{54\ W}$

(c) Power supplied $= \underline{108\ W}$
Power dissipated $= 18 + 36 + 54 = \underline{108\ W}$.   They are $\underline{equal}$.

Chapter 26

1. (a) $C = \dfrac{K}{4\pi k}\dfrac{A}{d}$, where $A = 60\ m \times 100\ m = 6000\ m^2$; $d = 3\ cm = 0.03\ m$;

$K = 1$ (air)

$C = \dfrac{(1)(6000\ m^2)}{4\pi(9 \times 10^9\ N\cdot m^2/C^2)(0.03\ m)} = 1.77 \times 10^{-6}\ F = \underline{1.77\ \mu F}$

(b) Increase the plate area, decrease the plate separation, put a dielectric of larger K between the plates.

(c) $PE_q = \tfrac{1}{2}CV^2 = \tfrac{1}{2}(1.77 \times 10^{-6}\ F)(1000\ V)^2 = \underline{0.885\ J}$

2. (a) $\underline{Charge\ remains\ constant}$; the plates are insulated, there is no way for the charge to escape.

(b) $C = KA/4\pi kd$, so as d increases, the $\underline{capacitance\ decreases}$.

(c) Since $V = q/C$ and q is constant, as C decreases, $\underline{voltage\ increases}$.

(d) $E = V/d$; d increases and V increases. Do they increase such that E remains constant?

$$V = \frac{q}{C} = \frac{q4\pi kd}{KA}, \text{ so } E = \frac{V}{d} = \frac{q4\pi kd}{KAd} = \underline{constant}$$

(e) $PE_q = \tfrac{1}{2}qV$; q is constant but V increases, so the $\underline{stored\ energy\ increases}$. (Where does this extra energy come from? From the work required to separate the plates: $W = \Delta PE_q$.)

3. (a) $q_0 = CV_0 = (6 \times 10^{-6}\ F)(100\ V) = \underline{6 \times 10^{-4}\ C} = \underline{600\ \mu C}$

(b) $i = i_0 e^{-t/RC}$; at $t = 0$, $i = i_0 = V_0/R = \dfrac{100\ V}{1 \times 10^6\ \Omega} = \underline{1 \times 10^{-4}\ A} = \underline{100\ \mu A}$

(c) $RC = (1 \times 10^6\ \Omega)(6 \times 10^{-6}\ F) = \underline{6\ s}$

(d) At $t = 6\ s$, $i = i_0 e^{-6/6} = (100\ \mu A)e^{-1} = (100\ \mu A)\dfrac{1}{e} = (100\ \mu A)\dfrac{1}{2.72} = (100$

$\mu A)(0.37) = \underline{37\ \mu A}$

(e) $V = iR = (37 \times 10^{-6}\ A)(1 \times 10^6\ \Omega) = \underline{37\ V}$

(f) $q = CV = (6 \times 10^{-6}\ F)(37\ V) = \underline{2.2 \times 10^{-4}\ C} = 220\ \mu C$
or $q = q_0 e^{-1} = (600\ \mu C)(0.37) = \underline{220\ \mu C}$

Chapter 27

1. (a) $\underline{vertically\ down}$;

(b) $\underline{into\ the\ page.}$ (c)

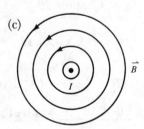

2. (a) By the right-hand rule, B is <u>into the page.</u>

(b) $\dfrac{mv^2}{r} = qvB$; so $mv = qrB = (3)(1.6 \times 10^{-19} \text{ C})(0.3 \text{ m})(0.1 \text{ T})$

$= \underline{1.44 \times 10^{-20} \text{ kg} \cdot \text{m/s}}$

(c) $T = \dfrac{2\pi r}{v}$; where $v = \dfrac{\text{momentum}}{m} = 1.44 \times 10^7 \text{ m/s}$

$T = \dfrac{2\pi(0.3 \text{ m})}{1.44 \times 10^7 \text{ m/s}} = 1.3 \times 10^{-7} \text{ s} = \underline{13 \ \mu\text{s}}$

or $T = \dfrac{2\pi r}{v}$; where $r = \dfrac{mv}{qB}$, so $T = \dfrac{2\pi mv}{qBv} = \dfrac{2\pi(10^{-27} \text{ kg})}{3(1.6 \times 10^{-19} \text{ C})(0.1 \text{ T})}$

$= 1.3 \times 10^{-7} \text{ s} = \underline{13 \ \mu\text{s}}$

3. $\bar{B}$ due to outer loop: $B_0 = \dfrac{\mu_0 I}{2a} = \dfrac{(4\pi \times 10^{-7} \text{ T} \cdot \text{m/A})(50 \text{ A})}{2(0.3 \text{ m})} = 10.5 \times 10^{-5} \text{ T}$ into

the page, $\otimes$

$\bar{B}$ due to inner loop: $B_i = \dfrac{\mu_0 I}{2a} = \dfrac{(4\pi \times 10^{-7} \text{ T} \cdot \text{m/A})(20 \text{ A})}{2(0.2 \text{ m})} = 6.3 \times 10^{-5} \text{ T}$ out of the

page, $\odot$

Resultant $\bar{B} + \bar{B}_0 = \bar{B}_i = B_\otimes - B_\odot = 10.5 \times 10^{-5} \text{ T} - 6.3 \times 10^{-5}$

$= \underline{4.2 \times 10^{-5} \text{ T into the page, } \otimes}$

Chapter 28

1. $|\mathcal{E}| = \dfrac{\Delta\Phi}{\Delta t} \ (N = 1); \ \Delta\Phi = \Phi - \Phi_0 = (2B)A \cos 30° - BA \cos 0°$, where $A = \pi r^2 = \pi(0.25 \text{ m})^2 = 0.2 \text{ m}^2$

$\mathcal{E} = \dfrac{2(0.2 \text{ T})(0.2 \text{ m}^2)(0.866) - (0.2 \text{ T})(0.2 \text{ m}^2)(1)}{0.15 \text{ s}} = \dfrac{0.069 \text{ T} \cdot \text{m}^2 - 0.04 \text{ T} \cdot \text{m}^2}{0.15 \text{ s}}$

$= \underline{0.193 \text{ V}}$

2. (a) External flux into the page increases, so induced field must oppose this by pointing *out of the page;* induced current is <u>counterclockwise.</u>

(b) In loop, $\bar{B}$ from wire is into the page by the right-hand rule. If the external $\bar{B}$ into the page decreases, the induced current opposes this by strengthening the field *into the page.* Therefore, the induced current is <u>clockwise</u> as seen from above.

(c) Solenoid $\bar{B}$ points to the left by the right-hand rule. If it *increases,* induced current must oppose this by creating a field *to the right.* Therefore, the induced current is <u>clockwise</u> as seen by an observer to the left.

3. (a) $\mathcal{E} = -L\dfrac{\Delta i}{\Delta t}$, where $|\Delta i| = 2 - (-2) = 4$ A

$|\mathcal{E}| = (50 \times 10^{-3} \text{ H})\dfrac{(4 \text{ A})}{0.02 \text{ s}} = \underline{10 \text{ V}}$

(b) For a transformer, $\mathcal{E}_2/\mathcal{E}_1 = N_2/N_1$, so $\mathcal{E}_2 = \dfrac{N_2}{N_1}\mathcal{E}_1 = \text{}^{60}\!/\!_{20}(10 \text{ V}) = \underline{30 \text{ V}}$

(c) $\mathcal{E}_2 = -M\dfrac{\Delta i_1}{\Delta t}$, so $M = \dfrac{\mathcal{E}_2 \Delta t}{\Delta i_1} = \dfrac{(30 \text{ V})(0.02 \text{ s})}{4 \text{ A}} = \underline{0.15 \text{ H}} = \underline{150 \text{ mH}}$

Chapter 29

1. (a) $\phi = 30°$; $\tan \phi = \dfrac{X - X_C}{R} = 0.577$; so $X_L - X_C = 0.577R$, where $X_L = 2\pi f L = 2260 \ \Omega$

$X_C = X_L - 0.577R = 2260 - (.577)(1000) = 1680 \ \Omega$

$C = \dfrac{1}{2\pi f X_C} = 1.6 \times 10^{-6} \text{ F} = \underline{1.6 \ \mu\text{F}}$

(b) $I = \dfrac{V}{Z}$, where $Z = \sqrt{R^2 + (X_L - X_C)^2} = \sqrt{(1000)^2 + (577)^2} = 1150\ \Omega$

$$I = \dfrac{140\ V}{1150\ \Omega} = \underline{0.12\ A}$$

(c) $V_L = IX_L = (0.12\ A)(2260\ \Omega) = \underline{270\ V}$

2. (a) $P = VI \cos \phi$; so $\cos \phi = \dfrac{P}{VI}$, where $I = \dfrac{V}{Z} = \dfrac{240\ V}{300\ \Omega} = 0.8\ A$

$$\cos \phi = \text{power factor} = \dfrac{130\ W}{(240\ V)(0.8\ A)} = \underline{0.68}$$

(b) $V = 0.707\ v_0$, so $v_0 = \dfrac{V}{0.707} = \dfrac{240\ V}{0.707} = \underline{340\ V}$

3. (a) $f = \dfrac{1}{2\pi \sqrt{LC}}$, so $L = \dfrac{1}{4\pi^2 f^2 C} = \dfrac{1}{4\pi^2 (200)^2 (20 \times 10^{-6})} = \underline{0.032\ H} = \underline{32\ mH}$

(b) For $R = 300\ \Omega$, $C = 40\ \mu F$, $L = 0.064\ H$

$$f = \dfrac{1}{2\pi \sqrt{LC}} = \dfrac{1}{2\pi \sqrt{(0.064)(40 \times 10^{-6})}} = \underline{100\ Hz}$$

(Note that the above calculations are *independent of R*.)

(c) At resonance, $X_L = X_C$, so $Z = \sqrt{R^2 + (X_L - X_C)^2} = R$; and $I = \dfrac{V}{Z} = \dfrac{40\ V}{150\ \Omega} = \underline{0.27\ A}$

Chapter 30

1. (a) $\dfrac{L}{L_0} = \left(1 - \dfrac{v^2}{c^2}\right)^{1/2} = 0.5$; squaring: $1 - \dfrac{v^2}{c^2} = 0.25$; so $\dfrac{v^2}{c^2} = 1 - 0.25 = 0.75$

$v^2 = 0.75\ c^2$; $v = \underline{0.866c} = \underline{2.6 \times 10^8\ m/s}$

(b) $\Delta t_0 = 8\ min$; $\Delta t = \dfrac{8\ min}{\left(1 - \dfrac{(0.866c)^2}{c^2}\right)^{1/2}} = \dfrac{8\ min}{(1 - 0.75)^{1/2}} = \dfrac{8\ min}{0.5} = \underline{16\ min}$

2. (a) $\underline{5}$ (You are not in relative motion with respect to yourself.)

(b) $\underline{3}$

(c) $\underline{3}$ (Einstein's second postulate of relativity predicts this.)

(d) $\underline{3}$

3. (a) $KE = qV$; $mc^2 - m_0 c^2 = qV$. Substituting the expression for relativistic mass gives:

$$m_0 c^2 \left[\dfrac{1}{\left(1 - \dfrac{v^2}{c^2}\right)^{1/2}} - 1 \right] = qV;$$

so $V = \dfrac{(9.1 \times 10^{-31}\ kg)(3 \times 10^8\ m/s)^2}{1.6 \times 10^{-19}\ C} \left[\dfrac{1}{\left(1 - \dfrac{(0.8c)^2}{c^2}\right)^{1/2}} - 1 \right]$

$V = (5.12 \times 10^5\ V) \left[\dfrac{1}{(0.36)^{1/2}} - 1 \right] = (5.12 \times 10^5\ V)(0.667)$

$= 3.4 \times 10^5\ V = 340,000\ V$

(b) $\dfrac{KE}{m_0 c^2} = \dfrac{mc^2 - m_0 c^2}{m_0 c^2} = \dfrac{m}{m_0} - 1 = \dfrac{m_0}{\left(1 - \dfrac{v^2}{c^2}\right)^{1/2} m_0} - 1 = (1.667 - 1)$

$= \underline{0.667} = \tfrac{2}{3}$

Chapter 31

1. E per photon $= hf = (6.63 \times 10^{-34} \text{ J}\cdot\text{s})(10^6 \text{ s}^{-1}) = 6.63 \times 10^{-28} \text{ J}$;

$$\frac{10^{-14} \text{ J/s}\cdot\text{m}^2}{6.63 \times 10^{-28} \text{ J/photon}} = \underline{1.5 \times 10^{13} \frac{\text{photons}}{\text{s}\cdot\text{m}^2}}$$

2. (a) $KE = hf - \phi = 5 \text{ eV} - 3 \text{ eV} = 2 \text{ eV} = 2 \text{ eV}(1.6 \times 10^{-19} \text{ J/eV}) = 3.2 \times 10^{-19} \text{ J}$

$$\text{Since } KE = \tfrac{1}{2}mv^2,\ v = \sqrt{\frac{2\,KE}{m}} = \sqrt{\frac{2(3.2 \times 10^{-19} \text{ J})}{9.1 \times 10^{-31} \text{ kg}}}$$
$$= \underline{8.4 \times 10^5 \text{m/s}}$$

(b) $E = \dfrac{hc}{\lambda}$; so $\lambda = \dfrac{hc}{E} = \dfrac{(6.63 \times 10^{-34} \text{ J}\cdot\text{s})(3 \times 10^8 \text{ m/s})}{(5 \text{ eV})(1.6 \times 10^{-19} \text{ J/eV})}$

$= 2.49 \times 10^{-7} \text{ m} = \underline{249 \text{ nm}}$ (ultraviolet)

3. (a) $E_A = 2E_B$; $E_A = hf_A = \dfrac{hc}{\lambda_A}$; so $\lambda_A = \dfrac{hc}{E_A}$, and $\lambda_B = \dfrac{hc}{E_B}$

Therefore, $\dfrac{\lambda_A}{\lambda_B} = \dfrac{hc/E_A}{hc/E_B} = \dfrac{E_B}{E_A} = \dfrac{E_B}{2E_B} = \dfrac{1}{2}$

(b) $p_A = \dfrac{h}{\lambda_A}$, and $p_B = \dfrac{h}{\lambda_B}$; so $\dfrac{p_A}{p_B} = \dfrac{h/\lambda_A}{h/\lambda_B} = \dfrac{\lambda_B}{\lambda_A} = \dfrac{2}{1}$

4. (a) $\lambda = \dfrac{h}{mv} = \dfrac{6.63 \times 10^{-34} \text{ J}\cdot\text{s}}{(9.1 \times 10^{-31} \text{ kg})(10^5 \text{ m/s})} = \dfrac{6.63 \times 10^{-34} \text{ J}\cdot\text{s}}{9.1 \times 10^{-26} \text{ kg}\cdot\text{m/s}}$

$= 7.3 \times 10^{-9} \text{ m} = \underline{7.3 \text{ nm}}$

(b) $\Delta p = \tfrac{1}{100}p = \tfrac{1}{100}mv = \tfrac{1}{100}(9.1 \times 10^{-26} \text{ kg}\cdot\text{m/s}) = 9.1 \times 10^{-28} \text{ kg}\cdot\text{m/s}$

$\Delta p\,\Delta x = \dfrac{h}{2\pi}$; so $\Delta x = \dfrac{h}{2\pi\,\Delta p} = \dfrac{6.63 \times 10^{-34} \text{ J}\cdot\text{s}}{2\pi(9.1 \times 10^{-28} \text{ kg}\cdot\text{m/s})} = 1.2 \times 10^{-7} \text{ m}$

$= \underline{120 \text{ nm}}$

Chapter 32

1. (a) $r \propto n^2$ (Eq. 32.9)

(b) $v \propto \dfrac{1}{n}$ (Eq. 32.8 and (a) above)

(c) $f \propto \dfrac{1}{n^3}\left[f = \dfrac{1}{T} = \dfrac{v}{2\pi r} \propto \left(\dfrac{1}{n}\right)\left(\dfrac{1}{n^2}\right)\right]$

(d) $E \propto -\dfrac{1}{n^2}$ (Eq. 32.11)

(e) $mvr \propto n$ (Eq. 32.8 or (a) and (b) above)

2. $E_n = -\dfrac{(13.6 \text{ eV})Z^2}{n^2}$; for $Z = 3$, $E_9 = -\dfrac{(13.6 \text{ eV})(3)^2}{(9)^2} = -1.51 \text{ eV}$

$E_4 = -\dfrac{(13.6 \text{ eV})(3)^2}{(4)^2} = -7.65 \text{ eV}$

$E_{\text{photon}} = -1.51 \text{ eV} - (-7.65 \text{ eV}) = \underline{6.14 \text{ eV}}(1.6 \times 10^{-19} \text{ J/eV})$

$= \underline{9.8 \times 10^{-19} \text{ J}}$

$\lambda = \dfrac{hc}{E} = \dfrac{(6.63 \times 10^{-34} \text{ J}\cdot\text{s})(3 \times 10^8 \text{ m/s})}{9.8 \times 10^{-19} \text{ J}} = 2.0 \times 10^{-7} \text{ m} = \underline{200 \text{ nm}}$

or $\dfrac{1}{\lambda} = RZ^2\left(\dfrac{1}{n_f^2} - \dfrac{1}{n_i^2}\right) = (1.097 \times 10^7 \text{ m}^{-1})(3)^2\left(\dfrac{1}{4^2} - \dfrac{1}{9^2}\right)$

$= 4.96 \times 10^6 \text{ m}^{-1}; \lambda = \underline{2.0 \times 10^{-7} \text{ m.}}$

Then $E = \dfrac{hc}{\lambda} = \dfrac{(6.63 \times 10^{-34} \text{ J}\cdot\text{s})(3 \times 10^8 \text{ m/s})}{2 \times 10^{-7} \text{ m}} = \underline{9.9 \times 10^{-19} \text{ J}}$

3. (a) $r_n = (0.53 \times 10^{-10} \text{ m}) \dfrac{n^2}{Z}$; for $Z = 2$, $n = 5$: $r_5 = (0.53 \times 10^{-10} \text{ m}) \dfrac{(5)^2}{2}$

$r_n = \underline{6.6 \times 10^{-10} \text{ m}}$

(b) 5 de Broglie wavelengths fit into the circumference of the $n = 5$ orbit:

$$5\lambda = 2\pi r; \text{ so } \lambda = \frac{2\pi r}{5} = \frac{2\pi(6.6 \times 10^{-10} \text{ m})}{5} = \underline{8.3 \times 10^{-10} \text{ m}}$$

(c) $mvr = \dfrac{5h}{2\pi} = \dfrac{5(6.63 \times 10^{-34} \text{ J}\cdot\text{s})}{2\pi} = \underline{5.3 \times 10^{-34} \text{ J}\cdot\text{s}}$

Chapter 33

1. (a) 4; (b) 3; (c) 4; (d) 2; (e) 1.

2. Xenon has 54 electrons: $1s^2 2s^2 2p^6 3s^2 3p^6 4s^2 3d^{10} 4p^6 5s^2 4d^{10} 5p^6$. The 54th electron completes the $5p$ subshell; all lower energy subshells are also completely filled. The xenon atom is electrically neutral and consists entirely of filled subshells; it is extremely stable and thus chemically inert.

3. (a) $\Delta E = E_{\text{photon}} = \dfrac{hc}{\lambda} = \dfrac{(6.63 \times 10^{-34} \text{ J}\cdot\text{s})(3 \times 10^8 \text{ m/s})}{0.153 \times 10^{-9} \text{ m}} = 1.3 \times 10^{-15} \text{ J}$

$= \underline{8.1 \text{ KeV}}$

(b) To see a transition from $n = 2$ to $n = 1$, an accelerated electron must have knocked out an electron from the $n = 1$ state. This requires an energy of

$$|E_1| \simeq \frac{(13.6 \text{ eV})Z^2}{n^2} = \frac{(13.6 \text{ eV})(29)^2}{(1)^2} = 11{,}400 \text{ eV}$$

Thus, $V \geqslant \underline{11{,}400 \text{ V}}$

(c) For $V = 20{,}000 \text{ V}$, $f_{\text{max}} = \dfrac{eV}{h} = \dfrac{(1.6 \times 10^{-19} \text{ C})(20{,}000 \text{ V})}{6.63 \times 10^{-34} \text{ J}\cdot\text{s}}$

$= \underline{4.8 \times 10^{18} \text{ Hz}}$

$\lambda_{\text{min}} = \dfrac{c}{f} = \dfrac{3 \times 10^8 \text{ m/s}}{4.8 \times 10^{18} \text{ s}^{-1}} = 6.2 \times 10^{-11} \text{ m} = \underline{0.062 \text{ nm}}$

Chapter 34

1. (a) $^{214}_{83}\text{Bi} \rightarrow {}^4_2\text{He} + {}^{210}_{81}\text{Tl}$

(b) $^9_4\text{Be} + {}^4_2\text{He} \rightarrow {}^1_0 n + {}^{12}_6\text{C}$

(c) $^{238}_{92}\text{U} + {}^1_0 n \rightarrow {}^{239}_{92}\text{U}$; $^{239}_{92}\text{U} \rightarrow {}^{\ 0}_{-1}e + {}^{239}_{93}\text{Np}$; $^{239}_{93}\text{Np} \rightarrow {}^{\ 0}_{-1}e + {}^{239}_{94}\text{Pu}$

2. $^{56}_{26}\text{Fe} \rightarrow {}^{55}_{25}\text{Mn} + {}^1_1\text{H}$; $m = (54.938046 + 1.007825) - 55.934939 = -0.010932$ u
$Q = (-0.010932)(931.5) = -\underline{10.2 \text{ MeV}}$ (*Note:* The total binding energy of iron is not pertinent here, as we are not separating the whole nucleus into all its constituent parts.)

3. (a) June 1–24 = 24 days = 3 half-lives; $N = \dfrac{N_0}{2^n}$, where $N = 3$ μg and $n = 3$

$N_0 = N(2^n) = (3 \text{ } \mu\text{g})(2^3) = \underline{24 \text{ } \mu\text{g}}$

(b) Activity $= \lambda N$, where $\lambda = 0.693/T = \dfrac{0.693}{(8 \text{ days})(8.64 \times 10^4 \text{ s/day})}$

$= 1 \times 10^{-6} \text{ s}^{-1}$

Activity $= (1 \times 10^{-6} \text{ s})(1.38 \times 10^{16} \text{ atoms}) = \underline{1.38 \times 10^{10} \text{ decays/s}} = \underline{0.37 \text{ Ci}}$

Chapter 35

1. $^7_3\text{Li} \quad + \quad {}^1_1\text{H} \quad \rightarrow \quad {}^4_2\text{He} + {}^4_2\text{He}$

Masses: 7.016005   4.002603
      $\underline{+ 1.007825}$ $\underline{+ 4.002603}$
       8.023830   8.005206

$\Delta m = 8.023830 - 8.005206 = 0.018624$ u; $Q = (0.018624)(931.5) = \underline{17.3 \text{ MeV}}$

2. $E = 5 \times 10^6$ tons TNT $\left(\dfrac{4.2 \times 10^9 \text{ J}}{1 \text{ ton TNT}} \right) = 2.1 \times 10^{16}$ J

$P = 1000 \times 10^6$ W $= 10^9$ W $= 10^9$ J/s

Solving $P = E/t$ for t; $t = \dfrac{E}{P} = \dfrac{2.1 \times 10^{16} \text{ J}}{10^9 \text{ J/s}} = \underline{2.1 \times 10^7 \text{ s}} \simeq \text{⅔ yr}$

(Small wonder that there is great interest in producing controlled fusion reactions.)

3. (a) charge; (b) strangeness; (c) baryon number; (d) strangeness; (e) baryon number.

4. $\cancel{2}E_\gamma = \cancel{2}m_n c^2$ (The photons have equal energy and opposite directions to conserve momentum. Their total energy equals the total annihilated rest mass to conserve energy.)

$E_\gamma = m_n c^2 = \underline{939.6 \text{ MeV}}$ (from Table 35.2 in textbook)